Silkworm Biofactory
Silk to Biology

Editors

Enoch Y. Park

Institute of Green Science & Technology
Shizuoka University
Shizuoka, Japan

Katsumi Maenaka

Hokkaido University
Sapporo, Japan

CRC Press
Taylor & Francis Group
Boca Raton London New York

CRC Press is an imprint of the
Taylor & Francis Group, an **informa** business
A SCIENCE PUBLISHERS BOOK

CRC Press
Taylor & Francis Group
6000 Broken Sound Parkway NW, Suite 300
Boca Raton, FL 33487-2742

Printed on acid-free paper
Version Date: 20180820

International Standard Book Number-13: 978-1-138-32812-9 (Hardback)

Visit the Taylor & Francis Web site at
http://www.taylorandfrancis.com

and the CRC Press Web site at
http://www.crcpress.com

Preface

Silkworms have been used for human needs for centuries. The silk derived from the silkworm has of great economic value and recently has used as value-added biomaterials for biopharmaceutical purpose. Silkworms have attracted the attention for expressing eukaryotic recombinant proteins, which require post-translational modifications. In 1985 Dr. Maeda demonstrated that silkworm larvae could produce a functional human α-interferon. Since then, to express recombinant proteins in silkworm various technologies have been developed. Baculovirus-based expression system (BES) comes foremost and has been used in for a long time. Baculovirus being the ideal vector carrying the desired cDNA of interest does not infect nor cause any known disease to humans or other livestock other than its host. The BES was suitable for expressing recombinant proteins in insect cells using *Autographa californica* multiple nucleopolyhedrovirus (AcMNPV). However, in the case of the silkworm, since *Bombyx mori* nucleopolyhedrovirus (BmNPV) has limited host cells and genetic manipulation kits, silkworm has not been easily used for protein production using BmNPV, but mainly silk production.

Today's biotechnology-based industry using silkworm faces a bottleneck towards further advancement owing to the lack of proper gene expression systems. However, with the development of BmNPV bacmid system, being capable of replicating in both *Escherichia coli* and *Bombyx mori*-derived cell lines or silkworm, silkworm larvae or pupae have been used for expression system for the recombinant protein production. This method utilizes the advantage of a bacmid it can be easily prepared and screened in *E. coli* to produce sufficient DNA for subsequent expression in silkworms. This would be a great breakthrough in the production of recombinant eukaryotic proteins and viruses, which will be a powerful tool in a new proteome era.

This volume provides comprehensive up to date review about recent and future trends of silkworm biotechnology and potential pharmaceutical applications, aiming at contributing in the field of life science and bioindustry. This volume consisted of three parts. In Part I Silkworm biology and physiology, genetics, breeding, posttranslational modification and glycoprotein pathway of silkworm are dealt in Chapters 1 to 6. In Part II Utilization of silkworm as a biofactory, new expression system, transgenic silkworm, expression of membrane proteins, virus-like particles and gene delivery are described in Chapters 7 to 13. In Part III Bioproducts from silkworm, bioproducts or the recombinant proteins from silkworm that nearby may be commercialized are introduced in Chapters 14 to 17. This volume begins with an overview of the fundamental research about the silkworm biofactory and extended to the wide applications for the large-scale expression of foreign genes.

Contents

Preface iii

Part I Silkworm Biology and Physiology

1. **Basic and Applied Genetics of Silkworms** 3
Toru Shimada

2. **Silkworm Breeding** 18
Yuyin Chen

3. **Focus on Physiological Vital Proteins from Silkworm *Bombyx mori*** 38
Raman Chandrasekar and *S. K. M. Habeeb*

4. **Metabolism of Molting Hormone and Juvenile Hormone in Silkworm** 62
Yungen Miao

5. **Post-translational Modification in Silkworm** 88
Sungjo Park, Enoch Y. Park and *Andre Terzic*

6. **Glycoprotein Biosynthesis of Silkworm** 107
Kazuhito Fujiyama

Part II Utilization of Silkworm as a Biofactory

7. **BacMam System in Silkworm** 129
Takashi Tadokoro

8. **Transgenic Silkworm** 143
Hideki Sezutsu

9. **Glycosyltransferase Expression in Silkworm and its Applications in Glycobiology** 163
Makoto Ogata, Taichi Usui and *Enoch Y. Park*

10. **Silkworm Expression of Cell Surface Molecules** 180
Shunsuke Kita

11. **Expression and Purification of Mitochondrial Membrane Protein OPA1 for Reconstitution of Membrane Fusion** 193
Tadato Ban and *Naotada Ishihara*

12. **Virus-like Particles Expression in Silkworms** 204
Vipin Kumar Deo and *Enoch Y. Park*

13. **Gene Delivery Based on *Bombyx mori* Nucleopolyhedrovirus (BmNPV)** 220
Tatsuya Kato and *Enoch Y. Park*

Part III Bioproducts from Silkworm

14. **Development of New Biomaterials from Insect and its Virus** 235
Eiji Kotani and *Hajime Mori*

15. **Production of Medical and Cosmetic Products using Transgenic Silkworms** 251
Masahiro Tomita

16. **Comparative Investigation of Influenza Virus-like Particles from Synthetic Chemical Substance and Silkworm** 266
Kuniaki Nerome, Kazumichi Kuroda and *Shigeo Sugita*

17. **Production of Feline and Canine Interferon from Silkworm** 287
Tsuyoshi Tanaka and *Takashi Tanaka*

Index **299**

Part I

Silkworm Biology and Physiology

CHAPTER 1

Basic and Applied Genetics of Silkworms

Toru Shimada

1. Introduction

Genomics is the basis for all researches, and no research in modern biology and biotechnology can be performed without genomic information. Currently the whole-genome sequences of several organisms are being made public and can be obtained from public databases for use in biological research.

The first draft of the genome sequence of *Bombyx mori* (strain p50T), the domesticated silkworm, was published in 2004 by the National Institute of Agrobiological Sciences in collaboration with other Japanese researchers (Mita et al. 2004). In the same year, a Chinese group independently published their draft sequence of *B. mori* strain Dazao, which was originally classified as the same strain as p50T. Later, both groups collaborated and exchanged raw sequence data in order to eventually produce a high-accuracy genome sequence containing new gene models (International Silkworm Genome Consortium 2008).

The silkworm genome analysis is a major contribution to basic biology. The silkworm is a representative insect of the order Lepidoptera (moths and butterflies), and the first species of the Lepidopterans to have a completely sequenced genome. Its genomic information is valuable not only for comparative genomic studies on insects and animals but also as the basis of studies on biological phenomena, such as morphogenesis, complicated pigmentation patterns and metamorphosis, unique to

Laboratory of Insect Genetics and Bioscience, Graduate School of Agricultural and Life Sciences, University of Tokyo, Yayoi 1-1-1, Bunkyo-ku, Tokyo 113-8657, Japan.
E-mail: shimada@ss.ab.a.u-tokyo.ac.jp

lepidopteran insects. In particular, the abundance of silkworm mutants and diverse traits in the races/strains are powerful tools for elucidating biological information coded in the genome.

The second advantage of having a sequenced silkworm genome is that researchers can utilize the genomic information for silkworm breeding and for the improvement of practical races in sericulture and insect biotechnology. Other model organisms, such as *Caenorhabditis elegans* and *Drosophila* with previously sequenced genomes, are not industrial animals, unlike the silkworm which is a beneficial and economically important insect raised for silk production worldwide. Therefore, several breeders are planning to use the silkworm genomic information in order to introduce beneficial traits suitable for silk and biomaterial production.

The third advantage of the sequenced silkworm genome is that it can be used to develop silkworms as model animals useful in the treatment of human diseases. *Drosophila* and *C. elegans* are also used in studies to determine the effects of pharmaceuticals and develop traits mimicking human diseases for the development of therapeutic methods. However, animals with a larger body size are more suitable as disease models. The silkworm is large enough that we can identify its organs with the naked eye, and its tissues and haemolymph can be easily distinguished and collected. Furthermore, chemicals can easily be injected into its body cavity or into the lumen of the digestive tract, and its developmental stages can be easily synchronized and accurately evaluated in terms of pharmacological effects. Considering these advantages and the availability of complete genomic information, the silkworm is an advantageous alternative experimental animal equivalent to mammals such as mice and rats. In recent years, the development of alternative experimental methods has been advanced from the viewpoint of animal welfare, and silkworms have become powerful candidates for this purpose.

In the silkworm *B. mori*, approximately 590 phenotypic mutants have been found thus far. Although these mutant phenotypes can be found during all four life stages, i.e., egg, larva, pupa/cocoon and adulthood, the phenotypic mutants are the most abundant in the larval stage. The most common larval trait is a variation in the pigment pattern and color in the integument. This abundance of larval mutants is opposite to that in *Drosophila* and the model coleopteran *Tribolium*, in which the mutant phenotypes are predominant during adulthood.

2. Mutations in Larval or Serosal Pigmentation and their Responsible Genes

The quantity and distribution of pigments such as melanin, ommochromes and pteridines on the integument of silkworm larva varies among different strains, and complicated pigmentation patterns differ depending on the developmental stages. Among these mutants, those with the melanin pigmentation phenotypes have been the most frequently studied for over 100 years. Typically, newly hatched larvae have entirely black integuments. However, the *chocolate* mutant (*ch*), first studied by Toyama (1909), exhibits a reddish-brown pigmentation in the newly hatched larval stage. One hundred years after the study by Toyama, Futahashi et al. (2008) revealed that the gene responsible for *ch* is an ortholog of *yellow* (*y*), which reduces melanin pigmentation in the adult *Drosophila*. Furthermore, it was elucidated that another pigmentation mutant *sooty* (*so*), which darkens the silkworm body color, is due to a mutation in the ortholog of the *Drosophila* pigmentation mutant, *ebony*. Recently, it has been revealed that the complex melanization pattern on the silkworm larval integument is controlled by the *p* locus coding for an Apontec-like transcription factor, the *Lunar* (*L*) locus coding for a signaling protein Wnt1 and the *Zebra* (*Ze*) coding for the Toll ligand Spätzle3 (Yamaguchi et al. 2013, Yoda et al. 2014, KonDo et al. 2017) (Fig. 1).

Several mutations related to ommochrome pigments are also present in silkworms. Although ommochromes are not present in vertebrates, they constitute the most important pigments in insects. Their color is expressed as purple or black and they are found on the serosa of diapause silkworm eggs and in the adult compound eyes. Ommochromes are synthesized from tryptophan and are localized in the cell cytoplasm as pigment granules. Several enzymes and transporters are involved in this pathway.

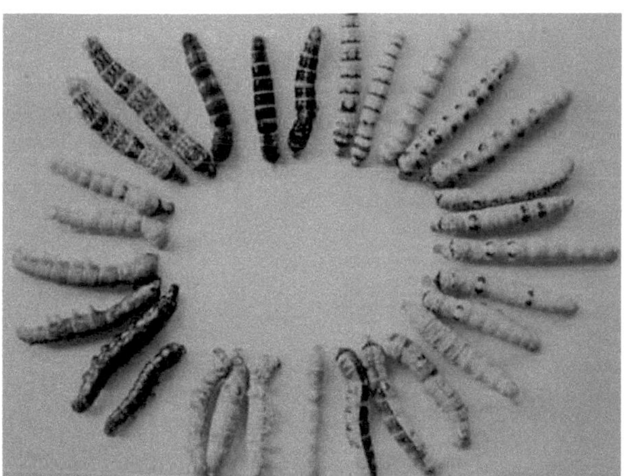

Fig. 1: The diversity of pigmentation patterns in the larval skin of *Bombyx mori*. The pigment is mainly melanin. The diversity can be partly explained as phenotypes of several pigmentation pattern genes, such as *p*, *Ze*, *L*, and the homeotic gene complex *E* (Ueno et al. 1992). Photograph courtesy of the National Bioresource Project Silkworm Newsletter No. 1, printed with permission of Professor Yutaka Banno, Kyushu University.

If the biosynthesis or accumulation of ommochromes is defective, then the "white egg" phenotype appears in the silkworm and the serosa and the adult compound eyes become colorless or white. Several strains produce white eggs, which is caused by different genetic mutations.

Of the white egg genes, *white egg 1* (*w-1*) encodes an enzyme that is involved in the biosynthesis of ommochrome BmKMO (Quan et al. 2007). *White egg 2* (*w-2*) and *white egg 3* (*w-3*) code for ABC transporters with similar structures. The genes responsible for *w-2* and *w-3* are orthologs of the *Drosophila* genes, *scarlet* (*st*) and *white* (w), respectively (Tatematsu et al. 2011). These orthologs control the uptake of ommochrome precursors into the pigment granules in the adult compound eyes. In silkworms, *w-2* and *w-3* are considered necessary for the accumulation of the ommochrome precursor, 3-hydroxykynurenine, in serosal cells of eggs and compound eyes, indicating their functional conservation between *Drosophila* and *Bombyx*. These white egg mutants are useful for transgenesis because it is preferable to use these in order to visualize transgenic fluorescence marker genes.

Although several commonalities in terms of phenotypes and genotypes exist between silkworm and *Drosophila* mutants, several mutants have phenotypes that are specific to the silkworm. An example is the *red blood* (*rb*) mutant, in which a large amount of xanthommatin, an ommochrome pigment, accumulates in the haemolymph and epidermis of the larva and produces a reddish appearance in both the epidermis and the haemolymph when removed from the haemocoel. This abnormality is due to a deficiency

Fig. 2: The diversity of *Bombyx mori* cocoons. The color, size and shape of cocoons are different among strains. The pigment is mainly carotenoids and flavonoids (see text for details). The genes controlling cocoon shape have not yet been identified. Photograph courtesy of the National Bioresource Project Silkworm Newsletter No. 20, printed with permission of Professor Yutaka Banno, Kyushu University.

in metabolizing 3-hydroxykynurenine to 3-hydroxyanthranilic acid in the ommochrome biosynthesis pathway. Silkworm genome analysis revealed that *BmKynu*, which encodes for kynureninase, is located at the proximal end of the chromosome 21 (Meng et al. 2009a). *BmKynu* in the *rb* mutant was observed to be disrupted, and the kynurenine hydrolytic activity of the recombinant protein was remarkably low. Interestingly, *BmKynu* had a low homology with eukaryotic kynureninases and a high homology to bacterial kynureninases, such as those of *Staphylococcus aureus*, which strongly suggests that *BmKynu* did not originate from insects but was acquired via a horizontal transfer from a prokaryote during the evolution of the silkworm or its ancestor.

3. Genes Controlling Metamorphosis and Wing Development

Molting and metamorphosis of silkworms are controlled by ecdysteroids and juvenile hormones (JH), as in other holometabolous insects. Several non-molting mutants of *B. mori* cannot enter the 1st or 2nd molt during the young larval stage. One of these mutants, *non-molting glossy* (*nm-g*), cannot synthesize ecdysteroids in its prothoracic glands because of an alteration in the gene encoding a short-chain dehydrogenase/reductase that is necessary for the synthesis of ecdysteroids (Niwa et al. 2010). Additionally, there is a genetic variation in moltinism, the number of larval molts, in *B. mori*. The *dimolter* (*mod*) mutant undergoes only two or three larval molts, whereas normal silkworms undergo four larval molts. The *mod* larvae become very tiny cocoons and pupae because of their precocious pupation. Daimon et al. (2012) revealed that a cytochrome P-450 gene coding for epoxidase, which catalyzes farnesoic acid to JH, is disrupted in the *mod* mutant. This expansion of knowledge and genetic resources may be useful not only for basic insect physiology but also for sericulture and biotechnology.

Lepidopteran insects are characterized by the presence of large and complex wings. Several mutants are related to the wing formation in silkworms, and one of them, *Vestigial* (*Vg*), is caused by a large deletion in the Z chromosome of the *Vg* strain in a silkworm ortholog of the *Drosophila* gene, *apterous* (*ap*) (Fujii et al. 2011a). The *Drosophila ap* mutant is also unable to form wings. ap^+ encodes a transcription factor containing a DNA-binding Lim-homeodomain and regulates the dorso-ventral compartmentalization of the wing disks. In the *ap* mutant, compartmentalization fails and the wing does not normally develop. The structural abnormality of the *ap* ortholog in the silkworm *Vg* mutant suggests that its ectopic expression in the wing disks leads to the deformed wing development. Other wing mutants such as *flügellos* (*fl*) and *crayfish* (*cf*) have been studied using a similar positional cloning method (Sato et al. 2008, Tong et al. 2015).

4. Genes for Sex Determination

B. mori is a dioecious organism, and its sex is determined by the presence or absence of the W chromosome and does not depend on environmental or endocrine factors. In vertebrates, sex expression is largely influenced by sex hormones and sexual differences are recognized as quantitative traits. Insects do not have sex hormones, and their sex is typically unambiguous. However, Hirokawa (1995) described a silkworm *Intersexuality* mutant, *Isx*, showing an intermediate sex. Later, Fujii et al. (2010a) discovered another intersexual line, the KG strain, which exhibits a male-like morphology in female adults. In this strain, the function of the female determinant on the W chromosome appears defective. Recently, Kiuchi et al. (2014) clarified the molecular status of the female determinant *Fem* on the W chromosome and revealed that *Fem* encodes a 29-base piRNA that interacts with the *Masculinizer* (*Masc*) gene on the Z chromosome. The KG strain produces reduced amounts of the *Fem* piRNA than the normal silkworm.

5. Genes for Disease Resistances

Various viral infections are known to occur in silkworms, including those due to nuclear polyhedrosis virus, cytoplasmic polyhedrosis virus, infectious flacharie virus and densonucleosis virus types 1 and 2. In any virus, the susceptibility of the host silkworm varies depending on genetic factors. Particularly, densonucleosis virus type 1 (BmDNV-1) and type 2 (BmDNV-2) show large differences in the virus susceptibility among races. These viruses are single-stranded DNA viruses, similar to human parvoviruses, which infect the columnar cells of the midgut of the silkworm and proliferate in their nuclei. The susceptibility to BmDNV-1 is controlled by two loci, namely *nsd-1* (chromosome 21) and *Nid* (chromosome 17), whereas that to BmDNV-2 is controlled by *nsd-2* (chromosome 17). Ito et al. (2008) positionally cloned *nsd-2*, which encodes a membrane protein that resembles an amino acid transporter. Genetic complementation using transgenic introductions of the wild-type gene ($+^{nsd\text{-}2}$) into resistant individuals of *nsd-2/nsd-2* using the *piggyBac* vector restored the susceptibility to BmDNV-2 and resulted in larval midgut infection, confirming *nsd-2* as the responsible gene. Recently, *nsd-1* was also isolated and revealed to code for a mucin-like membrane protein (Ito et al. 2018). These achievements on densovirus resistance genes are typical examples that succeeded in the positional cloning of beneficial traits using genomic information. The resistance gene to *Bacillus thuringiensis* (*Bt*) toxin Cry1AB is another example of a beneficial gene that was isolated by analyzing genetic variations among silkworm races (Atsumi et al. 2015).

6. Genes for Food Preferences

Of all the genetic factors that are responsible for silkworm behavior, those responsible for food preference are critical because they are related to the adaptability to the artificial diet. Normal silkworms feed only on mulberry leaves, whereas some limited strains containing the "polyphagous" trait feed on the leaves of non-mulberry plants; thus, this trait has been introduced to the commercial races for rearing on an artificial diet. We have been studying two polyphagous mutants, *beet feeder* (*Bt*) and *soft and pliable* (*spli*). The *Bt* mutant was identified by Ohnuma and Tajima (1986) after X-ray irradiation, and *Bt* silkworms can feed on the leaves of beet (*Beta vulgaris*). Although the *spli* mutant was originally discovered as a mutant larva with a soft body trait, Fujii et al. (2011b) found that *spli* is genetically allelic to *Bt* and, thus, displays the polyphagous trait. Genetic analysis of the Z chromosome of the *spli* strain identified that the coding region of *Bmacj6*, an ortholog of *Drosophila acj6* (abnormal chemosensory jump 6), was largely deleted. *Drosophila acj6* is a mutant that decreases the repelling behavior of flies to chemical substances and has a decreased expression of several olfactory receptor genes. Both *acj6* and *Bmacj6* encode transcription factors that include a POU-homeobox domain and that putatively regulate the olfactory receptor genes. It is unknown why *Bt* and *spli* mutant silkworms exhibit polyphagous behavior. Interestingly, male *spli* moths are incorrectly attracted to bombykol (the aldehyde form of bombykol) and not to the real sex pheromone, bombykol, suggesting that the *spli* mutants have altered chemical sensing.

7. Genes Controlling Cocoon Coloration

The main product of the silkworm is silk. Recently, the sericin proteins and the pigments associated with sericin have been of interest to applied scientists because of new functions and utility values of the sericin layer in silk fiber. There is a specific class of silk-associated pigment flavonoids derived from "Sasa-mayu" cocoons that includes the fluorescent green pigment, which was identified by Daimon et al. (2010) to be the quercetin 5-*O*-glucoside compound (Q5G). They identified that *Green-b* (*Gb*) that encodes a UGT glucosyltransferase (Q5GT), which glucosylates the quercetin at position 5. Q5GT is deleted and non-functioning in most normal strains, and, thus, these strains do not show the green fluorescence in cocoons.

Q5G emits strong fluorescence under ultraviolet irradiation and has an outstanding ultraviolet absorption among the quercetin derivatives. *B. mandarina*, a wild species of the silkworm, contains Q5G in cocoons to possibly protect the pupa in the cocoon from damage due to ultraviolet light. The powerful ultraviolet absorbing activity of the Sasa-mayu pigments has been used to develop commercial products, including those for cosmetics and functional silk fiber.

Carotenoid pigments in cocoons are more commonly recognized and utilized to discriminate races and sometime sexes. The inheritance of the yellow cocoon was studied by Toyama, which became the first evidence of the Mendelian inheritance in animals (Toyama 1906). The yellow carotenoid pigment in cocoons is lutein, which is derived from mulberry leaves and is transferred to the middle silk gland first through the midgut and then through the haemolymph. This transport is mainly controlled by the *Y* (*Yellow blood*) gene, which encodes a carotenoid binding protein in the midgut cells (Sakudoh et al. 2007), and the *C* (*Yellow cocoon*) gene, which encodes a membrane protein "Cameo2" in the silk gland (Sakudoh et al. 2010).

8. Silkworm Mutants as Models of Human Diseases

Some phenotypic genes isolated by positional and candidate cloning are orthologous to genes responsible for human genetic diseases, and silkworms could potentially function as models for these human diseases.

skunk (*sku)* is a mutant that dies in the last larval instar. A large amount of isovaleric acid is produced in the larval body and feces of *sku*, which has a distinct odor. This build-up of isovaleric acid is similar to that in the human disorder, isovaleric acidemia (IVA). Urano et al. (2010) have revealed that *sku* is caused by an amino acid substitution in the enzyme isovaleryl-CoA dehydrogenase (IVD), which is essential in leucine metabolisms. This gene is also responsible for human IVA. Because of the similarities between silkworm *sku and human IVA*, *sku* can possibly be used as a disease model for IVA, potentially for drug discovery, because IVA does not currently have any therapeutic treatment.

In addition, as described above, *rb* (*red blood*) shows a low activity of kynureninase, which is necessary for tryptophan metabolism, and is similar to the genetic disease xanthurenic aciduria in humans. Meng et al. (2009a) have revealed that *rb* is caused by a decrease in the enzyme activity due to a single amino acid substitution of kynureninase. Thus, *rb* may also be used as a therapeutic model for human disease.

The translucent larval skin is a phenotype specific to the silkworm that is controlled by over 30 loci. They are also models of human diseases described below. In the epidermal cells of normal silkworm larvae, uric acid crystallizes to form urate granules, which make the integument opaque. In silkworms, uric acid is synthesized from xanthine as a final metabolite of nitrogen in the fat body. A part of the uric acid is excreted from the malpighian tubules and a part is transported from the fat body to the epidermal cells via the haemolymph and taken into the cytoplasm to form urate granules. Larval translucent mutants cannot form urate granules because of an abnormality in the synthesis, transport, or granulation of uric acid.

A larval translucent mutant, *distinct translucent* (*od*), has an abnormality in the formation of urate granules and has a highly translucent integument. The *od* mutation has been utilized as a marker on the Z chromosome because it is easy to discriminate. Fujii et al. (2010b) have reported that the gene responsible for *od* is *BmBLOS2*, homologous to *BLOS2*, which encodes a sub-unit of the protein complex BLOC-1, essential for the formation of lysosome-related organelles (LROs) in mammalian cells. Fujii et al. (2010b) rescued the *od* phenotype by transgenic complementation using the wild-type *BmBLOS2* gene, which produced silkworms with partially white skin. *BmBLOS2* is also utilized for a marker of genome editing, as described below.

Following studies on *od*, larval translucent mutants, such as the *Aojuku translucent* (*oa*) and *mottled translucent of Var* (*ov*), were also identified to be caused by defects of the genes involved in the formation of LRO. Because human HPS syndromes are caused by abnormalities in the vesicular transport within a cell, the deficiency of uric acid granules in the larval translucent silkworm may also be caused by the same anomaly. Larval translucency is unique to silkworms, is not found in *Drosophila* or other model organisms, and could potentially be utilized as models for studies on vesicular transport and membrane traffic in metazoans.

Some mutants may be rescued by drug administration. For example, *lcm*^{l} (*lcmon-lethal*) is a mutant that shows yellow pigmentation in the integument at the beginning of the 2nd larval instar and dies without feeding. Meng et al. (2009b) identified the sepiapterin reductase gene *Spr* as the causal locus in *lem*^{l} mutants. SPR is an enzyme that synthesizes the coenzyme tetrahydrobiopterin (BH4), which is essential for the synthesis of dopamine and serotonin. *Spr* could potentially serve as a model for Parkinson's disease in humans. Thus, an oral administration of BH4 or dopamine can rescue *lem*^{l} individuals and facilitate longer survival.

9. Transgenic Technology and Genome Editing

Tamura et al. (2000) have established a method for producing transgenic silkworms using the transposon *piggyBac* as a vector. This technology can be used to develop silkworms that produce materials of high commercial value, such as functionalized silk (such as fluorescent silk) and useful pharmaceuticals and cosmetics. Targeted gene knockouts can be generated in silkworms using either transcription activator-like effector nucleases (TALENs) (Takasu et al. 2013) or the clustered regularly interspaced palindromic repeats-associated (CRISPR/Cas9) system (Daimon et al. 2014, Yuasa et al. 2017). Furthermore, knock-ins can be generated in silkworms using a new "PITCh (Precise Integration into Target Chromosome)" technique with TALEN or CRISPR (Nakade et al. 2014). The effectiveness of these techniques in silkworms shows that they can function as an excellent model system that is capable of investigating genetic functions and gene networks. In addition, by gaining scientific knowledge, a new era has emerged that uses these transgenic technologies and genome editing techniques in order to promote applied research, such as in the development of new fibers and biomaterials from silkworms.

The conventional mutants can also contribute to the new technologies and effectively serve as genetic resources. For example, for the secretion and recovery of proteins in the sericin layer, it is more efficient to utilize a line of naked pupa because it lacks the fibroin filament. The naked pupa used by Inoue et al. (2005) is a deletion mutant of the fibroin L chain gene and the causative allele is $Nd\text{-}s^{D}$. In addition, non-diapausing eggs are necessary for the egg microinjection, and therefore either tropical polyvoltine strains (N4, etc.) or the *pigmented and non-diapause* (*pnd*) mutants are used. Genes controlling the non-diapause trait have not been identified; however, identifying these genes in the future would lead to the elimination of strain restrictions for microinjection, making the technique more convenient.

Disease resistance is critical to all silkworms, although the prevention of pathogen contamination and infection is much more crucial during the production of useful substances by genetically modified silkworms. Because of the importance of gene resistance in silkworms, techniques have been developed in order to acquire resistance to nuclear polyhedrosis virus (BmNPV) using a transgenic CRISPR/Cas9 cassette (Tan et al. 2017), which is a promising method.

10. Genetics of the Wild and Non-mulberry Silkworms

B. mandarina is a wild mulberry silkmoth inhabiting East Asia and is genetically the most closely related species to the domesticated silkworm *B. mori*. They can easily hybridize with each other and the resulting progeny are viable and fertile. In sericulture, *B. mandarina* is not only a mulberry pest but also a potential mating partner for the genetically modified *B. mori* that could lead to the introgression of transgenes into the wild (Yukuhiro et al. 2012, Yukuhiro et al. 2017). However, *B. mandarina* is also a potential genetic resource for *B. mori* breeding.

The Eri silkworm *Samia ricini* (Lepidoptera: Saturniidae) is a fully domesticated species closely related to *S. canningi*, which is a wild species distributed in southeast Asia. It is commercially cultivated in the tropical and subtropical regions of Asia and is also grown for basic research in the temperate countries. *S. ricini* is not only utilized for silk production but also has potential as a new model for molecular biology and biotechnology research because its body size is bigger than *B. mori*. It can be easily cultivated on various non-mulberry plants such as the castor bean (*Ricinus communis*) and cassava (*Manihot esculenta*) and is tolerant of bacterial diseases. The whole-genome analysis of *S. ricini* is currently in progress in Japan. Furthermore, the technique for the genome editing of *S. ricini* has been recently developed (Lee, in submission), and genetically modified *S. ricini* could become a new system for biotechnology. In addition, a baculovirus-based system for foreign gene expression in *S. ricini* has already been established (Wu et al. 2013).

The Japanese oak silkmoth *Antheraea yamamai* inhabits the Japanese islands and is cultivated by farmers for the production of special silk with a beautiful green color. The Chinese oak silkmoth *A. pernyi* is native to China and is cultivated in the East Asian countries for silk production. *A. mylitta*, the Indian tussar moth, is grown in India. The Muga silkmoth *A. assamensis* is reared in northeast India in order to produce Muga silk, which has a golden yellow color. Among these silkmoth species, the whole genome of *A. yamamai* has been recently sequenced by Kim et al. (2018). The TALEN-based genetic modification could be applicable to not only *S. ricini* but also *Antheraea* species. Further, the baculovirus expression vector system has been developed in *A. pernyi* nucleopolyhedrosis virus (Huang et al. 2002).

11. Availability of Strains and Mutants of Silkworms

Silkworm strains, including more than 500 mutants and geographic races from different origins, are preserved in the Kyushu University, Japan, under the support of the National Bioresource Project (NBRP), Silkworm. Several wild silkworm species are maintained in the Shinshu University. These resource libraries of domesticated and wild silkworms are available through NBRP (Banno et al. 2010). The information can be retrieved at "SilkwormBase" (http://shigen.nig.ac.jp/silkwormbase/).

12. Conclusion

The genomic information for *B. mori* has been accurately analyzed and annotated, whereas genome analyses of non-mulberry silkworms

(*Antheraea* and *Samia*) are ongoing. In *B. mori*, forward genetics is advancing using both conventional mutants and genomic information-based gene isolations. Reverse genetics is also promoting basic and applied sciences in *B. mori* using piggyback-based transgenesis as well as TALEN and CRISPR-based genome editing. Recently, a non-mulberry silkworm, *S. ricini*, has become a new model insect in which TALEN-based genome editing is functional. There is an approximately 6000-year history of sericulture in China and approximately 2000-year history in Japan, and silkworms are still important agricultural and economic insects. The history of domestication and breeding is engraved in the genomic information as well as target traits to be improved in the future. The development of special commercial materials using silkworms and the production of useful substances are becoming industrialized, and genomic information will be critically useful.

References

Atsumi, S., K. Miyamoto, K. Yamamoto, J. Narukawa, S. Kawai, H. Sezutsu, I. Kobayashi, K. Uchino, T. Tamura, K. Mita, K. Kadono-Okuda, S. Wada, K. Kanda, M. R. Goldsmith and H. Noda. 2012. Single amino acid mutation in an ATP-binding cassette transporter gene causes resistance to Bt toxin Cry1Ab in the silkworm, *Bombyx mori*. Proc. Natl. Acad. Sci. USA. 109: E1591–E1598.

Banno, Y., T. Shimada, Z. Kajiura and H. Sezutsu. 2010. The silkworm—an attractive BioResource supplied by Japan. Exp. Anim. 59: 139–146.

Chen, S., C. Hou, H. Bi, Y. Wang, J. Xu, M. Li, A. A. James, Y. Huang and A. Tan. 2017. Transgenic clustered regularly interspaced short palindromic repeat/Cas9-mediated viral gene targeting for antiviral therapy of *Bombyx mori* nucleopolyhedrovirus. J. Virol. 91: 5–16.

Daimon, T., C. Hirayama, M. Kanai, Y. Ruike, Y. Meng, E. Kosegawa, M. Nakamura, G. Tsujimoto, S. Katsuma and T. Shimada. 2010. The silkworm *Green b* locus encodes a quercetin 5-*O*-glucosyltransferase that produces green cocoons with UV-shielding properties. Proc. Natl. Acad. Sci. USA. 107: 11471–11476.

Daimon, T., R. Kozaki, R. Niwa, I. Kobayashi, K. Furuta, T. Namiki, K. Uchino, Y. Banno, S. Katsuma, T. Tamura, K. Mita, H. Sezutsu, M. Nakayama, K. Itoyama, T. Shimada and T. Shinoda. 2012. Precocious metamorphosis in the juvenile hormone-deficient mutant of the silkworm, *Bombyx mori*. PLoS Genet. 8: e1002486.

Daimon, T., T. Kiuchi and Y. Takasu. 2014. Recent progress in genome engineering techniques in the silkworm, *Bombyx mori*. Dev. Growth Differ. 56: 14–25.

Fujii, T., H. Abe and T. Shimada. 2010. Molecular analysis of sex chromosome-linked mutants in the silkworm *Bombyx mori*. J. Genet. 89: 365–374.

Fujii, T., T. Daimon, K. Uchino, Y. Banno, S. Katsuma, H. Sezutsu, T. Tamura and T. Shimada. 2010. Transgenic analysis of the *BmBLOS2* gene that governs the translucency of the larval integument of the silkworm, *Bombyx mori*. Insect Mol. Biol. 19: 659–667.

Fujii, T., H. Abe, S. Katsuma and T. Shimada. 2011. Identification and characterization of the fusion transcript, composed of the *apterous* homolog and a putative protein phosphatase gene, generated by 1.5-Mb interstitial deletion in the vestigial (*Vg*) mutant of *Bombyx mori*. Insect Biochem. Mol. Biol. 41: 306–312.

Fujii, T., T. Fujii, S. Namiki, H. Abe, T. Sakurai, A. Ohnuma, R. Kanzaki, S. Katsuma, Y. Ishikawa and T. Shimada. 2011. Sex-linked transcription factor involved in a shift of sex-pheromone preference in the silkmoth *Bombyx mori*. Proc. Natl. Acad. Sci. USA. 108: 18038–18043.

Futahashi, R., J. Sato, Y. Meng, S. Okamoto, T. Daimon, K. Yamamoto, Y. Suetsugu, J. Narukawa, H. Takahashi, Y. Banno, S. Katsuma, T. Shimada, K. Mita and H. Fujiwara. 2008. *Yellow* and *ebony* are the responsible genes for the larval color mutants of the silkworm *Bombyx mori*. Genetics. 180: 1995–2005.

Hirokawa, M. 1995. Control of the sex in the silkworm, *Bombyx mori*. Bull. Fukushima Seric. Exp. Sta. 28: 1–104 (in Japanese).

Huang, Y. J., J. Kobayashi and T. Yoshimura. 2002. Genome mapping and gene analysis of *Antheraea pernyi* nucleopolyhedrovirus for improvement of baculovirus expression vector system. J. Biosci. Bioeng. 93: 183–191.

Inoue, S., T. Kanda, M. Imamura, G.-X. Quan, K. Kojima, H. Tanaka, M. Tomita, R. Hino, K. Yoshizato, S. Mizuno and T. Tamura. 2005. A fibroin secretion-deficient silkworm mutant, *Nd-s*^{D}, provides an efficient system for producing recombinant proteins. Insect Biochem. Mol. Biol. 35: 51–59.

International Silkworm Genome Consortium. 2008. The genome of a lepidopteran model insect, the silkworm *Bombyx mori*. Insect Biochem. Mol. Biol. 38: 1036–1045.

Ito, K., K. Kidokoro, H. Sezutsu, J. Nohata, K. Yamamoto, I. Kobayashi, K. Uchino, A. Kalyebi, R. Eguchi, W. Hara, T. Tamura, S. Katsuma, T. Shimada, K. Mita and K. Kadono-Okuda. 2008. Deletion of a gene encoding an amino acid transporter in the midgut membrane causes resistance to a *Bombyx* parvo-like virus. Proc. Natl. Acad. Sci. USA. 105: 7523–7527.

Ito, K., K. Kidokoro, S. Katsuma, H. Sezutsu, K. Uchino, I. Kobayashi, T. Tamura, K. Yamamoto, K. Mita, T. Shimada and K. Kadono-Okuda. 2018. A single amino acid substitution in the *Bombyx*-specific mucin-like membrane protein causes resistance to *Bombyx mori* densovirus. Sci. Rep. (in press).

Kiuchi, T., H. Koga, M. Kawamoto, K. Shoji, H. Sakai, Y. Arai, G. Ishihara, S. Kawaoka, S. Sugano, T. Shimada, Y. Suzuki, M. G. Suzuki and S. Katsuma. 2014. A single female-specific piRNA is the primary determiner of sex in the silkworm. Nature. 509: 633–636.

KonDo, Y., S. Yoda, T. Mizoguchi, T. Ando, J. Yamaguchi, K. Yamamoto, Y. Banno and H. Fujiwara. 2017. Toll ligand Spätzle3 controls melanization in the stripe pattern formation in caterpillars. Proc. Natl. Acad. Sci. USA. 114: 8336–8341.

Meng, Y., S. Katsuma, K. Mita and T. Shimada. 2009. Abnormal red body coloration of the silkworm, *Bombyx mori*, is caused by a mutation in a novel kynureninase. Genes Cells. 14: 129–140.

Meng, Y., S. Katsuma, T. Daimon, Y. Banno, K. Uchino, H. Sezutsu, T. Tamura, K. Mita and T. Shimada. 2009. The silkworm mutant *lemon* (*lemon-lethal*) is a potential insect model for human sepiapterin reductase deficiency. J. Biol. Chem. 284: 11698–11705.

Mita, K., M. Kasahara, S. Sasaki, Y. Nagayasu, T. Yamada, H. Kanamori, N. Namiki, M. Kitagawa, H. Yamashita, Y. Yasukochi, K. Kadono-Okuda, K. Yamamoto, M. Ajimura, G. Ravikumar, M. Shimomura, Y. Nagamura, T. Shin-I, H. Abe, T. Shimada, S. Morishita and T. Sasaki. 2004. The genome sequence of silkworm, *Bombyx mori*. DNA Res. 11: 27–35.

Nakade, S., T. Tsubota, Y. Sakane, S. Kume, N. Sakamoto, M. Obara, T. Daimon, H. Sezutsu, T. Yamamoto, T. Sakuma and K. T. Suzuki. 2014. Microhomology-mediated end-joining-dependent integration of donor DNA in cells and animals using TALENs and CRISPR/Cas9. Nat. Commun. 5: 5560.

Niwa, R., T. Namiki, K. Ito, Y. Shimada-Niwa, M. Kiuchi, S. Kawaoka, T. Kayukawa, Y. Banno, Y. Fujimoto, S. Shigenobu, S. Kobayashi, T. Shimada, S. Katsuma and T. Shinoda. 2010. *Non-molting glossy/shroud* encodes a short-chain dehydrogenase/reductase that functions in the 'Black Box' of the ecdysteroid biosynthesis pathway. Development. 137: 1991–1999.

Ohnuma, A. and Y. Tajima. 1986. Further studies on non-preference mutations in the silkworm. 3. A new sex-linked mutant that shows defective food-preference. Rep. Silk Sci. Res. Inst. 34: 17–25.

Quan, G.-X., I. Kobayashi, K. Kojima, K. Uchino, T. Kanda, H. Sezutsu, T. Shimada and T. Tamura. 2007. Rescue of *white egg 1* mutant by introduction of the wild-type *Bombyx* kynurenine 3–monooxygenase gene. Insect Sci. 14: 85–92.

Sakudoh, T., H. Sezutsu, T. Nakashima, I. Kobayashi, H. Fujimoto, K. Uchino, Y. Banno, H. Iwano, H. Maekawa, T. Tamura, H. Kataoka and K. Tsuchida. 2007. Carotenoid silk coloration is controlled by a carotenoid-binding protein, a product of the *Yellow blood* gene. Proc. Natl. Acad. Sci. USA. 104: 8941–8946.

Sakudoh, T., T. Iizuka, J. Narukawa, H. Sezutsu, I. Kobayashi, S. Kuwazaki, Y. Banno, A. Kitamura, H. Sugiyama, N. Takada, H. Fujimoto, K. Kadono-Okuda, K. Mita, T. Tamura, K. Yamamoto and K. Tsuchida. 2010. A CD36-related transmembrane protein is coordinated with an intracellular lipid-binding protein in selective carotenoid transport for cocoon coloration. J. Biol. Chem. 285: 7739–7751.

Sato, K., T. M. Matsunaga, R. Futahashi, T. Kojima, K. Mita, Y. Banno and H. Fujiwara. 2008. Positional cloning of a *Bombyx* wingless locus *flügellos* (*fl*) reveals a crucial role for fringe that is specific for wing morphogenesis. Genetics. 179: 875–885.

Takasu, Y., S. Sajwan, T. Daimon, M. Osanai-Futahashi, K. Uchino, H. Sezutsu, T. Tamura and M. Zurovec. 2013. Efficient TALEN construction for *Bombyx mori* gene targeting. PLoS One. 8: e73458.

Tatematsu, K., K. Yamamoto, K. Uchino, J. Narukawa, T. Iizuka, Y. Banno, S. Katsuma, T. Shimada, T. Tamura, H. Sezutsu and T. Daimon. 2011. Positional cloning of silkworm *white egg 2* (*w-2*) locus shows functional conservation and diversification of ABC transporters for pigmentation in insects. Genes Cells. 16: 331–342.

Tong, X., S. He, J. Chen, H. Hu, Z. Xiang, C. Lu and F. Dai. 2015. A novel laminin β gene *BmLanB1-w* regulates wing-specific cell adhesion in silkworm, *Bombyx mori*. Sci. Rep. 5: 12562.

Toyama, K. 1906. Studies on the hybridology of insects, I. On some silkworm crosses, with special reference to Mendel's law of heredity. Bull. Coll. Agric. Tokyo Imp. Univ. 7: 259–393.

Toyama, K. 1909. Studies on the hybridology of insects. II. A sport of the silk-worm, *Bombyx mori* L., and its hereditary behavior. J. Coll. Agric. Tokyo Imp. Univ. 2: 85–103.

Ueno, K., C. C. Hui, M. Fukuta and Y. Suzuki. 1992. Molecular analysis of the deletion mutants in the E homeotic complex of the silkworm *Bombyx mori*. Development. 114: 555–563.

Urano, K., T. Daimon, Y. Banno, K. Mita, T. Terada, K. Shimizu, S. Katsuma and T. Shimada. 2010. Molecular defect of isovaleryl-CoA dehydrogenase in the *skunk* mutant of silkworm, *Bombyx mori*. FEBS J. 277: 4452–4463.

Wang, L., T. Kiuchi, T. Fujii, T. Daimon, M. Li, Y. Banno, S. Kikuta, T. Kikawada, S. Katsuma and T. Shimada. 2013. Mutation of a novel ABC transporter gene is responsible for the failure to incorporate uric acid in the epidermis of *ok* mutants of the silkworm, *Bombyx mori*. Insect Biochem. Mol. Biol. 43: 562–571.

Wu, Y., Y. Huang, D. Lei, Y. Wu, M. Li, J. Kobayashi and X. Wang. 2013. Research for the production of recombinant human epidermal growth factor using *Samia cynthia ricini* pupae bioreactor. Sheng Wu Yi Xue Gong Cheng Xue Za Zhi. 30: 136–140 (in Chinese with English summary).

Yamaguchi, J., Y. Banno, K. Mita, K. Yamamoto, T. Ando and H. Fujiwara. 2013. Periodic *Wnt1* expression in response to ecdysteroid generates twin-spot markings on caterpillars. Nat. Commun. 4: 1857.

Yoda, S., J. Yamaguchi, K. Mita, K. Yamamoto, Y. Banno, T. Ando, T. Daimon and H. Fujiwara. 2014. The transcription factor Apontic-like controls diverse colouration pattern in caterpillars. Nat. Commun. 5: 4936.

Yuasa, M., T. Kiuchi, Y. Banno, S. Katsuma and T. Shimada. 2016. Identification of the silkworm quail gene reveals a crucial role of a receptor guanylyl cyclase in larval pigmentation. Insect Biochem. Mol. Biol. 68: 33–40.

Yukuhiro, K., H. Sezutsu, T. Tamura, E. Kosegawa, K. Iwata, M. Ajimura, S. H. Gu, M. Wang, Q. Xia, K. Mita and M. Kiuchi. 2012. Little gene flow between domestic silkmoth *Bombyx mori* and its wild relative *Bombyx mandarina* in Japan, and possible artificial selection on the CAD gene of *B. mori*. Genes Genet. Syst. 87: 331–340.

Yukuhiro, K., H. Sakaguchi, N. Kômoto, S. Tomita and M. Itoh. 2017. Three single nucleotide polymorphisms indicate four distinctive distributions of Japanese *Bombyx mandarina* populations. J. Insect Biotech. Sericol. 86: 77–84.

CHAPTER 2

Silkworm Breeding

Yuyin Chen

1. Introduction

Silkworms were first domesticated in China over 5,000 years ago, since then, the silk production capacity of the species has increased nearly ten-fold through artificial selection and an intensive breeding process. The silkworm is an organism wherein the principles of genetics and breeding are applied in order to harvest the maximum output, being the second following maize in exploiting the principles of heterosis and cross breeding. Traditional silkworm breeding has been focused on achieving higher production with better quality while improving survival, reproduction and physiological characteristics. Silkworm breeding has made a great contribution to the improvement of silk quality and to the sustainable development of sericulture in the past and still today.

Traditionally, the sericulture is divided into distinct activities, including egg production, silkworm rearing, cocoon reeling and silk weaving. Correspondingly, silkworm breeding aims for the overall improvement of silkworms from a commercial point of view. The major objective is to improve fecundity, the health of larvae, disease resistance, and the quantity of cocoon and silk production. With the development of molecular biotechnology, and the establishment and wide application of *Bombyx mori* nucleopolyhedrovirus (BmNPV) as protein expression vector and later genetic technology to generate transgenic silkworms by using the piggyBac transposon derived from the lepidotpteran *Trichoplusia ni*, the silkworm has been well known as a good biofactory for the production

College of Animal Sciences, Zhejiang University, 866 Yu-Hang-Tang Road, Hangzhou, 310059, P. R. China.
E-mail: chenyy@zju.edu.cn

of high value functional biomaterials. However, the characteristics of silkworm strains used for such purposes are different from those in commercial silk production. This chapter will briefly introduce the silkworm germplasm resources and traditional silkworm breeding techniques, demonstrate examples of silkworm breeding for special utilization and integrate the new trend of the silkworm as biofactory.

2. Silkworm Germplasm Resources

Silkworm germplasm resource, also known as genetic resource or gene resource, is the important material for silkworm breeding and genetic research. As a silkworm breeder, the richness of silkworm resources on hand, the width and depth of research on the maintained silkworm strains, and the proper selection of silkworm strains as parental breeding materials greatly determine the success and efficiency of new silkworm variety breeding. There are several huge silkworm germplasm resource-maintaining centers in the world, such as China, Japan and India, and the largest one with an online search function is in China. As mentioned above, the specific characters of silkworm strains required for biofactory are different, and there is currently not much knowledge on the differences. It is necessary to introduce some related information on the silkworm germplasm resources as they are useful either for the material of biofactory, or for the special needs of the transgenic silkworm studies.

2.1 Types of Silkworm Germplasm Resources and their Utilization

There are thousands of silkworm strains and genetic mutations maintained in the pool of different research institutes in the world. Since spreading to the world through China's silk road, the silkworm has differentiated into several geographical races and formed four major races with three types of voltinisms, including (1) univoltine European race, (2) univoltine, biovoltine, and diapaused multivoltine Chinese race, (3) univoltine and biovoltine Japanese race, and (4) the non-diapause multivoltine tropical race. Traditionally, the silkworms are divided into local varieties, improved varieties, genetic mutations, and closely related species based on the origination (Xiang 1994). For example, the representative local varieties are Lanxi, Mianyanghong, Dazhao from China, Akajuku from Japan, Blance pur from France, Ascoli from Italy, Szegrad from Austria, Bilianzhou from Vietnam and Mysore from India. The improved varieties are those which make hybrids eggs used for commercial cocoon production or those with special characteristics as silkworm breeding materials. Commercial varieties used in sericulture of different countries are significantly different in larval healthiness, cocoon yield and silk quality, which depends on the local climate condition, silkworm rearing technical level and silkworm breeding level. Special breeding materials are those with outstanding characteristics, such as particularly tolerant to hot and humid climates (e.g., Nistari 306, Dong 34); completely resistant to *Bombyx mori* nucleopolyhedrovirus (Qiufen N); especially short larval duration (Lan 5, Lan 10); significantly high cocoon shell weight (010, 610, BC9); exclusively high cocoon shell ratio (TB, 124); particularly long filament length (MK, CH8L); excellent silk reelability (Qiuhua, Dongfei, Su 12) and so on. Genetic mutations are mostly developed by artificial radiation treatment and only few are selected from natural mutation. Most of these silkworm strains are good for genetic research but could not be used for commercial silkworm breeding due to some severe defects in economic traits. The best samples of the induced mutation useful for commercial silkworm breeding are the sex-limited larval marker strains and the sex-linked balanced lethal system. Among several strains with sex-limited larval marker, the strains of female with normal marker, male with plain skin have been successfully used in practical breeding in order to develop the new varieties with easy sex separation in hybrid production. The sex-linked balanced lethal system was established by Russian scientist Strunikow (1995) who irradiated γ-ray to female pupae. This will be explained in detail in the section of male-only silkworm breeding. Figure 1 shows some distinguishing traits of different developmental stages from egg to moth of *B. mori*.

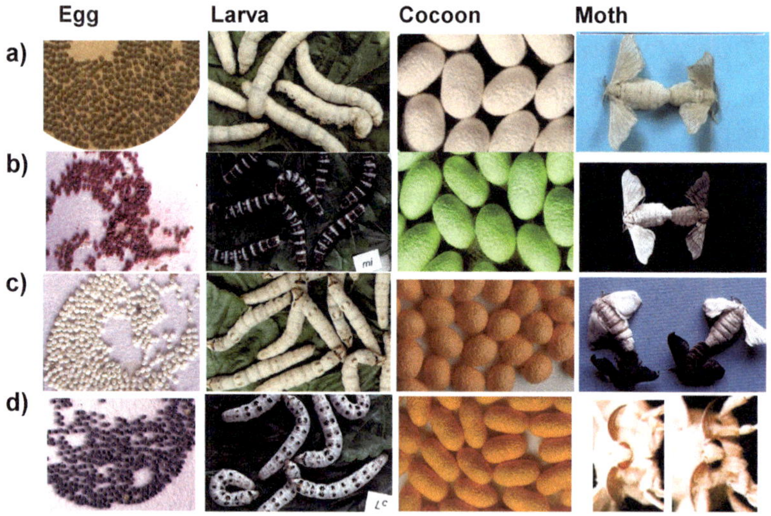

Fig. 1: Examples of distinguishing traits from egg to moth of silkworm *Bombyx mori* maintained in germplasm resources. (a) from left Normal brown, Plain skin, Normal white and Normal white; (b) from left Red mutation, White mold, Green mutation and Grey mutation; (c) from left White mutation, Normal marker, Pink mutation and Black mutation; (d) from left Black mutation, Multi star, Golden yellow mutation and Red or black eyes.

2.2 Conservation and Innovation of Silkworm Germplasm Resources

2.2.1 Maintenance of Original Character

The most important task for conservation of silkworm germplasm resources is to maintain the original genetic structures and to prevent the genetic drift caused by improper selection of seed stock or natural divergence over time. The principles to follow are: (1) preventing inbreeding and maintaining the original traits by taking 20 parent moth-laying, cutting eggs from one corner of each laying and putting together for incubation, brushing 0.5–1.0 g of newly hatched larvae and mixing rearing according to the standard silkworm rearing techniques. At least two sheets of 28 egg-laying on paper are required to be produced, one for the next generation rearing and one for backup of the seed. (2) rearing univoltine and bivoltine varieties in the spring season once a year, diapause multivoltine varieties from spring to autumn 2–3 times a year, and non-diapause multivoltine varieties year around; and (3) maintaining some local varieties in original place, or at least similar geographical conditions, for example, maintaining tropical silkworm varieties in southern China such as Guangzhou Province and maintaining univoltine and bivoltine varieties in Zhejiang and Jiangsu Province in China.

2.2.2 Innovation of Silkworm Germplasm Resources

Innovation of silkworm germplasm resources is regarded as the material foundation in super silkworm breeding. This could be achieved by systematic selection, induced mutation and molecular gene editing.

The classic example of systematic selection for the creation of special breeding materials is the basic strains MK with the longest silk length. Its longest individual silk length of a single cocoon reached as high as 3300 m in the 49th generation of systematic selection in 1976. Induced mutation breeding technology has become the most efficient way to create novel features of silkworm in the last century since its discovery. The pioneer of induced mutation breeding in silkworms is Japanese breeder Tajima, who used X-rays to treat silkworm pupae and created the sex-limited marker strains in 1938. Since then, several induced mutation strains have been bred and used for commercial silkworm breeding, such as the sex-limited egg color strains, sex-limited cocoon color strains and the sex-linked balanced lethal system. With the rapid development and wide application of molecular biology, genomics, transgene and genome editing, the innovation of silkworm genetic resources has entered a new era: from traditional cross breeding or induced mutation to molecular design. Though detailed research progresses on the transgenic silkworm are also presented in other chapters, particular attention is given here to the transgenic technology useful for the improvement of commercial silk production. Ma et al. (2011) reported that Ras1CA overexpression in the posterior silk gland improves silk yield by 60%; Teule et al. (2012) reported that silkworms transformed with chimeric silkworm/spider silk genes spin composite silk fibers with improved mechanical properties; Jiang et al. (2014) reported that silkworm molecular breeding for disease resistance; and Wang et al. (2016) reported that the transgenic silkworm yielded silk with calcium-binding activity. Molecular design no doubt gives renewed impetus to the creation of completely new silkworm strains to enrich the silkworm germplasm resources.

2.3 Information System of Silkworm Germplasm Resources

The database for silkworm germplasm resources has been established in each maintained institute with the assistance of computer information searching technology, although earlier systems were not online searchable. For example, the germplasm resource maintained in the Institute of Sericulture Research in China (Zhang et al. 2004) uses the Chinese Crop Germplasm Resource Information System (CGRIS) software, users can carry out retrieval of 81 characters of silkworm simply by browsing the pictures of different stages in the retrieval result. The recently established Silkworm Germplasm Resource Database and Information Exchange and Sharing Platform in Zhejiang Province of China (http://zjjc.zaas.ac.cn) has an online searching service (Meng 2014). It offers resource catalog, screening, retrieval and pedigree of 830 silkworm varieties. Unfortunately, there is still no international sharing information database available for silkworm germplasm resources.

3. Traditional Silkworm Breeding

3.1 Set up the Breeding Goals

Silkworm breeding has been focused on the commercial cocoon production in order to meet the needs of the silk industry. The general principles for setting up the breeding goals in traditional silkworm breeding are: (1) satisfying the consumer's needs for silk consumption, (2) being adaptable to the local climate conditions and technical level of silkworm rearing farmers, and (3) balancing the benefit among hybrid producers, cocoon producers, silk reeling and weaving factories through the biological and economic efficiency of new hybrids. Therefore, silkworm breeding is considered to be the systematic engineering and the art that manipulates new silkworm varieties at the highest level on the balanced development to all related sectors of sericulture.

The requirements for making silkworm as the biofactory are different from those of the traditional silkworm breeding, and the breeding goals should be rehabilitated. However, so far there has been no specific research that deals with the principles of setting up breeding goals. Some ordinary questions are addressed below for discussion. First, what type of silkworm strains should be selected when the BmNPV is used as the protein expression vector. It is well-known that some commercial silkworm varieties are very sensitive to BmNPV, and that this is one of the bad traits for commercial cocoon production but being a good feature when BmNPV is used as an efficient protein expression vector when silkworm used as the biofactory. However, there are silkworm strains that are completely resistant to BmNPV, such as Qiufeng N and Baiyu N, and there are sensitive strains such as Quifeng and Jinsong. This is an example of opposite direction required for different utilization. Second, what type of silkworm strain is more suitable when the piggyBac transposon is used for research on transgenic silkworm. Various transgenic expression systems based on the piggyBac transposon have been established for the production of exogenous proteins in the silk of transgenic silkworms. These transgenic resources make it possible to create genetically modified silk materials with improved mechanical properties and new biofunctionalities. When the constructed plasmid is microinjected into preblastoderm embryos of the silkworm eggs, an important prerequisite condition to successfully obtain transgenic silkworm is that eggs must be non-diapause. That is the reason why earlier studies on the transgenic silkworm usually chose Nistari as the experimental material for its non-diapause multivoltine feature. The drawback of Nistari in the transgenic research is its non-diapause trait, meaning that when the researchers successfully get the transgenic strains, they have to be reared year around. In case of accidental failure, the valuable transgenic silkworm strain would be lost completely. If there were diapause eggs of the transgenic silkworm, there would be backup eggs for new batch of silkworm to grow. It was reported that some researchers chose silkworm strains with diapause multivoltine features like Dazhao, Lan 10 as the research materials (Peng et al. 2014, Wang et al. 2016). This is because such strains of silkworm produce majority non-diapause eggs but small percentage of diapause eggs in the first crop of a year, then produce majority diapause eggs in the third crop. The transgenic silkworm eggs would be able to be stored in refrigerator in the winter and earlier spring, and also there are backup eggs for accidental loss during the rearing period. Third, the adaptability of silkworm larvae to an artificial diet. All silkworm varieties have good ingestion and adaptability to their natural host mulberry leaves, which is the result of long-term natural evolution and artificial selection. It was found that the adaptability of silkworms to artificial diet shows significant differences due to the type of silkworm varieties. The silkworm strains with excellent artificial diet adaptability might be put into consideration when and where mulberry leaves are not available or germ-free rearing condition is required as bioreactor.

3.2 Selection of Parental Materials for Cross Breeding

Three most important traditional methods of silkworm breeding are systematic selection, cross-breeding and induced mutation breeding, among which the systematic selection method is a fundamental technique and the similar procedure is used in the mid and late stage of all breeding programs. The major difference is the method of acquiring genetic variation at the beginning of the breeding project. New silkworm varieties for commercial cocoon production are mostly bred by cross-breeding, therefore, its common procedure will be discussed.

Systematic selection was widely used in the early age of silkworm breeding when the general level of silkworm varieties used for practical production was relatively low. This method now is used only for purification and rejuvenation of the used varieties, improvement of single trait or elimination of single defect trait. Induced mutation breeding was a very effective means to obtain totally different genetic traits and had created many novel strains of silkworm in the last century. Of them, the sex-limited strains and sex-linked balanced lethal strains are particularly used as the basic materials of cross-breeding today.

As for cross-breeding, generally two or more silkworm strains with different characteristics are selected as the parental materials to create a new variety with comprehensive superiority. It has been used for breeding most commercial varieties in sericulture, and is still a common method today. In silkworm breeding, combining ability and F_1 heterosis of two parents in a mating must be considered in beginning of the breeding plan because the hybrids are used in cocoon production. The past experience showed that the highest heterosis is the cross between Chinese race and Japanese race, that is, why the cross-breeding is usually to breed a pair of new varieties at same time, one Chinese race and the other Japanese race. Therefore, the new varieties are either fixed from all Chinese, or all Japanese, or Japanese-European, sometimes Chinese-Tropical, or Japanese-Tropical.

The first step for cross-breeding is the selection of parental materials, which is the key to success in the implementation of the breeding project. Five factors should be comprehensively considered: (1) The breeding goal, for example, if the new varieties to be bred are reared in the spring of southeastern China with good climate and skilled silkworm rearing farmers, the priority for consideration of parental materials is in the cocoon yield and silk quality. In contrast, if the new variety is bred in the tropical region with relatively poor rearing facility and less skilled farmers, the priority should shift to the robustness of silkworm larvae with balanced quantitative and quality characteristics. When two or more strains are selected for parental materials, in addition to all traits meeting the requirements for the breeding goal, each strain should carry distinct characteristics, such as one is particularly high in yielding and one is excellent in silk quality, or other special characteristics. (2) All selected parental materials should be comprehensively good in economic characters and genetic stability. At present, silkworm hybrids are required in such a high quality that any defect of a single parental material could cause unwanted features in the newly bred variety. It is important to carefully investigate the selected material when searching from the silkworm germplasm resource database. (3) The advantages and disadvantages of two parent traits should complement each other. There is no strain of silkworm that has all around excellent characteristics, by the complementation of two parent traits and proper breeding and selection of filial generations, the new variety with combined advantages of two parents could be expected. (4) The blood relationship of parental materials should be close. As mentioned above, the pair of new varieties to be used for hybrid production must have good combinability so as to show superior heterosis. For this purpose, the parental material for each of the paring varieties must be very close in blood relationship, being all Chinese, all Japanese, or all European. When multivoltine race is needed in order to breed the high resistant varieties, usually the Chinese multivoltine strains are used to cross Chinese parental material, and Vietnamese multivoltine strains are used to cross Japanese parental material from the consideration of blood closeness. (5) The general combining ability of all selected parental materials should be good so as to ensure the high combining ability of the new bred varieties. Early special combining ability should be tested for the selection of the best matched pairing varieties. Usually the existing commercial varieties with excellent combining ability are chosen as one of the parental strains for the new pair in the breeding plan, respectively, so that high special combining ability could be expected.

3.3 Crossing Patterns

The single cross, multiple cross and back cross patterns are all used in silkworm breeding as outlined in Fig. 2a and b. The cross pattern should be adopted in a specific breeding program, depending on the breeding goals. When single cross could satisfy the breeding goal, the first consideration for this is how to most quickly stabilize the genetic characters in the filial generations. However, it is very hard to meet the breeding goal from two basic strains nowadays, three or more parental materials are usually used. Back-crossing is widely used for the improvement of single weak traits in commercial varieties or transfer of single special traits into targeted varieties, for example, when the sex-limited marker trait is required to

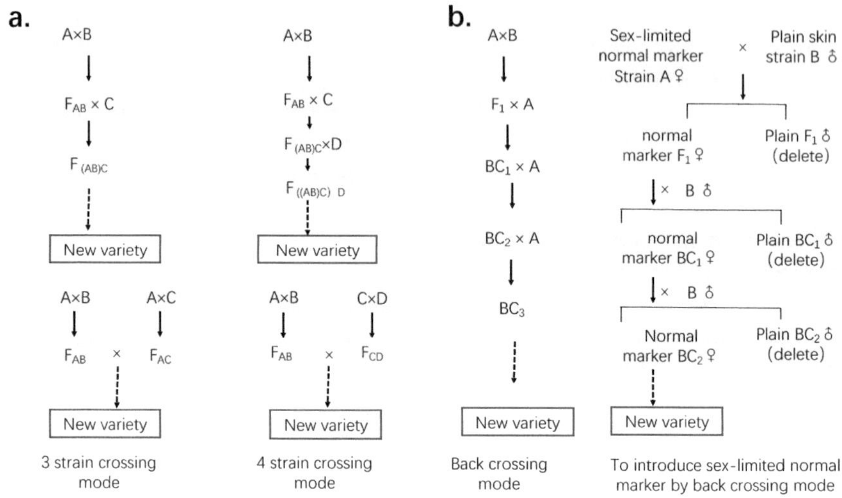

Fig. 2: (a) Outline of multiple cross patterns used in silkworm breeding, (b) Outline of back cross patterns used in silkworm breeding.

be introduced into commercial variety, the back-crossing strategy will be adopted as shown in Fig. 2b.

3.4 Set up of Breeding Environmental Conditions

The phenotype is the interaction of genotype and environmental condition. The full expression of a genotype, especially some economic characters, requires the suitable environmental condition. The correct selection of seed cocoons to meet the breeding goal greatly depends on the environmental conditions set up for the filial generation of a particular breeding program. After cross of parental materials, there are significant differences among individuals or batches in the filial generations due to gene segregation and recombination, the breeding environment should be set up to ensure maximum expression of desired genotypes carried by individuals. In the next segment, two typical cases are discussed.

3.4.1 For the Breeding of Silkworm Varieties with High Cocoon Yield and Super Silk Quality

Spring is the best season for silkworm rearing since summer is hot and humid and leaves in autumn and winter are of relatively low quality. To breed the new varieties in the spring season with the goal of high yield and excellent silk quality, the breeding condition should be set up for the maximum expression of cocoon and silk characters, such as relatively large breeding scale in the spring, relatively high temperature (26–28°C, RH 80–85%) in the young larval stage, and lower temperature in grown up larval stage (23–24°C, RH 70–75%), feeding good quality mulberry leaves until fully satiated, strict disinfection of rearing environment and careful nurturing. In the spring rearing season, the moth batch of best performance in developmental uniformity, high cocoon yield and silk quality as seed batch is first selected, and then the best individuals within the seed batch are selected so that silk quantity-related genes can be gradually accumulated generation by generation, and finally stabilized. In the summer and autumn rearing seasons, equal weighting is given to the selection of larval healthiness and quantitative characters. Breeding practice can be carried out year-round in the early breeding generations, and then mainly in the spring season and at a smaller scale in the late autumn season.

3.4.2 Breeding of Robust Varieties with High Yield

Strong tolerance to high temperatures and humidity as well as strong resistance to diseases are important in ensuring stable cocoon yield in harsh climates and seasons or tropical areas. In this case, the early breeding stage is usually carried out in the summer and autumn rearing season and fed with low quality leaves, so as to ensure full expression of healthy characters. A general procedure consists of first selecting the moth batch with shorter larval duration and high vitality as seed batch, and then selecting the individual cocoons with high shell weight, shell ratio and silk quality from the seed batch to reproduce next generation. Artificial climate with high temperature and humidity should be adopted if breeding in the spring season and laying more weight on the selection of cocoon yield and quality characters. This is a mature breeding technique for balancing the larval vitality and cocoon quantitative and quality characters at the highest level. The temperature and humidity are preferably set up at 28–30°C and RH 85–90%, usually being higher in the grown-up silkworm or early and middle breeding stage and being moderate in the young larvae and late breeding generation. It is important to note that excessively high temperatures could cause a rapid decline in the cocoon quantitative and quality characters, resulting in high mortality or even extinction of breeding strains.

3.5 Selecting Techniques and Mating Control

3.5.1 Selection Principles in Different Breeding Generations

Correct selection of seed cocoons is the key to breeding the outstanding new varieties. Selection in silkworm breeding is carried out by three levels of individual, moth batch and strain. These three levels of selection should be incorporated properly at different stages of the breeding procedure and should complement each other. The general principles for cross-breeding are given below.

3.5.2 First Filial Generation (F_1)

According to genetic principals, individuals in this generation are highly homozygous with uniform phenotype when two pure parental strains are crossed. Therefore, selection is not carried out in the individual level. Mixed rearing of 0.3 g newly hatched larvae from over 20 laying of each breeding material is adopted. There are many cross combinations of breeding materials to start, for example, Chinese × Chinese, Japanese × Japanese, or other form. By focusing on the observation among different combinations, first select the combinations with better performance as further breeding strain and eliminate the undesirable combinations because F_1 has the combined genetic constitute of two parents. If the F_1 performance is not desirable, there is a very low chance to breed a good new variety through the late generation selection. This is selection at the strain level.

3.5.3 Segregating Generation (F_2–F_3)

Filial generation of high segregation of parental genotypes is the key stage of individual selection. Mixed rearing of 0.5 g newly hatched larvae from egg laying with high number per laying and high hatchability is adopted. Selection starts among the combinations by selecting the combinations with comprehensively good performance as seed strains for further breeding and eliminating more combinations with lower performance. Then, the individuals with intermediate traits of two parents that meet the breeding goals from the seed combinations are selected as the seed cocoons. The number of seed cocoons is 15–20 females and 10–20 males.

3.5.4 Mid-stage (F_4–F_7)

Genetic traits of breeding strain are gradually stabilized by selection after the segregation and recombination of parental traits. Individual moth batch rearing and all three levels of selection are adopted in this stage. Five to ten egg laying with best egg quality, high egg number per laying, and high hatching rate are selected from the seed breeding strains for individual moth batch rearing. The number depends on the importance of breeding strains, and sometimes 15–20 batches are the excellent breeding strains. There are significant genetic differences among moth batches in this stage, some batches show comprehensively good performance, some carry one or two distinctive characters while other characters are average. It is the time to establish multiple lines within the same strain, particularly important for the high silk yielding varieties, for these types of varieties usually lay less eggs. F_1 hybrid egg production can be greatly increased by crossing between different lines in order to produce hybrid parent eggs so as to insure the profit of silkworm egg farms. Selection starts at the strain level, further eliminating undesirable strains, and then focuses on the moth batches of the remaining breeding strains in order to determine the seed batches of each strain after evaluation of the comprehensive performance such as larval developmental uniformity, larval-pupae vitality, and cocoon and silk quantitative and quality characteristics. Generally, 20% of seed moth batches are selected, more seed batches might be selected if the strain shows particularly excellent performance. Next, 50–100 female and male cocoons are respectively selected from each seed moth batch and a serial number is placed on each cocoon, each cocoon and cocoon shell and reel silk of individual cocoon with live pupae is weighed if necessary, then all related data is calculated and arranged from high to low order. Finally, the best cocoons of 20–25 females and 15–20 males are selected based on the overall data of cocoon and silk quantitative and quality test. The male and female seed cocoons are lined up from high to low according to their respective performance, then the male and female are mated in the order from top down correspondingly. The cocoon number in each laying is marked in each laying for next generation rearing.

3.5.5 Late Stage (From F_8 after)

The merit genes tend to be homozygous after strict selection and mating within the moth laying through $F_{3\text{–}7}$, now the difference between the lines of the same breeding strain increased, but the difference between the moth batches of same line decreased, and the genotype of individuals within the moth batch was very close (any difference in phenotype is mostly caused by environmental variation), therefore, selection is made mainly on the strain and moth batch level. Selection on the strain level is conducted by paying attention to difference among different lines of the same strain and maintaining a moderate difference of one or two traits between the lines of a strain so that there is moderate heterosis when crossed for hybrid parent. Selection on moth batch level is done by choosing the moth batches with similar performance within the line or strain. 15–20 moth batches (or more for important strains) are raised for each strain (or line), and 20–40% of seed moth batches are selected. Individual selection within moth batch is done as follows: cocoon weight just above average but not the highest, cocoon shape with uniform size and typical racial feature, and the seed cocoon numbers are the same as the mid-breeding stage. Mating between moth batches is adopted in this stage so as to insure the robustness of the new varieties because mating of individuals within the moth batch is in favor of homozygosity of gene and stabilization of traits, but unfavored of vitality. Combinability test for early predication between the Chinese and Japanese strains is done when the breeding strains are relatively stabilized so that any short comings can be further improved in the late breeding stage.

3.5.6 Selection of Important Traits

The important traits for silkworm breeding are cocoon quality, larval robustness and silk quality. Cocoon quality, such as cocoon weight, shell weight and shell ratio, is easily affected by the environmental and nutritional conditions. It is very important to ensure that all the breeding strains are reared under same conditions and treatments. There is a negative relationship between cocoon quality and larval robustness. In order to manipulate the new strain at the best balance based on the breeding goal, more focus must be given to cocoon quantitative and quality selection for high yielding strains and robustness in the breeding of strains used for adverse environmental conditions. The index of daily increase of cocoon shell weight in 5th instar (Cocoon shell weight (g)/Duration of fifth instar(d)) is very useful for unifying the contradiction of quantitative and robust traits; the index of larvae-pupal survival rate (survival rate of larvae × (1-dead worm cocoon rate)) is most important for the evaluation of robustness; and the index of cocoon and cocoon shell yield of ten thousand larvae (Cocoon yield of the testing block(kg)/real larval number, Cocoon yield of ten thousand larvae × cocoon shell ratio (%)) is used for the hybrid productivity evaluation of the newly bred varieties.

Cocoon reeling ability is an important silk quality trait and easily affected by cocooning environmental conditions. Cocoon reeling ability test should be done before the final determination of seed moths. Silk neatness is another important silk quality trait and hardly affected by environmental conditions. It should be strictly selected based on the test in early and mid-breeding stage, otherwise it will be very hard to be improved upon in the late breeding stage.

Selection of morphological traits, such as larval marker, cocoon shape, cocoon color, cocoon wrinkle, abnormal cocoon shape, etc., is relatively simple and efficient as these traits are generally controlled by major genes or few multiple alleles, therefore, these traits should be strictly selected in the early and mid-breeding stage.

4. Breeding of Silkworm Variety with Special Traits

The traditional silkworm breeding reached its peak in late 1980s and recently faced a bottleneck. With the rapid development of molecular biotechnology, silkworm breeding is shifting from traditional hybridization to molecular design. In the past 10 years, molecular technology applied on silkworm research has accumulated a lot of genetically modified materials in terms of yield, quality and resistance, and opened new applications of silkworm in bioreactors, pest control and other fields (Ma et al. 2017). Silkworm expression chip, genetic variation map, proteomics and metabolomics have obtained, thus might break through the limits of traditional resources in the near future. Given the challenges facing sericulture today, a much higher diversity of silkworm varieties with special traits is required. Therefore, it is worth the words to introduce some recently developed silkworm varieties with special traits.

4.1 Breeding of only Male Silkworm Hybrid with High Silk Yield

It has been noticed that male silkworm larvae are more resistant to adverse environmental conditions, are able to spin better quality silk and have a higher food efficiency (no nutrition utilized for egg forming). Male-only silkworm hybrid for cocoon production has been considered to be cost effective for producing the best quality silk with less food consumption and easier care. Silkworm breeders have been working on this study since last 70s, as mentioned earlier, the sex-linked balanced lethal system was established by the Russian scientist Strunikov (1975). By irradiating γ rays on female pupae two days before eclosion, he isolated the translocation strain of a small piece of Z chromosome attached to W chromosome. He also irradiated γ rays on the male pupae to induce several recessive lethal mutations and then isolated the strains of lethal mutation gene locus matched with the small piece of Z chromosome translocated to W chromosome. The outline of this procedure is showed in Fig. 3.

If the male of this system crosses any normal female, the female hybrids die at egg stage, leaving all hatched larvae as male only. If they mated inside the system, 100% female and 50% male survive, so that the system will be maintained. The creation of this system made the male-only hybrid for cocoon production possible, however, this system has many deficiencies due to the low-quality silk of original material, by-effect of chromosome translocation and gene mutation by radiation treatment. It was not practically used until early this century after improved strains were established by Chinese breeders through crossing and back-crossing with practical strains one after another. The original system was established on the European race, therefore, high quality varieties of Japanese race (they

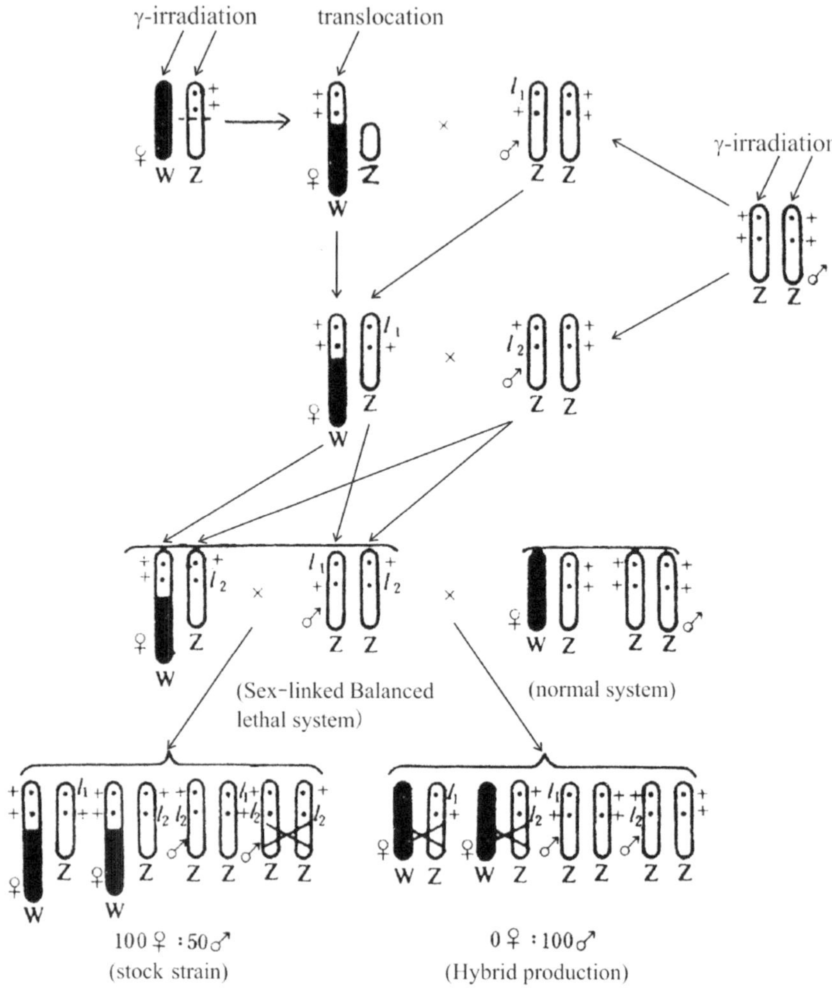

Fig. 3: Production of male-only silkworm hybrid.

have relatively closer blood relationship) were used for its improvement and all the improved strains are near Japanese race. Thus, the vigorous heterosis of hybrids for cocoon production can be achieved when crossed with practical varieties of the Chinese race. A series of improved strains with different characteristics of robustness, cocoon and silk quality, such as Ping 6, Ping 28, Ping 30 and Ping 48, were used for hybrid egg production and tested in farmers' rearing condition. In 2002, taking the high silk yield trait as the main breeding goal, Zhu et al. (2014) used the improved sex-linked balanced lethal silkworm strain Ping 8 as the acceptor parent and common silkworm variety 872 as gene donor for excellent economic traits. By crossing, self-crossing and back-crossing with 872 in turn for 5 generations, and then following the breeding procedure of mid and late stages, a new sex-linked balanced lethal strain Ping 72 with excellent economic characters was bred. Meanwhile, among several breeding strains of the Chinese race, a sex-limited marking silkworm variety with high silk yield (Huajing) was selected for its high combinability with Ping 72 and comprehensively excellent economic traits. The hybrids of Huajing × Ping 72 are all male silkworm. This pair of silkworm varieties are officially authorized and released into the field in 2013 after its hybrid has been tested jointly in five laboratories for two years, followed by farmers' trial rearing for two years in Zhejiang Province. The hybrid has been widely distributed in China for its excellent comprehensive economic traits, increased profits for silkworm farmers and best silk quality for silk industry.

4.2 Establishment of Silkworm Parthenogenetic Clones

Parthenogenesis (also known as asexual reproduction) refers to the phenomenon of female reproductive cells developed directly to individual under the unfertilized situation. The occurrence of silkworm parthenogenesis is rare in natural condition. However, a very effective, simple and low cost artificial treatment technique was developed by Strunnikov (1975) in Russia by soaking unfertilized eggs in hot water at 46°C for 18 min. Parthenogenetic Silkworm clones are all female with identical genotype, they are good materials for the study of the heterosis mechanism, quantitative traits and epigenetic analysis, and also specific value for silkworm breeding.

In the case of male-only silkworm hybrid for cocoon production, there is a dilemma: because the sex-separation cannot be done before the pupation, only half of the parental pupae, females of maternal variety and males of the sex-linked balanced lethal variety, can be used for hybrid egg production, thus, the production cost of hybrid eggs is increased. Therefore, the benefit of breeding the practical maternal variety from parthenogenesis used for hybrid production is a reduction in the required amount of labour. In this way, the sex-separation of female pupae of the maternal variety can be avoided, and only half the amount of silkworm larvae of paternal variety is needed for the same amount of hybrid egg production by taking the advantage of male moth for multiple mating. As a result, the cost of hybrid egg production can be greatly reduced.

The parthenogenic rate was not high enough to meet the requirement of practical production, the hatching rate of parthenogenic eggs was very low, sometimes lower than 10%, and deformed silkworm larvae were frequently observed when the technique was developed. Studies on the genetic mechanism of parthenogenetic characters, construction of practical parthenogenetic clones and its application on silkworm breeding have rapidly progressed since 1996. The silkworm breeding group led by Wang (2010) in China, has been focusing on the technical improvement, selection and breeding of parthenogenic silkworm varieties. They have established a good size of bivoltine silkworm parthenogenetic clones with an averaged parthenogenetic rate of over 90% and hatching rate of over 80%: including 11 Chinese, 11 Japanese and 3 hybrid parthenogenetic clones. Based on these parthenogenesis clones, a practical new variety Female 35 has successfully bred, and matched Ping 28 with high combinability. The new combination "Female 35 × Ping 28" was officially authorized in 2016 in China after joint identification in five labs for two years and then trial rearing of farmers for two years. This combination shows not only the high potential of lowering hybrid egg production cost, but also higher cocoon yield and excellent silk quality. It will gradually become the major varieties for silk production in China.

4.3 Breeding of Silkworm for Silk with Ca-binding Activity

Silk is an ideal biomaterial for clinical uses. Recent researches (Sofia et al. 2001, Yang et al. 2012) found that *B. mori* fibroin with predominantly β-sheet protein has considerable potential for use in strong and tough implantable biomaterials and scaffolds for tissue engineering, and the capability of mineralization of natural silk fibroins mineralizing with the bone mineral hydroxyapatite indicates that silk fibroins may have potential to be used as the bone graft substitute materials (Kong et al. 2004).

Silkworm breeding group led by Professor Chen in Zhejiang University, China, successfully bred a series of transgenic silkworm strains and used them in order to produce silk fibers containing the Ca-binding sequence [(AGSGAG)6ASEYDYD DDSDDDDEWD]2 (referred to as CABP) through the systematic transformation of silkworm to produce silk fibroin-based biomaterial with Ca-binding activity in large scale at low cost. The group established the transgenic strains in 2013 and found that there was a significant increase in the Ca-binding activity and mineralization while having similar or decreased mechanical properties of transgenic silk as compared with the normal silk. Moreover, significant differences were observed in mechanical properties and Ca-binding activity of the silk fibroin from various transgenic strains with different insertion sites. A breeding design was set up based on the goal of creating silkworm hybrid spinning transgenic silk with strong Ca-binding activity and excellent silk mechanical property. Among 17 original transgenic strains, 4 strains, namely Sa2, SCa3, SCa8 and SCa10, were selected as the breeding donor strains based on the overall performance of Ca-binding ability examined *in vitro* and silk mechanical properties. A pair of silkworm varieties, Qiufeng and Baiyu, being the two most popularly used for silk production in China, were selected as the recipient strains. The cross design is shown as Fig. 4. By back-crossing to the recipient strains for eight generations and through properly selecting, strictly monitoring and adequate rearing techniques throughout the whole breeding process, the final transgenic hybrid silkworms show four inserted sequences in different chromosome sites having super high Ca-binding activity, similar silk mechanical properties, larval robustness and silk productivity as compared with the Qiufeng × Baiyu. Further research on the application of this transgenic silk as the bone repair materials is ongoing.

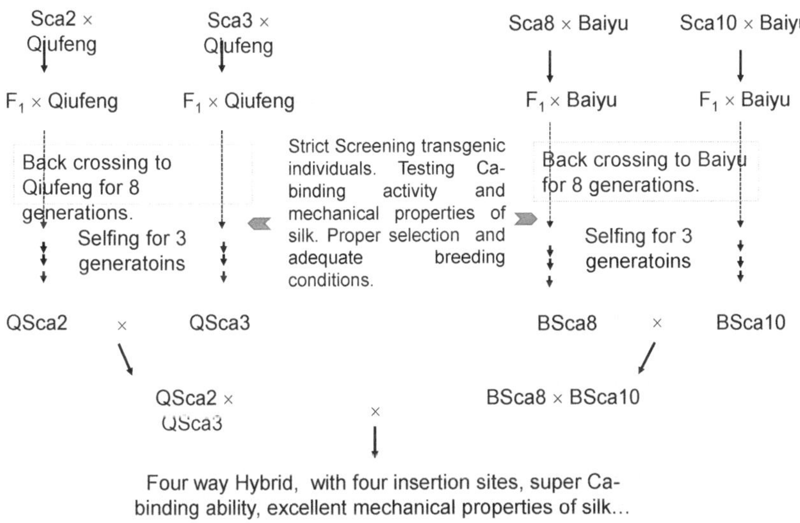

Fig. 4: Illustration of breeding silkworm to spin silk with super high Ca-binding activity and excellent mechanical property.

4.4 Silkworm Variety with Artificial Diet Adaptability

Sericulture has been maintained in the form of traditional labor-intensive agriculture and is vulnerable to weather conditions, whereas most other agriculture sections have been modernized and are well-equipped to deal with adverse conditions. With the rapid industrialization of agriculture, the comparative efficiency of sericulture in Japan has dropped significantly since the 70s. To stabilize sericulture, the best solution is to use artificial feed for the establishment of an year-round silkworm rearing system with labor-saving technology, so as to increase the production efficiency and stability. A series of researches have been carried out and a well-equipped system for silkworm rearing with artificial diet was established in the 1980's (Cui et al. 2016). However, it did not rescue Japan's sericulture because the cost of cocoon production still was not competitive with China where the labor was so cheap at that time. Today, China is also facing the same problem, though it is still the major sericulture country with the capability to produce near 80% of the world's raw silk. Establishing a factory-farmed system for sericulture is no doubt a preferential solution, and breeding of silkworm varieties with artificial diet adaptability should be ahead of other related programs. Moreover, with the rapid progress in the research of silkworm as a biofactory, it is predicted that the need for experimental silkworm with a good artificial diet adaptability will increase rapidly.

Genetic analysis on artificial diet adaptability of over 600 silkworm strains has shown that this trait is controlled by multiple genes in a diversified heredity pattern, and that there are significant differences among silkworm strains. In breeding of new silkworm varieties with good artificial diet adaptability, the parent strains for breeding materials must carry some related genes. Based on the breeding experiences, the most efficient way of selecting artificial diet adaptable silkworm varieties for practical cocoon production is to test the adaptability of many combinations of silkworm varieties popularly used for hybrid production and select the combinations that are better than the average performance. Both parents of each combination are used as the breeding materials in the breeding pairs correspondingly. Directive selection and breeding by artificial diet feeding should start at the young larval feeding in the early breeding stage, and gradually to the whole larval feeding, thus, the adaptability will be significantly improved generation by generation as the related genes accumulate. With this breeding strategy, the newly bred combination will not only be completely adaptable to artificial diet, but also have high combining ability and comprehensively good economic traits.

5. Concluding Remark

Silkworm breeding is a discipline studying the silkworm genetic structure, the theory, method and technology of breeding new silkworm varieties for the need of continuous changing silk consumers traditionally, and now also of the growing requirement as a biofactory. Great progress has been made in theoretical research not only in the aspects introduced above but also in those not yet touched upon in this chapter. Technical knowledge on silkworm breeding is by no means complete but this chapter offers a brief outline on the current and near future developing trends.

References

Cui, W. Z., S. X. Zhang, Q. X. Liu, Y. W. Wang, H. L. Wang, X. L. Liu and Z. M. Mu. 2016. An overview on research and commercialization of artificial diet for silkworm (*Bombyx mori*) in China. Science of Sericulture. 42(1): 0003–0015.

Jiang, L., P. Zhao and Q. Y. Xia. 2014. Research progress and prospect of silkworm molecular breeding for disease resistance. Science of Sericulture. 40(4): 571–575.

Kong, X., F. Cui, X. Wang, M. Zhang and W. Zhang. 2004. Silk fibroin regulated mineralization of hydroxyapatite nanocrystals. Journal of Crystal Growth. 270: 197–202.

Ma, S. Y. and Q. Y. Xia. 2017. Genetic breeding of silkworms: from traditional hybridization to molecular design. Hereditas. 39(11): 1025–1032.

Ma, L., H. F. Xu, J. Q. Zhu, S. Y. Ma, Y. Liu, R. J. Jiang, Q. Y. Xia and S. Li. 2011. Ras1CA overexpression in the posterior silk gland improves silk yield. Cell Res. 21(6): 934–943.

Meng, Z. Q. 2014. Zhejiang Silkworm Germplasm Resources. Hangzhou, China.

Peng, P. Zhao and Q. Y. Xia. 2014. Advanced silk material spun by a transgenic silkworm promotes cell proliferation for biomedical application. 10: 4947–4955.

Sofia, S., M. B. McCarthy, G. Gronowicz and D. L. Kaplan. 2001. Functionalized silk-based biomaterials for bone formation. J. Biomed. Mater. Res. 54: 139–148.

Strunnikov, V. A. 1975. Sex control in silkworms. Nature. 255(5504): 111–113.

Strunnikov, V. A. 1995. Control over Reproduction Sex and Heterosis of the Silkworm. Harwood Academic Publishers, Moscow, Russia.

Tajima, Y. 1964. The Genetics of the Silkworm. London, UK.

Teule, F., Y. G. Miao, B. H. Sohn, Y. S. Kim, J. J. Hull, M. J. Jr Fraser, R. V. Lewis and D. L. Jarvis. 2012. Silkworms transformed with chimeric silkworm/spider silk genes spin composite silk fibers with improved mechanical properties. Proc. Natl. Acad. Sci. USA. 109(3): 923–928.

Wang, F., H. F. Xu, Y. C. Wang, R. Y. Wang, L. Yuan, H. Ding, C. N. Song, S. Y. Ma, Z. X. Wang, S. H., Y. Y. Zhang, M. Y. Yang, L. P. Ye, L. Gong, Q. J. Qian, Y. J. Shuai, Z. Y. You, Y. Y. Chen and B. X. Zhong. 2016. Characterization of transgenic silkworm yielded biomaterials with calcium-binding activity. PLoS One. 11(7): e0159111.

Wang, Y. Q., X. R. Zhu, K. R. He, Y. T. Yao, J. R. Cao, J. Q. Zhou, Y. F. Huang, X. J. Liu, X. L. He and Z. Q. Meng. 2010. The breeding and application of new male silkworm varieties by using female silkworm parthenogenetic clones. 36(2): 0268–0273.

Xiang, Z. H. 1994. Genetics and Breeding of Silkworm *Bombyx mori*. Beijing, China.

Yang, M., Y. K. Shuai, W. He, S. J. Min and L. J. Zhu. 2012. Preparation of porous scaffolds from silk fibroin extracted from the silk gland of *Bombyx mori* (*B. mori*). Int. J. Mol. Sci. 13: 7762–7775.

Zhang, Y. H., A. Y. XU, M. W. Li, C. X. Hou and P. J. Sun. 2004. Construction of database for silkworm germplasm resource. Science of Sericulture 30(3): 296–299.

Zhu, X. R., K. R. He, X. J. Liu, Y. Q. Wang, S. Chen and Z. Q. Meng. 2014. Breeding of new male silkworm variety Huajing × Ping 72 with high silk yield. Science of Sericulture. 40(2): 0248–0253.

CHAPTER 3

Focus on Physiological Vital Proteins from Silkworm *Bombyx mori*

Raman Chandrasekar[1,*] and *S. K. M. Habeeb*[2,3]

1. Introduction

The silkworm is a useful model for biochemical, genetic and physiological studies, as in insect model systems it occupies the position next to the fruit fly, *Drosophila melanogaster*, but is the model system of choice among the Lepidoptera order. It was the first insect within the order Lepidoptera for which a draft genome sequence was completed. This was accomplished in 2004 by the International Silkworm Genome Consortium (http://silkworm.genomics.org.cn/). It is an important model species for investigations of development and metamorphosis, reproduction, pheromone & olfaction, insect plant-interaction, immunity, host-pathogen interaction, enzymes, and many other aspects of RNAi approaches. Entomo-informatics, a term coined by Habeeb (Habeeb and Chandarsekar 2014), is a new trend for studying an insect's global effects on genome and drug discovery, comparing molecular evolution of proteins/enzymes and comparing structure analysis of physiological important proteins. Hence, it has been the purpose of this review to highlight this change of emphasis

[1] Department of Biochemistry and Molecular Biophysics, Kansas State University, Manhattan 66502, KS, U.S.
[2] Department of Genetic Engineering, School of Bioengineering, SRM University, TN, India.
[3] Entomoinformatics Lab, Department of Genetic Engineering, School of Bioengineering, SRM University, TN, India.
E-mail: biochandrus@yahoo.com
* Corresponding author: chandbr@ksu.edu

and raise awareness of the silkworm physiological vital proteins (i.e., Storage Proteins, Vittellogenin, Microvitalin, Apolipophorin-III, Ferritin, Lipocalin) and their molecular evolution. In addition, the manner in which structural studies of these proteins and an increasing understanding of their molecular recognition properties gives insights into their biological functions is also discussed.

2. Physiological Vital Proteins

2.1 Storage Proteins

The most abundant proteins in the larvae of holometabolous insects are Arylphorins. These proteins are members of the hexamerin family, which primarily act as storage proteins used as a source of energy during metamorphosis. Arylphorin storage proteins are of two major classes in lepidopterans, SP1 (sex specific proteins or methionine rich > 4%) and SP2 (common for both sexes and enriched in aromatic amino acids > 15%). Silkworm pupae do not feed during metamorphosis, and depend on nutrients that have already been accumulated in the fat body during the larval period. The fat body is a relatively large organ, distributed throughout the insect body, preferentially located underneath the integument and surrounding the gut and reproductive organ. The silkworm fat body is comprised of complex multifunctional tissues that participate in multiple biochemical and physiological functions. It is a major site for synthesis and storage of carbohydrates, lipids, proteins and nitrogenous components (Arrese and Soulages 2010, Chandrasekar et al. 2007, Keeley 1985, 2009, Ryu et al. 2009). Due to the existence of fat body heterogeneity and from previous correlation of their structure and functions in *Bombyx mori* (Chandrasekar et al. 2008, Tojo et al. 1980), these storage proteins are synthesized abundantly during "active feeding larval period" in the peripheral fat body (PF) tissue and subsequently release proteins into the haemolymph (Fig. 1). Our previous electron micrograph studies during the spinning stage provide direct evidence for the sequestering ability and huge accumulation of SP hexameric-nano granules crystalline, which was observed in previsceral fat body tissues (PVF) of silkworm (Fig. 2A–C). The PF is mainly responsible for larval protein biosynthesis and SP-1, whereas the PVF (Fig. 2B, C) is a specialized storage organ (Chandrasekar et al. 2012, Sumithra et al. 2009).

Pietrzyk and co-workers' (2013) first X-ray crystallographic study made a significant contribution towards revealing the SP2–SP3 hetero-hexamer molecular structure in the silkworm. Insect storage proteins are evolutionary related to the arthropod haemocynin family. Although these proteins have similar amino acid sequences and structures, they display different functions. *B. mori* 3D model was rebuilt by using (*A. pernyi* PDB ID: 3GWJ and PDB: 4L37) PyMOL program (Fig. 2D–E). The model of hexamer (SP2, SP3) can be divided into three distinct sub-domains: the haemocynin_N domain (residues 17–156), haemocynin_M domain (residues 157–441) and the haemocynin_C domain (residues 442–675). Although *B. mori* arylphorin (SP2) shares 67% sequence similarity with *A. pernyi* arylphorin, *B. mori* SP2 contains only one glycosylation site (N211), corresponding to *A. pernyi* arylphorin residue N196, with the chemical formula of Glc1Man9GlcNAc2. The glycosylation site and the chemical formula of the oligosaccharide chain in *B. mori* SP2 are identical to the oligosaccharide chain buried inside the hexamer in *A. pernyi* arylphorin (Kim et al. 2003). However, hemocyanins contain di-copper centers, where two copper atoms are coordinated by six histidine residues involved in oxygen transport throughout arthropod bodies (van Holde and Miller 1995). By contrast, the corresponding residues of *B. mori* SP2 and SP3 are primarily made up of tyrosine residues, which are not capable of copper coordination.

Why does the fat body first secrete storage protein into the haemolymph before sequestering them, apparently unchanged? It has been proposed that a functional shift of the fat body takes place at the end of the last larval stage, from biosynthetic organ to storage organs. A physiological link between amino acids of storage proteins (SP) or smart nano-particles and those amino acids used in vitellogenesis may be common in insects

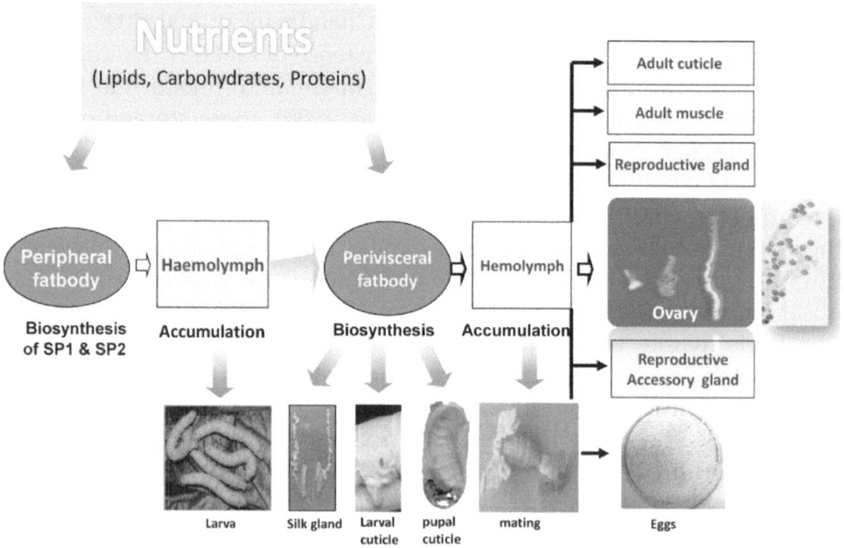

Fig. 1: Schematic diagram showing storage protein bio-synthesis, and its distribution in the silkworm, *Bombyx mori*.

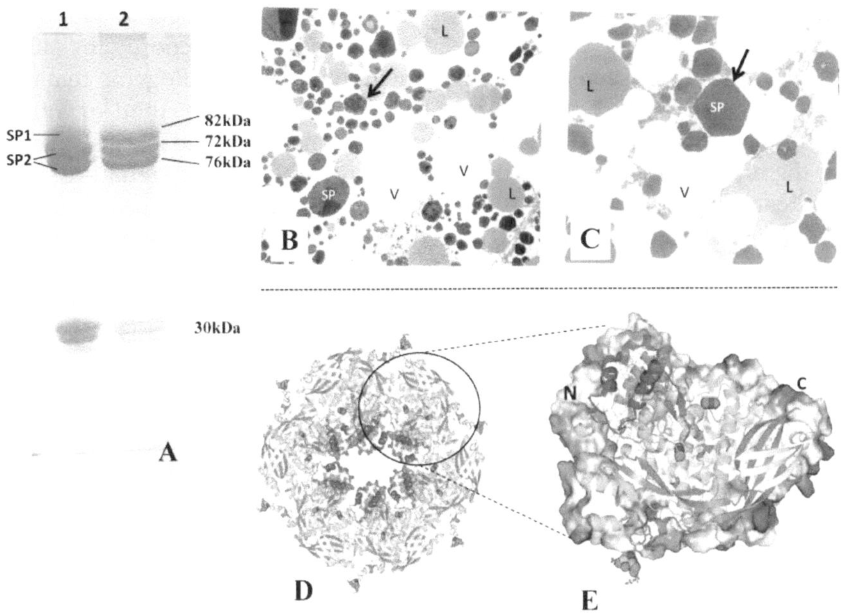

Fig. 2: (A) Accumulation of storage proteins in the haemolymph from the final instar stage (Lanes 1 and 2, Day 4 and Day 5) by one-dimensional electrophoresis indicating bands (SP1 & SP2) Lane 1, 2nd Day. (B, C) Electron micrograph illustrating crystalline hexameric storage protein granules (arrow heads) in the pervisceral fat body tissue (S0) of silkworm. S0-spinning day 0; V-vacuole; L-lipid droplet; SP-storage proteins; scale bar –0.5 mm. Gel and EM photograph adapted from Chandrasekar. (D) The 3D model structure of Storage protein of *Bombyx mori* by comparative modeling using PDB: 3GWJ and PDB: 4L37 template to design with the help of PyMOL program. (E) Superimposition (electrostatic surface with helix) of the 3D structure of SP and three distinct regions of N domain (residues 17–156 in dark blue color), M domain (residues 157–441 in cyan color) and the C domain (residues 442–675 in yellow and magenta color β-sheets).

(Figs. 1 and 2). The dynamics between sequestration and release of amino acids from storage protein in the transition from pupa to adult stage may be an important feature, enabling many of their diverse strategies of reproduction (Chandrasekar et al. 2007–2009, 2012).

2.2 *Vitellogenin Proteins*

The developing embryo of oviparous animals and insects draws practically all of its requisite nutrients from a cache of proteins, lipids and carbohydrates stored within the egg as yolk (Hagedorn and Kunkel 1979, Sappington and Raikhel 1998). Yolk is the major internal food supply, upon which most embryos depend. *Bombyx mori* Vg are encoded by mRNAs of 6–7 kb and translated as primary products of ~ 203 kDa, which are cleaved into sub-units (apoprotein) ranging from 90 to 110 kDa. After extensive co- and post-translational modifications, vitellogenin sub-units form high molecular weight oligomeric phospholipoglycoproteins (400–550 kDa) that are secreted into the haemolymph of vitellogenic females (Fig. 3A). The BmVg sequence has homology with that of other lepidopteran species. The evolutionary link between these insects species is, therefore, of considerable interest to some conservative signature such as ligand-binding domain (Cytoplasmic tail, epidermal growth factor precursor domain (YWXD & Cys-rich repeat) and O-linked sugar domain (Tufail and Takeda 2008). As the phospholipid affinity of Vg-domain is now reported for phylogenetically distant proteins as human apoB and BmVg, it is tempting to interpolate this property to another LDL receptor family (Lin et al. 2013, Meenakshi et al. 2007, Yano et al. 1994). However, the relatively well-conserved Vg-domain has great variation among species at two sites (Fig. 3B).

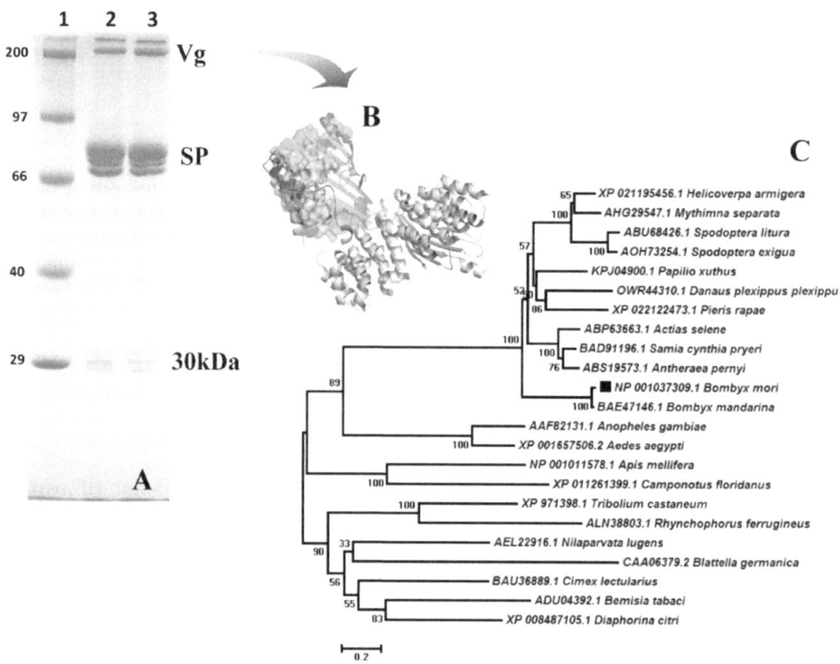

Fig. 3: (A) Protein profile of silkworm haemolymph of the late-pupal stage (Lane 1: protein marker; Lane 2: day 9 of pupa; Lane 3: day 10 of pupa). (B) The Three-dimensional model of the BmVg. A cartoon representation is depicted, generated on the basis of sequence similarity with the PDB: 1LSH as a template. The colouring of the chain graduates from dark red (N-terminus), conserved residues (Cys-rich repeat and O-linked sugar domain) in sky blue and green, cytoplasmic and C-terminus region in cement color. (C) The molecular phylogenetic tree constructed based on the sequence of vitellogenin (Vg) from 24 insects (different insect orders) by using MEGA 7 program.

In order to examine the evolutionary relationship of insect Vg with other insect groups, a phylogenetic analysis of 24 Vg families available in the literature and Gene Bank database was carried out. Using a neighbor-joining tree construction program with MEGA version 7, based on distances of amino acid sequences, a tree was obtained, as shown in Fig. 3C. The result revealed that lepidopteran Vg groups are more closely related to other insect groups, and a unique pattern of amino acids sequence of C-terminal domains was observed. The strict regulation of its utilization is essential for providing nutrients to the developing tissues at the right time, and to ensure survival of the embryo until it becomes a free-living organism (Fagotto 1995). Major yolk proteins may be divided into four categories (Raikhel and Dhadialla 1992): (i) Vitellin (Vtn), found in most insect species, is derived from vitellogenin (Vg), the precursor from which it is synthesized in the fat body cells (PF and PVF) in a female-specific manner. They are secreted into the haemolymph and sequestered into developing oocytes. (ii) Yolk proteins (YPs) of the higher Diptera order, such as *Drosophila melanogaster*, are the products of genes expressed both in the female fat body cells and ovarian follicle cells, incorporated into developing oocytes. (iii) Proteins such as the egg-specific proteins (ESP) of *Bombyx mori*, are produced in the ovarian follicle cells and incorporated into developing oocytes. (iv) A group of low molecular weight lipophorin proteins (~ 30 kDa), which has been designated as micro-vitellogenin, seen in *B. mori*, is produced in the fat body cells and secreted into the haemolymph, but in a sex non-specific manner. Numerous studies of electron-microscopy and immunocytochemistry clearly demonstrate the internalization of yolk proteins (Vg) via coated pits and coated vesicles eventually delivered into late endosomes or transitional yolk bodies (TYBs). They are subsequently delivered to a storage compartment, the mature yolk body (MYB), in which the storage of the vitellins and other yolk proteins occur (Meenakshi et al. 2007, Raikhel and Dhadialla 1992, Snigirevskaya et al. 1997).

The insect oocyte provides an excellent system for studying receptor-mediated endocytosis because of the high intensity of protein uptake, probably occurring through four processes: (a) endogenous synthesis of elements carried out by the oocytes; (b) synthesis of elements by follicular cells, followed by transport into the oocytes by endocytosis; (c) selective uptake of extra-ovarian material by follicular cells, followed by transport to the oocytes; and (d) transport of extra-ovarian material into the oocytes through intercellular spaces of the follicular epithelium (Caperucci and Mathias 2007, Han et al. 2017, Meenakshi et al. 2007, Raikhel and Dhadialla 1992, Telfer and Takeda 2008). Further insilco analysis of the 3D model of BmVgR (PDB: 1LSH was used as a template) indicated that the N-terminal region was recognized as the binding site of BmVg and assumed a conserved β-sheet structure. The rest of the protein would serve to enhance the affinity for the BmVg receptor. Understanding the mechanism of SP/SPr or Vg/VgR interaction in invertebrates might provide new insight into animal reproduction. The successful reproduction in insects encompasses the synthesis of yolk protein (Vg) and its deposition into the ovary (Fig. 1).

2.3 Lipophorin

The silkworm *Bombyx mori* is exploited both as a powerful biological model organism and also as a tool for converting leaf proteins into silk. In previous work, a group of structure-related proteins with a molecular mass of approximately 30 kDa was shown to accumulate in a stage-dependent fashion in the haemolymph of the silkworm, *Bombyx mori* (Britto et al. 2012, Sumithra et al. 2009). These proteins, referred to as 30K proteins, belong to the lipoprotein family and several authors hypothesize that they are involved in multiple functions, such as the transport of sterols and hormones, immunity, reproduction, programmed cell death, etc. (Fig. 4A). However, their physiological functions still remain largely unresolved.

Our previous study found that 30 kDa protein synthesis during the larval-pupal as well as pupal-adult metamorphosis occurred in the peripheral fat body tissue (PF) during larval stage, followed by subsequent transport into specific tissues (Chandrasekar et al. 2008, Sumithra et al. 2009). The SDS-PAGE revealed that the 30 kDa-like protein had three sub-units, 26 kDa, 29 kDa and 31 kDa, and was predominately expressed in the late larval (fifth instar larva), pupal and decline adult stage. Further purification of each sub-unit was confirmed by using Vg-specific antibody (Fig. 4B). Based on their sequence identity homology with lipoprotein family (Chandrasekar et al. 2011), it has a signal peptide in the N-terminal region (1–20) with a degrading site of Ser and repeated Ser-Ala region. It is speculated that the 30 k protease A hydrolyse the 30 K proteins between Ser and Ala, producing different products involved in different functions. The recent draft of the silkworm genomic sequence revealed the presence of 46 genes for 30 kDa lipophorins and shows the three sub-families of 30 kDa protein, based on the binding properties such as typical 30 kDa LPs (BmLP1-24), serine/threonine-rich LPs (BmLP25-36) and ENF peptide-binding proteins (BmLP37-46) (Sun et al. 2007, Zheng et al. 2012). Recent crystal structures of 30 kDa (BmLP-3: PDB ID: 4IY9; BmLP-7: PDB ID: 4EFP) provide the insightful information required to understand this important event at the molecular level (Pietrzyk et al. 2013, 2014, Yang et al. 2011). It has a signal peptide and two distinct domains for all α-helix N-terminal domain, followed by all-β trefoil C-terminal domain (Fig. 4C). Furthermore, the phylogenetic tree (Fig. 4D) was constructed as

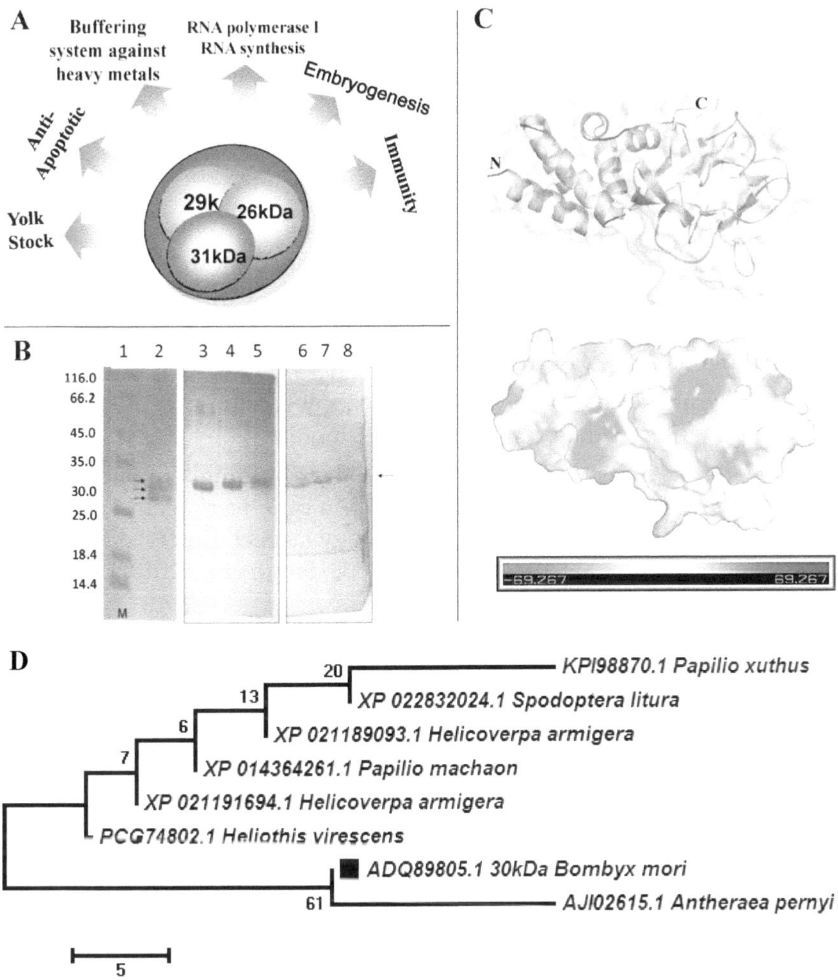

Fig. 4: (A) Schematic diagram shown the various functions of microvitellin (30 kDa) in the silkworm, *Bombyx mori*. (B) Purified 30 kDa proteins from haemolymph by SDS-PAGE (Lane 1: protein marker, Lane 2: purified 30 kDa; Lane: 3, 4, and 5 purified individual subunit) and stained with Coomassie Brilliant Blue and confirmed by immunoblot analysis (Lanes 6, 7, 8) by using specific Vg antibody. (C) Three Dimensional Molecular model of *Bombyx mori* 30 kDa protein. The 3D-predict protein structure could be modeled at 60% sequence identity using the multiple-template (BmLP-3: PDB: 4IY9; BmLP-7, PDB: 4EFP). The topology of the visible colored secondary structure shown as the molecular surface model (top): N-terminal signal peptide followed with helix region (sky blue); β-sheets (yellow) and electrostatic surface model (bottom) generated using PyMOL program. (D) Phylogenetic tree was constructed within lepidopteron group insects by using MEGA 7 program. The neighbor-joining method with a Poisson correction model is based on the alignment of the micro-vitellogenin from related sequence found in Accession number from NCBI Data base.

two branches using the 30 kDa amino acid sequences of *B. mori* and other lepidopteran (*Antheraea pernyi, Heliothis virescens, Helicoverpa armigera, Papilio Machaon, Spodoptera litura, Papilio Xuthus). B. mori* 30 kDa and their homologs in other species may have originated from a common ancestor, and then increased in number through gene duplication.

2.3a Functions

(a) *B. mori* 30 K proteins have sequence similarity with Vg and cross-react with Vg specific antibody. It suggests that major source yolk proteins, as a major source of nutrition for growing embryos, i.e., micro-vitellogenin.
(b) The overall sequence analysis of 30 kDa shows serine-rich glycoproteins and suggests a broad range of binding targets.
(c) Due to the presence of a putative sugar-binding domain, Bm 29 kDa plays a crucial role in the protection of *B. mori* against invading pathogens and acts as an immunology aspect (Britto et al. 2012, Chandrasekar et al. 2011, Kim et al. 2002, Singh et al. 2013, Ujita et al. 2005).
(d) During insect metamorphosis, the larval tissues are degraded. This decomposition involves programmed cell death (PCD) triggered by ecdysteroids (Chandrasekar et al. 2009, Hou et al. 2010).
(e) Potential functions of 30 kDa are prolonging life and inhibiting programmed cell death in *B. mori* (Park et al. 2003, Sumithra et al. 2009).
(f) In general, 30 kDa proteins are multifunctional and permanently required by insects. Therefore, they exist at various regions throughout the life cycle of the silkworm.

2.4 Ferritin

Iron is essential for almost all organisms, fulfilling a variety of biological functions, because of its important role in growth, differentiation, oxygen transport, storage, energy production, cell proliferation, a range of catalytic processes, RNA/DNA synthesis and the cell cycle (Fig. 5A). Iron can promote the formation of toxin-free radicals in the presence of oxygen and water, and is safely accumulated in living tissues in the special protein, ferritin (Nichol and Locke 1996, Nichol et al. 2002).

The ferritin molecule consists of an approximately spherical protein of 24 sub-units of two types, H (21 kDa) and L (19 kDa), which assemble to form a hollow protein shell, capable of encapsulating up to 4500 Fe atoms in the form of a hydrous ferric oxide-phosphate complex (Cozzi et al. 2000, Hamburger et al. 2005). In vertebrates, ferritin exists in two

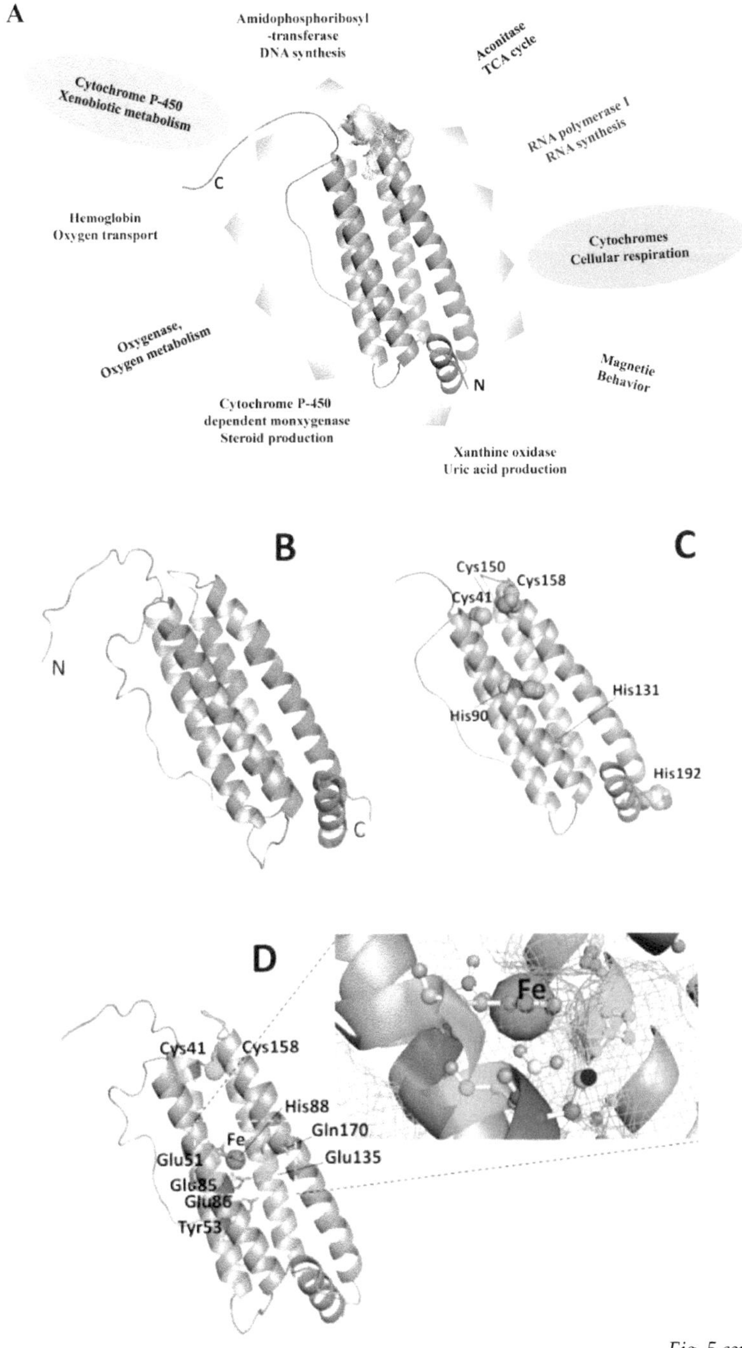
A
Amidophosphoribosyl
-transferase
DNA synthesis
Aconitase
TCA cycle
Cytochrome P-450
Xenobiotic metabolism
RNA polymerase I
RNA synthesis
C
Hemoglobin
Oxygen transport
Cytochromes
Cellular respiration
Oxygenase,
Oxygen metabolism
Magnetic
Behavior
N
Cytochrome P-450
dependent monxygenase
Steroid production
Xanthine oxidase
Uric acid production
B
N
C
C
Cys150
Cys158
Cys41
His131
His90
His192
D
Fe
Cys41
Cys158
His88
Gln170
Fe
Glu51
Glu135
Glu85
Glu86
Tyr53

Fig. 5 contd....

...Fig. 5 contd.

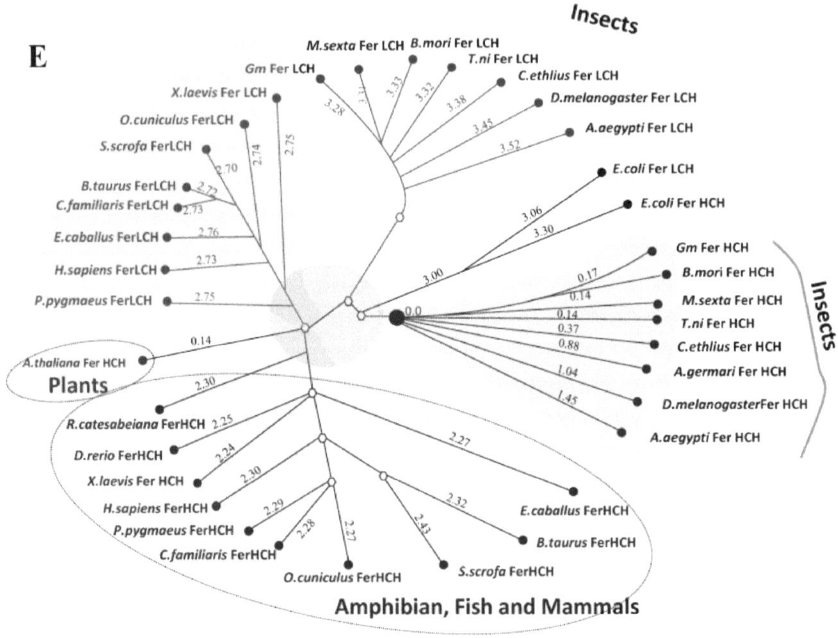

Fig. 5: (A) Schematic diagram indicating that the silkworm ferritin (iron) is essential for a variety of biological functions. Homology modeling analysis of Bm ferritin showed five α-helices (A-E) with a loop connecting helices B and C. (B) Bm Ferritin predicted three-dimensional ribbon model obtained using two templates from Human ferritin HCH (PDB 2FHA) and *T. ni* ferritin HCH (PDB 1Z60) and conserved residues shown in red color. (C) Metal binding key residues highlighted in sphere shape Cys41, His90, His192, His131, Cys150, Cys158. (D) Residues at the ferroxidase (Fe^{3+}) center are highlighted in red color and following Glu51, Glu85, Glu86, Tyr53, His88, Gln170, Glu135 residues showing coordinating binding ferroxidase (Fe^{3+}) molecules and close up view of the binding pocket. Cysteine residues (Cys41, Cys158) involved in disulfide bonds are marked green and blue. (E) Phylogenetic analysis of *B. mori* Ferritin LHC and HCH with other organisms. Amino acid sequences for mature ferritin polypeptides were aligned using Clustalw, and an unrooted phylogenetic tree was constructed using the Minimum Evolution Method from Phylip package (MEGA 7 software). Accession numbers for H and L chain ferritins protein sequences obtained from NCBI (Table 1). Lengths of branches along the axis are proportional to evolutionarily distances calculated from the pairwise amino acid identity matrix with high bootsrap value of 1000. All positions containing gaps and missing data were eliminated.

forms, the circulatory (serum) and cytosolic (intracellular) form. The latter exists in greater abundance. Our current knowledge and understanding of the ferritin structure from studies in vertebrates as well as few insects suggest that ferritin has two types of sub-unit called H and L polypeptide chains encoded by distinct genes (Hamburger et al. 2005, Proudhon et al. 1996). Ferritin genes have been described in several insects species (Yu et al. 2017), followed by amphibians, fish, mammals and birds. Thus, gene

Table 1: Ferritin proteins used for Amino acid sequence alignments.

Species (Strain)	Sub-unit Name	Mature Peptide (kDa)	Accession #	Ferroxidase Resides	Signal Peptide
Bombyx mori	HCH	24.8	XP_004927855	+	+
	LCH	23.3	NP_001037580	–	+
Galleria mellonella	HCH	23.7	AAG41120	+	+
	LCH	26.5	AAL47694	–	+
Manduca sexta	HCH	23.9	AAK39636	+	+
	LCH	26.5	AAF44717	–	+
Tricho	HCH	21.8	AAX94728	+	+
	LCH	23	AAX94729	–	+
Calpodes ethilus	HCH	24.3	AAD50238	+	+
	LCH	26.1	AAD50240	–	+
Apriona germari	HCH	24.2	AAM44043	+	+
Drosophilla mellogaster	HCH	23.1	AAB70121	+	+
	LCH	25.3	AAF07879	–	+
Aedes agypti	HCH	23.8	AAL85607	+	+
	LCH	25	AAO41698	–	+
Pongo pygmaeus	HCH	21	CAH91913	+	+
	LCH	20	CAH93128	–	+
Canis familiaris	HCH	21.3	BAD96176	+	+
	LCH	20.1	BAD96179	–	+
Oryctotagus cuniculus	HCH	19.2	AAA31247	+	+
	LCH	20.1	P09451	–	+
Equus caballus	HCH	21.3	AAM51631	+	+
	LCH	20	BAD96182	–	+
Bos taurus	HCH	21.1	AAW82097	+	+
	LCH	20	BAA24819	–	+
Sus scrofa	HCH	21	BAA03666	+	+
	LCH	18.3	AAG16228	–	+
Xenopus xevis	HCH	20.9	AAQ10928	+	+
	LCH	20.5	AAQ10929	–	+
Homo spiens	HCH	21.2	AAH11359	+	+
	LCH	20	CAG32996	–	+
Rana catesbeiana	HCH	20.5	1MFR	+	+
Escherichia coli	HCH	19.4	1EUM	+	+
	LCH	19.3	1SQ3		+
Danio rerio	HCH	20.7	AAG37837	+	+
Arabidopsis thaliana	HCH	20.9	BAD43342	+	+
Gallus gallus	HCH		CAA75004	+	+
	LCH		NP_989714	–	+

sequencing in the Gene Bank has increased (NCBI data bank). However, molecular structural information on insect ferritins is still limited to a few insect species. Accurate identification of a specific group of proteins by their amino acid sequence is an important goal in genome research (Atchley et al. 2005). Based on available crystal structure studies, the iron binding domain is discussed in terms of its functional consequences (Hamburger et al. 2005). Hence, silkworm studies focus on a computational analysis of the Bm ferritin structure. Molecular evolutionary aspects are discussed in depth. A brief survey of the insect ferritin sub-unit mRNA suggests that the IRE is likely involved in translational control of ferritin synthesis in a manner similar to that found in vertebrates. Comparison of the *Bm* ferritin sequence homology in insects to humans, animals and plants shows that it may exist in order to maintain a particular intern/exon pattern within ferritin genes (Fig. 5B, C). The iron-responsive element (IRE) sequence with a predicted stem-loop (CAGUGN) structure is present in the 5′-UTR of ferritin H Chain mRNA and was highly conserved among all species (Chandrasekar et al. 2007, Hamburger et al. 2005, Hong et al. 2014, Nichol et al. 2002, Yu et al. 2017, Zhang et al. 2001). Other common characteristics of insect Ferritin HCH and LCH is the conservation of an integrin attachment site consisting of Glu51 (E), Glu85 (E), Glu86 (E), Tyr53 (Y), His88 (H), Glu135 (E) and Gln170 (Q) in most insects' ferritin and also in many other species (Fig. 5D). The Bm Ferritin also includes inter-subunit disulfide bonds connecting the H chain residue (Cys41-N-terminal helix and Cys158-C-terminal helix) and it makes contact with an L chain. Generally, the conservation of key metal binding residues (Fig. 5C), cysteine (Cys), involved in inter-subunit disulfide bonds, (BmHCH: His35, Cys38, His57, His73, His89, Cys107, His152, Cys153, His174, His187 and His190; BmLCH: Cys23, Cys41, His90, His131, Cys150, His158, His192) suggested that Bm ferritin is common in lepidopterans and other insect groups. Recent preliminary data (quantitative polymerase chain reaction) also strongly suggest that Bm Ferritin is predominately expressed in the fat body tissues, midgut, malpighian tubule, ovarian tissues and egg tissues for ROS and oxidative stress (Chandrasekar unpublished data). Also, it is involved in the role of ferritin in the immune response, anti-oxidation and the regulation of iron homeostasis (Pham et al. 2010, Yu et al. 2017).

We traced the evolutionary histories of the Bm HCH/LCH ferritin protein-coding genes using orthologous delineation and observed a close relationship with mammals, amphibians, fish, birds, plants and *Escherichia coli* (Fig. 5E). The presence of heteromeric ferritin in animals suggested that there is an adaptive value in altering the ferritin function by mixing two types of sub-unit. The existence of ferritin genes in humans with structure similar to that in lower animals, and plants or *vice versa* has yet to be identified. The possibility of changes in the severity of selective constraints during evolution with weaker constraints could possibly have allowed variation after the plant and animal lineages diverged, followed by stronger constraints which froze the ferritin gene organization into the contemporary forms. Silkworm ferritin has a dynamic role in various physiological and biological functions, including detoxification, developmental regulation and immunity (signaling pathway).

2.5 ApoLipophorin-III

Apolipophorin (BmApoLp-III) is a haemolymph protein whose function is to facilitate lipid transport in an aqueous medium. ApoLp-III is synthesized in a variety of tissues apart from the fat body cells, including hemocytes, midgut epithelial cells, ovaries and testes. The differential tissue-specific expression levels of ApoLp-III in insects suggest their additional functions in immune responses against various pathogens. ApoLp-III has been abundantly characterized from insect haemolymph and has been shown to mediate insect immune responses in various species such as *Galleria mellonella* (wax moth), *Hyphantria cunea* (fall webworm), *Heliothis virescens* (tobacco budworm), *Locusta migratoria* (locust), *Anopheles gambiae* (mosquito), *Thitarodes pui* (ghost moth), and more recently in beetles *Tribolium castaneum* and *Tenebrio molitor*.

ApoLp-III has been extensively investigated with regards to the molecular details of the binding interactions with cell wall components such as lipopolysaccharide (LPS), peptidoglycan (PGN) and β-1,3 glucan (Chandrasekar et al. 2005, Leon et al. 2006, Patnaik et al. 2014, Weers and Ryan 2006). Interestingly, silkworm ApoLp-III studies on insect apoLp-III have focused on sequence-structure relationships and their evolutionary positions within the insect orders. The identities of apoLp-III at the primary sequence level were found to be less striking within the orthologous groups (Fig. 6A), although they fold to similar tertiary structures (Chandrasekar et al. 2005, Patnaik et al. 2014, Wang et al. 2002). The structural homology of BmApoLp-III was identified using a known 3D structure (PDB-1EQ1 template of *M. sexta* structure) by PSI-BLAST. Structural analysis reveals that BmApoLp-III molecules consist of five long, amphipathic alpha helical bundles with short loops connecting the helix 1 (N-terminal helix) and helix 2, 3 and 5 (C-terminal) to the center of helix 4 (Fig. 6B). In the predicted model of BmApoLp-III, the hydrophobic faces oriented towards the center of the bundle, whereas the hydrophilic faces interacted with the solvent. This molecular architecture resembles that of the four-helix, 22 kDa N-terminal domain of human apolipoprotein E. Another distinctive feature of apoLp-III helix bundle is 3rd helix that is predicted to initiate lipid binding (Leu109) through one end of the bundle (Patnaik et al. 2014). The sequence and structure similarities found

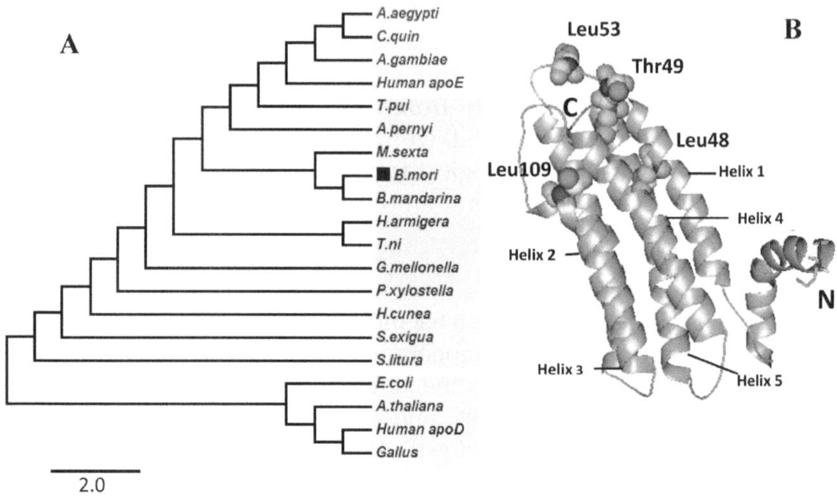

Fig. 6: (A) Phylogenetic tree analysis of the aligned BmApoLp-III protein sequences. The analyses were conducted using the branch and bound parsimony algorithm and single un-rooted phylogram; The length of the branch lines indicates the extent of divergence according to the scale 2.0 bar. The Gene Bank accession numbers of the ApoLp-III protein sequences obtained from NCBI. (B) Deduced 3D structure of *Bombyx mori* (BmApoLp-III) predicted using *Manduca sexta* apoLp-III model (PDB ID: 1EQ1) as a template. Leu48, Thr49, Leu53, Leu109 are actively involved in the binding affinity towards the LDL.

between BmApoLp-III and other organisms (*E. coli*, Plants, Human apoE, apoD) suggest that they originate from a common ancestor, and that they possibly share the same function of carrying hydrophobic ligands in a hydrophilic medium.

2.6 Bombyrin

Lipocalins are typically small (165–200 amino acids) extracellular proteins that have limited similarity in primary sequence, but share a highly conserved tertiary structure, 8-stranded anti-parallel β-barrel. More recently evolved lipocalins tend to (a) show greater rates of amino acid substitutions, (b) have more flexible protein structures, (c) bind smaller hydrophobic ligands, and (d) increase the efficiency of their ligand-binding contacts (Ganfornina et al. 2006, 2007). Thus, attention has focused on the silkworm, *Bombyx mori* and how to increase the economic character by using lipocalin protein. The 27 kDa protein encoded a protein of 186 amino acids with theoretical pI of 6.05. *Bombyx mori* lipocaline proteins are designated as Bombyrin (Sakari et al. 2001), because they have 70% homology with *Galleria mellonella* (Gallerin). We used the selective template structure of *E. coli* (PDB: 1QWD) to predict the structure of Bombyrin. The protein has a highly symmetrical, all β-structure, dominated by a single eight-stranded anti-parallel β-sheet closed back on itself to form a continuously hydrogen-bounded β-barrel. Interestingly, Bombyrin had a characteristic three motif (Fig. 7A) region (Motif 1: Asn30 – Phe47 (blue color); Motif 2: Val115 – Tyr129 (red color); Motif 3: Leu147 – Val161 (magenta color)) with similarity to other lipocaline family proteins. Lipocalins usually bind to small lipophilic ligands within the hydrophobic cavity of the β-barrel and transport them inside a cell via receptor-mediated endocytosis.

Table 2, shows the specific residues that co-ordinate the binding pocket (Fig. 7B) of the Bombyrin in comparison with other lipocalin family proteins [Gallerin (*G. mellonella*), Bilin-Binding Protein (*Pieris rapae*), Insecticyanin (*M. sexta*), Neural lazarillo (*D. melanogaster*), *Arabidopsis thaliana* (plant), Lipoprotein (*E. coli*) and Apo-D (*Homo sapiens*)]. To accommodate ligands of different size and shape, the binding sites of different lipocalins can be quite varied. These roles, not related to the binding pocket of lipocalins, seem to have evolved independently many times along the evolution of the family. This illustrates the versatility of the lipocalin folding combined with flexibility for change and allows Bombyrin to interact with different ligands in their binding pocket (Fig. 7C) in order to make contact with different molecules throughout the protein surface. It was found that Bombyrin has a relatively high sequence identity with lipocalins of other species such as *G. mellonella* (70%), *Pieris rapae* (53%), *M. sexta* (46%), *Homo sapiens* (37%), *Arabidopsis thaliana* (30%), *E. coli* (33%), and *D. melanogaster* (28%) the evolution of this protein family (Fig. 7D).

The known structure of Blc, the first of a bacterial lipocalin, exhibits a classical fold formed by a β-barrel and a α-helix similar to that of the moth bilin-binding protein (Filippov et al. 1995). The empty and open cavity, however, is too narrow to accommodate bilin but the alkyl chains of two fatty acids or of a phospholipid could be readily modeled inside the cavity. Blc was reported to be expressed under stress conditions such as starvation or high osmolarity, during which the cell envelope suffers and requires maintenance. These data, together with our structural interpretation, suggest a role for Blc in storage or transport of lipids necessary for membrane repair or maintenance (Campanacci et al. 2004). Insect lipocalins primarily act in invertebrate coloration, such as bilin-binding protein from the cabbage white butterfly (*Pieris brassicae*), the closely related *Manduca sexta* (tobacco hornworm) protein insecticyanin, and the lobster protein crustacyanin (Holden et al. 1987, Schmidt and Skerra 1994). Like other members of the family, they bind small molecules, and gain their colorant properties from interaction with their ligands. In *Hyphantria cunea*, it has been hypothesized to play an important role in the repair process of brain (Kim et al. 2005). The lipocalin apoD is a mammalian plasma lipoprotein that has long been thought to function

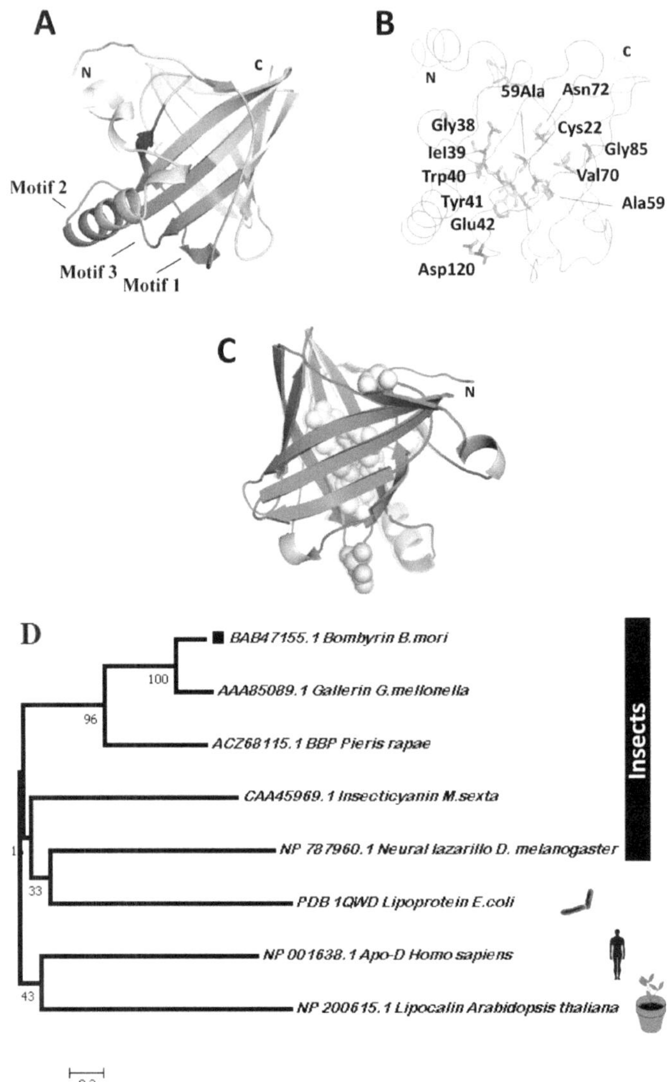

Fig. 7: (A) The three-dimensional predicted structure of silkworm Bombyrin. Bacterial Lipocalin (PDB: 1QWD) was used as a template to build *Bombyx mori*, bombyrin 3D-structure. The three distinct motif 1: N30-F47 (blue), motif 2: 115V-Y129 (magenta), motif 3: 147L-V161 (red) signature depicted as shown. (B) Multi-species conserved sequences identity (Cys22, Gly38, Iel39, Trp40, Tyr41, Glu42, Ala59, Val70, Asn72, Gly85) represented in line model Bombyrin. (C) The binding mode of ligands in the binding pocket of bombyrin depicted as sphere shape (white color). Structure figures were generated by using PyMOL (http://pymol.sourceforge.net). (D) Phylogenetic tree constructed from Bombyrin protein sequences with other lipocalin protein sequences. The possible classifications of different groups (lipocalin protein family) were used to construct phylogenetic tree accordingly to sequence similarity.

Table 2: Sequence characteristics and a hydrophobic environment formed by a small pocket adjacent to the main cavity of lipocalin variations present in Bombyrin and other organisms. Those conserved hydrophobic cavities consisting of α & β residues presented in the lipid binding protein family of lipocalins. β strands shown in the numbers.

Organisms	–1	1	α	2	α	3	4	5	6	7	8	9
Blc		V28,V31	F34,Y39	G41,W43 Y44,E45	F53,E54 L57	A62,Y64 L66	L72,V74	A93	V106,F108	Y116,A120	Y125,A128 C131	W139,I140 L141
Hyphantrin	C23		F32	G39, W41 Y42, E43	E52	Y62,C57			V104	Y117,L120 T122,Y124	Y126,C132 A127	W144,I145 L146,S147
ApoD	C28		F37,Y142	G44,W46 Y47,E48	E57	C61,A64 Y66,L68	V76	A94	L105,V107 F109	Y118,L121 A122,T123	Y128,A129 Y132,C134	W147,I148 L149
Bombyxrin	C23	V29	F32,Y37	G39,W41 Y42,E43	E52	C57,A60 Y62,L64	V71	E85 L100	V102,F104 F106	V117,L118 A119,T120	Y123,A124 Y125,C127	W141,I142 L143,S144
Dmlazarillo	C33	V36	F42,Y47	G49,W51 Y52,E53	E62,F61	C67,A70 Y72,L74			L99,V111 F113	Y124,L127 T129	Y134,A135 Y138,C140	W153,I155 L153
Gallerin	C22	V28	F31,Y37	G39,W41 Y42,E43	E52	C57,A60 Y62,L64	E85,A88		L100,V102 Y122	A119, T120 Y129,C131	Y125,A126	W145,I146 L147
BBP	C23	V26,V29	F32,Y37	G39,W41 Y42,E43	E52	C57,A60 Y62		E85 V121		L117,T119 C130	Y124,A128	W144,V145 L146
Arabidopsis		V9,V10	Y18	G20,W22 Y23,E24	F32	A42,Y44 L46	V54	E68 A74	L83,V85 F87	Y102,L105 L115	Y111,A114	W125,I126 L127,S128
Insecticyanin	C23	V29	F35	E46,W44 G42	E55	C60,A63 Y65		E88	F111	L123,T125 Y127	Y130,A131 Y134,C136	W150,I151 L152,S153

in cholesterol metabolism, although its precise physiological role remains unclear. In insects, lipocaline as a multifunctional protein that is used in different cellular or physiological contexts such as retinol transport, cryptic coloration, olfaction, pheromone transport, the prevention of the harmful local accumulation of endogenous (immunomodulation), enzymatic synthesis and the regulation of cell homeostasis (Flower 1996, Ganfornina et al. 2006, 2007, Sanchez et al. 2003).

3. Conclusion

The continuing post-genomic research in gene and protein expression patterns in the various physiological, developmental and environmental contexts of silkworm has provided a vast amount of information. In conclusion, the present study clearly demonstrates that the storage function of the fat body is essential to the life of holometabolous insects since it distributes major nutrients (SP, Vg) during their developmental periods. Lipophorin protein has important biological functions such as immunity, regulation of growth and development of embryogenesis for the life-span of silkworm. In addition, other groups of low molecular weight proteins (ApoLp-III, Ferritin, and Bombyrin) are also essential in the normal and pathophysiological condition of silkworm for carrying various nutrients throughout the body and acting as signals that coordinate biological functions between the different organs. Three-dimensional structural models are useful because they can provide detailed information about the nature of molecular interactions giving rise to the functional properties of a molecule. But there are still many interesting questions that remain to be answered for individual proteins, such as: co-factor affecting the protein-protein interactions, post-translation modification, identification of associated proteins/protein net-work complex, etc.). The recent demand for research on comparative evolution, molecular biology/visualizing and an understanding of complex biological processes demands integrated efforts of entomo-informatics (structure and functional) which would open new vista of research in Seri-biotechnology that will not only improve the economy of the silk industry but also human health and welfare.

Abbreviations

SP	–	Storage protein
Vg	–	Vitellogenin
VgR	–	Vitellogenin Receptor
LDL	–	Low-density lipoprotein
ESP	–	Egg-specific proteins
ApoLp	–	Apo lipoprotein

PF	–	Peripheral fat body tissues
PVF	–	Perivisceral fat body tissues
TYBs	–	Transitional yolk bodies
Bm	–	*Bombyx mori*
LPs	–	Lipopolysaccharide
PDB	–	Protein Data Bank
HCH	–	Heavy chain homolog
LCH	–	Light chain homolog
IRE	–	Iron responsive element
BLAST	–	Basic local alignment search tool
UTR	–	Untranslated region
SDS-PAGE	–	Sodium dodecyl sulphate-polyacrylamide gel electrophoresis
kDa	–	Kilo Dalton
Blc	–	Bacterial Lipocalin
S0	–	Spinning stage day 0

Acknowledgement

We would like to thank Prof. S. Tojo, Saga University, Japan, and Dr. Hiroaki Abe, Tokyo University of Agriculture and Technology, Japan, for providing invaluable support in this project. The author thanks APSERI-08 Nagoya University, Japan, APMC9-South Korea, for providing the travel grant for Oral & Poster presentation of this project data.

References

Arrese, E. L. and J. L. Soulages. 2010. Insect fat body: energy, metabolism, and regulation. Ann. Rev. Entomol. 55: 207–225.

Atchley, W. R. and A. D. Fernandes. 2005. Sequence signatures and the probabilistic identification of proteins in the Myc-Max-Mad network. Proc. Natl. Acad. Sci. USA. 102(18): 6401–6406.

Britto, C. P., N. K. Singh, M. Krishnan and S. König. 2012. A proteomic view on the developmental transfer of homologous 30 kDa lipoproteins from peripheral fat body to perivisceral fat body via haemolymph in silkworm, *Bombyx mori*. BMC Biochem. 13: 5.

Burmester, T. 2001. Molecular evolution of the arthropod hemocyanin superfamily. Mol. Biol. Evol. 18: 184–195.

Campanaccim, V., D. Nurizzom, S. Spinelli, C. Valencia, M. Tegoni and C. Cambillau. 2004. The crystal structure of the *Escherichia coli* lipocalin Blc suggests a possible role in phospholipid binding. FEBS Lett. 562(1-3): 183–188.

Caperucci, D. and M. I. Camargo Mathias. 2007. Female reproductive system of the sugarcane spittlebug *Mahanarva fimbriolata* (Auchenorrhyncha): vitellogenesis dynamics and protein quantification. Micron. 38(1): 65–73.

Chandrasekar, R., R. Dhanalakshmi, M. Krishnan, H. J. Kim and S. J. Seo. 2005. Computational analysis of apolipohorin-III in Hyphantria cunea. Inter. J. Indus. Economic Entomol. 10: 25–33.

Chandrasekar, R. 2006. Molecular evolution and structural analysis of silkworm LCH ferritins. Vth International Conference on Bioinformatics and Biotechnology, InCoB-2006, The Ashok Hotel, Kautilya Marg, Chankyapuri, New Delhi, 18–20.

Chandarsekar, R., S. J. Seo and M. Krishnan. 2007. Expression and localization of storage protein 1 (SP1) in differentiated fat body tissues of red hairy caterpillar, *Amsacta albistriga* Walker. Archi. Insect Biochem. Physiol. 69(2): 70–84.

Chandrasekar, R., P. Sumithra, S. J. Seo and M. Krishnan. 2008. Sequestration of storage protein 1 (SP1) in differentiated fat body tissues of the female groundnut pest *Amsacta albistriga* (Lepidoptera: Arctiidae). Intl. J. Trop. Insect Sci. 28(2): 78–87.

Chandrasekar, R., M. Meenakshi and M. A. Luiz Paulo. 2009. Insect Hexamerin Storage Proteins: Biosynthesis, Utilization and Evolution. Short Views on Insect Molecular Biology, (Ed.) R. Chandrasekar, International Book Mission, South India. Chapter 3: 49–20.

Chandrasekar, R., A. V. Ananad, P. G. Brintha and B. K. Tyagi. 2011. Impact of enterobacter sp. infection in phenotypic changes and haemolymph protein profiling in silkworm, *Bombyx mori*, L. Chapter 11. pp. 1–13. *In*: B. K. Tyagi and Vijay Veer (eds.). Entomology: Ecology and Biodiversity, Scientific Publisher, India.

Chandrasekar, R. 2012. Electron microscopy and immunogold labelling analysis of smart nano-particles in insects. *In*: A. Méndez-Vilas (ed.). Current Microscopy Contributions to Advances in Science and Technology. Vol. (1): 168–178. Formatex publisher, Spain.

Cozzi, A., B. Corsi, S. Levi, P. Santambrogio, A. Albertini and P. Arosio. 2000. Overexpression of wild type and mutated human ferritin H-chain in HeLa cells: *in vivo* role of ferritin ferroxidase activity. J. Biol. Chem. 275: 25122–25129.

Fagotto, F. and F. R. Maxfield. 1994. Changes in yolk platelet pH during *Xenopus laevis* development correlate with yolk utilization. A quantitative confocal microscopy study. J. Cell Sci. 107: 3325–3337.

Filippov, V. A., M. A. Filippova, D. Kodrík and F. Sehnal. 1995. Two lipocalin-like peptides of insect brain. pp. 35–43. *In*: A. Suzuki, H. Kataoka and S. Matsumoto (eds.). Molecular Mechanisms of Insect Metamorphosis and Diapause. Tokyo: Industrial Publishing and Consulting, Inc.

Flower, D. R. 1996. The lipocalin protein family: structure and function. Biochem. J. 318: 1–14.

Ganfornina, M. D., H. Kayser and D. Sanchez. 2006. Lipocalins in arthropoda: Diversification and functional explorations. pp. 49–74. *In*: B. Åkerström, N. Borregaard, D. R. Flower and J-. Ph. Salier (eds.). Lipocalins. Landes Bioscience.

Ganfornina, M. D., G. Gutiérrez, M. Bastiani and S. Diego. 2007. A Phylogenetic analysis of the lipocalin protein family. Mol. Biol. Evol. 17(1): 114–126.

Habeeb, S. K. M. and R. Chandrasekar. 2014. Entomo-informatics: A prelude to the concepts in bioinformatics. *In*: Short Views on Insect Biochemistry and Molecular Biology, Vol. 2: 621–662.

Hagedorn, H. H. and J. G. Kunkel. 1979. Vitellogenin and vitellin in insects. Ann. Rev. Entomol. 24: 475–505.

Han, C., E. Chen, G. Shen, E. Peng, Y. Xu, H. Zhang, H. Liu, Y. Zhang, J. Wu, Y. Lin and Q. Xia. 2017. Vitellogenin receptor selectively endocytoses female-specific and highly-expressed haemolymph proteins in the silkworm, *Bombyx mori*. Biochem. Cell Biol. 95(4): 510–516.

Hamburgerac, A. E., A. P. West Jr, Z. A. Hamburgera, P. Hamburger and J. P. Bjorkmana. 2005. Crystal structure of a secreted insect ferritin reveals a symmetrical arrangement of heavy and light chains. J. Mol. Biol. 349: 558–569.

Higgins, D. and W. Taylor. 2000. Bioinformatics: sequence, structure and data bank: A practical approach. Oxford University Press, New York.

Holden, H. M., W. R. Rypniewski, J. H. Law and I. Rayment. 1987. The molecular structure of insecticyanin from the tobacco hornworm *Manduca sexta* L. at 2.6 A resolution. EMBO J. (6): 1565–1570.

Hong, S. M., H. Mon, J. M. Lee and T. Kusakabe. 2014. Characterization and recombinant protein expression of ferritin light chain homologue in the silkworm, *Bombyx mori*. Insect Sci. 21: 135–146.

Hou, Y., Z. Yong, F. Wang, J. Gong, X. Zhong, Q. Xia and P. Zhao. 2010. Comparative analysis of proteome maps of silkworm haemolymph during different developmental stages. Proteome Sci. 8: 45.

Hou, Y., J. Li, Y. Li, Z. Dong, Q. Xia and Y. A. Yuan. 2014. Crystal structure of *Bombyx mori* arylphorins reveals a 3:3 heterohexamer with multiple papain cleavage sites. Protein Sci. 23(6): 735–746.

Kawooya, J. K., E. Osir and J. H. Law. 1986. Physical and chemical properties of microvitellogenin. A protein from the egg of the tobacco hornworm moth, *Manduca sexta*. J. Biol. Chem. 261: 10844–10849.

Keeley, L. L. 1985. Physiology and biochemistry of the fat body. pp. 211–248. *In*: G. A. Kerkut and L. I. Gilbert (eds.). Comprehensive Insect Physiology, Biochemistry and Pharmacology. New York, Pergamon. v. 3.

Kim, S., S. K. Hwang, R. A. Dwek, P. M. Rudd, Y. H. Ahn, E. H. Kim, C. Cheong, S. I. Kim, N. S. Park and S. M. Lee. 2003. Structural determination of the N-glycans of a lepidopteran arylphorin reveals the presence of a monoglucosylated oligosaccharide in the storage protein. Glycobiology. 13: 147–157.

Kim, T. Y., Y. Choi, H. S. Cheong and J. Choe. 2002. Identification of a cell surface 30 kDa protein as an candidate receptor for *Hantann virus*. J. Gen. Virol. 83: 767–773.

Kim, H. J., H. J. Je, H. M. Cheon, S. Y. Kong, J. H. Han, C. Y. Yun, Y. S. Han, I. H. Lee, Y. J. Kang and S. J. Seo. 2005. Accumulation of 23 kDa lipocalin during brain development and injury in *Hyphantria cunea*. Insect Biochem. Mol. Biol. 35: 1133–1141.

Leon, L. J., H. Idangodage, C. L. Wan and P. M. M. Weers. 2006. Apolipophorin III: Lipopolysacharide binding requires helix bundle opening. Biochem. Biophys. Res. Commun. 348: 1328–1333.

Lin, Y., Y. Meng, Y. X. Wang, J. Luo, S. Katsuma, C. -W. Yang, Y. Banno, T. Kusakabe, T. Shimada and Q. -Y. Xia. 2013. Vitellogenin receptor mutation leads to the oogenesis mutant phenotype "scanty vitellin" of the Silkworm, *Bombyx mori*. J. Biol. Chem. 288(19): 13345–13355.

Liu, L., Y. Wang, Y. Li, Z. Lin, Y. Hou, Y. Zhang, S. Wei, P. Zhao and H. He. 2016. LBD1 of vitellogenin receptor specifically binds to the female-specific storage protein SP1 via LBR1 and LBR3. PLoS ONE. 11(9): e0162317.

Meenakshi, P. M., R. Chandrasekar and M. Krishnan. 2007. Molecular Characterization of the Vitellogenin and Vitellogenin Receptor in the Cotton Pest, *Spodoptera litura* Dimensions of Molecular Entomology, Universities Press, India Chapter 9: 122–137.

Nichol, H. and M. Locke. 1990. The localization of ferritin in insects. Tissue and Cell. 22: 767–777.

Nichol, H., J. H. Law and J. J. Winzerling. 2002. Iron metabolism in insects. Ann. Rev. Entomol. 47: 535–559.

Park, H. J., E. J. Kim, Y. K. Tai and H. P. Tai. 2003. Purification of recombinant 30 k protein produced in *Escherichia coli*, and its anti-apoptotic effect in mammalian and insect cell systems. Enzym. Microb. Technol. 33: 466–471.

Patnaik, B. B., R. Chandrasekar and Y. S. Han. 2014. Molecular expression and structure-function relationship of apolipophorin III in insects with special references to innate immunity. *In*: R. Chandrasekar, B. K. Tyagi et al. (eds.). Short Views on Insect Biochemistry and Molecular Biology. International Book Mission, Academic Publisher, Chapter 29, Vol. (2): 663–684. ISBN No. 978-1-63315-205-2, printed in Manhattan, KS, U.S.

Pham, D. Q. D. and J. J. Winzerling. 2010. Insect ferritins: typical or aptypical? Biochem. Biophys. Acta. 1800(8): 824–833.

Pietrzyk, A. J., A. Bujacz, J. Mueller-Dieckmann, M. Jaskolski and G. Bujacz. 2013. Crystallographic identification of an unexpected protein complex in silkworm haemolymph. Acta. Crystallogr. Sect. D. 69: 2353–2364.

Pietrzyk, A. J., A. Bujacz, M. Jaskolski and G. Bujacz. 2014. Crystal structure of *Bombyx mori* lipoprotein 6: Comparative structural analysis of the 30-kDa lipoprotein family. PLos One. 9(11): e108761.

Proudhon, D., J. Wei, J. -F. Briat and E. C. Theil. 1996. Ferritin gene organization: Differences between plants and animals suggest possible kingdom-specific selective constraints. J. Mol. Evolution. 42: 325–336.

Raikhel, A. S. and T. S. Dhadialla. 1992. Accumulation of yolk proteins in insect oocytes. Ann. Rev. Entomol. 37: 217–251.

Ryu, K. S., J. O. Lee, T. H. Kwon, H. H. Choi, H. S. Park, S. K. Hwang, Z. W. Lee, K. B. Lee, Y. H. Han, Y. S. Choi, J. H. Jeon, C. Cheong and S. Kim. 2009. The presence of monoglucosylated N196-glycan is important for the structural stability of storage protein, arylphorin. Biochem. J. 421: 87–96.

Sakai, M., C. Wu and K. Suzuki. 2001. Nucleotide and deduced amino acid sequences of a cDNA encoding a lipocalin protein in the central nervous system of *Bombyx mori*. J. Insect Biotechnol. Sericol. 70: 105–111.

Sanchez, D., M. D. Ganfornina, G. Gutierrez and A. Mar. 2003. Exon-intron structure and evolution of the lipocalin gene family. Mol. Biol. Evol. 20(5): 775–783.

Sappington, T. W. and A. S. Raikhel. 1998. Ligand-binding domains in vitellogenin receptors and other LDL-receptor family members share a common ancestral ordering of cysteine-rich repeats. J. Mol. Evol. 46: 476–487.

Schmidt, F. S. and A. Skerra. 1994. The bilin-binding protein of *Pieris brassicae*. cDNA sequence and regulation of expression reveal distinct features of this insect pigment protein. Eur. J. Biochem. 219(3): 855–863.

Singh, N. K., B. C. Pakkianathan, M. Kumar, T. Prasad, M. Kannan, S. König and M. Krishnan. 2013. Vitellogenin from the silkworm, *Bombyx mori*: An effective anti-bacterial agent. PLoS One. 8(9): e73005.

Snigirevskaya, E. S., A. R. Hays and A. S. Raikhel. 1997. Secretory and internalization pathways of mosquito yolk protein precursors. Cell Tissue Res. 290: 129–142.

Sumithra, P., R. Chandrasekar and M. Krishnan. 2009. Autophagic programmed cell death in the peripheral fat body tissues of the silkworm, *Bombyx mori* L. (strain, Tamil Nadu-NB4D2). pp. 159–174. *In*: R. Chandrasekar (ed.). Short Views on Insect Molecular Biology. International Book Mission, South India. Vol. (1).

Sun, Q., P. Zaho, Y. Lin, Y. Young, G. Y. Xia and Z. H. Xiang. 2007. Analysis of the structure and expression of the 30 K protein genes in silkworm, *Bombyx mori*. Insect Sci. 14: 5–14.

Tojo, S., M. Nagata and M. Kobayashi. 1980. Storage proteins in the silkworm, *Bombyx mori*. Insect Biochem. 10(3): 289–303.

Tufail, M. and M. Takeda. 2008. Insect vitellogenin/lipophorin receptors: Molecular structures, role in oogenesis, and regulatory mechanisms. J. Insect Physiol. 55: 88–104.

Ujita, M., Y. Katsuno, I. Kawachi, Y. Ueno, Y. Banno, H. Fujii and A. Hara. 2005. Glucan-binding activity of silkworm 30-kDa apolipoprotein and its involvement in defense against fungal infection. Biosci. Biotechnol. Biochem. 69(6): 1178–1185.

van Holde, K. E. and K. I. Miller. 1995. Hemocyanins. Adv. Protein Chem. 47: 1–81.

Wang, J., B. D. Sykes and R. O. Ryan. 2002. Structural basis for the conformational adaptability of apolipophorin III, a helix bundle exchangeable apolipoprotein. Proc. Natl. Acad. Sci. USA. 99: 1188–1193.

Weers, P. M. and R. O. Ryan. 2006. Apolipophorin III: role model apolipoprotein. Insect Biochem. Mol. Biol. 36: 231–240.

Yang, J. P., X. X. Ma, Y. Xing, H. Wei, F. Li, Y. Kang, R. Bao, Y. Chen and C. Z. Zhou. 2011. Crystal structure of the 30 K protein from the silkworm *Bombyx mori* reveals a new member of the β-trefoil superfamily. J. Struct. Biol. 175(1): 97–103.

Yano, K. I., M. T. Sakurai, S. Izumi and S. Tomino. 1994. Vitellogenin gene of the silkworm, *Bombyx mori*: Structure and sex-dependent expression. FEBS Lett. 356(2-3): 207–211.

Yu, H. Z., S. Z. Zhang, Y. Ma, F. Dong-Qiong, L. Bing, Y. Li-Ang, W. Jie, Y. Zhang, Z. Dong, S. Liu, Q. Yang, P. Zhao and Q. Xia. 2012. Identification of novel members reveals the structural and functional divergence of lepidopteran-specific Lipoprotein_11 family. Funct. Integr. Genomics. 12(4): 705–715.

Zhang, D., C. Ferris, P. Kohlhepp and J. J. Winzerling. 2001. *Manduca sexta* IRP1: Molecular characterization and *in vivo* IRP1/IRE binding activity in response to iron. Insect Biochem. Mol. Biol. 32: 85–96.

Zhen, L., M. Azharuddin and J. P. Xu. 2017. Molecular characterization and functional analysis of a Ferritin heavy chain subunit from the eri-silkworm, *Samia cynthia ricini*. Intl. J. Mol. Sci. 18: 2126.

CHAPTER 4

Metabolism of Molting Hormone and Juvenile Hormone in Silkworm

Yungen Miao

1. Introduction

The studies of basic phenomena of silkworm growth and development, and their regulatory mechanisms are of instructive significance in establishing a rational rearing method for producing high quality cocoons and silks in large quantities (Zhou and Lv 1980).

Growth is usually understood as an increase of body weight and size, while development is the process of tissue differentiation, molting, metamorphosis and maturation of reproductive organs. Growth is the basis of development. Development creates the precondition for further growth.

As soon as silkworms are hatched from eggs, they feed on mulberry leaves, resulting in an uptake of nutrition and, thereby, an increase in body weight and size. All these processes represent the growth of silkworm until they attain adult stage, at which point the development of sexual organs takes place.

The growth of larval organs and tissues of the silkworm can be divided into three types: (1) Cell division: For instance, the number of blood cells increases as the larvae grow, but there is little difference between the size of blood cells of young silkworm and that of mature silkworm.

College of Animal Sciences, Zhejiang University, 866 Yuhangtang Road, Hangzhou 310058, China.
E-mail: miaoyg@zju.edu.cn

(2) Cell enlargement: cell division in the silk glands, Malpighian tubules and salivary glands, etc. occurs only during the embryonic stage. There is no cell division during the larval stage. Their growth is the result of an increase in cell size. (3) Cell division and enlargement: The growth of dermal cells, epithelial cells of the midgut and the fat body cell, etc. is a result of cell division and enlargement (Wu et al. 1981).

The development of silkworm mainly involves the molting during larval stages and metamorphosis during the pupa stages. Like other insects, the growth of silkworm larvae is discontinuous. The larvae must molt several times because of their chitin exoskeleton, and be divided into several instars. For normal varieties, the larvae molt 4 times and are divided into 5 instars.

2. Metamorphosis

The silkworms belong to holometabolan. There is a great difference in external morphology and structure of internal organs between the larval stage and adult stage. There is a transition stage called pupa stage. Metamorphosis in pupa is the morphological and structural changes of insects during their development from larvae to adults. During this stage, the tissues and organs are decomposed and destroyed, even degenerated or eliminated completely. This process is called histolysis. At the same time, undifferentiated embryonic tissues or cells, which exist inside the larval body in the form of adult imaginal buds, develop rapidly, and differentiate to tissues or organs of adults. This process is called histogenesis. The morphological changes of tissues or organs of silkworms during pupal stage can be divided into five types. Briefly they are as follows:

(1) Decomposition and disappearance: The further existence of organs which only larval life needs, such as silk glands, salivary glands, peritracheal glands, small intestine, colon, mandibles and abdominal legs, etc. will impede the development of adult organs and their roles. These organs are degenerated and eliminated by means of histolysis. The products of histolysis are utilized during the formation of adult organs and provide energy.

(2) Recombination and reconstruction: Adult life still needs the tissues or organs, such as midgut, fat body, most musculature, tracheae, maxillae, ocelli and thoracic legs, etc., but their morphology and function are different from those in larvae. These tissues and organs are reconstructed from differentiation of adult-type cells. For example, the larval maxillae have the function of gestation, but after reconstruction, the adult maxillae have different morphology and function. Their function is mainly the secretion of cocoonase. Together with enzymes produced in the midgut of pupae, cocoonase acts on the emergence of adult moths. The morphology

of the same organ changes during pupal stage. This organ has function which differs from that of larvae. The foregut and hindgut of silkworm larvae are differentiated from the ectoderm, while the midgut originates from the endoderm. The epithelium of the larval midgut mainly consists of cylindrical cells, goblet cells, and undifferentiated regenerative cells. During metamorphosis, the epithelial cells of larvae midgut drop out. The regenerative cells differentiate into epithelium of pupae and adults, which mainly consists of cylindrical cells with different functions. Different from midgut, the adult foregut is differentiated from cells at the larval pharynxes, the crop comes from cells at the boundary of esophagus and midgut, the hindgut from cells at the end of pyloric valve, and the rectal sac from cells at the centrally shrunk part of the colon. All these cells are properties of imaginal buds.

(3) Function retaining: Between larvae and adults, the functions of the nervous system, Malpighian tubules and reproductive glands are basically the same. Hence, there are little morphological changes during pupal stage. There are only some cell-level transformations and they differentiate uninterrupted from larval stage to adult stage, and finally from the reproductive organ. At the end of 5th instar, the spermatogonia and oogonia differentiate to spermatocytes and oocytes. During metamorphosis, they differentiate gradually to sperm and eggs.

(4) Differentiation and development: This type belongs to adult-type tissues and organs. During the larval stage, they exist in different parts in the form of imaginal buds, carrying genetic information which will differentiate into adult tissues and organs. However, during the larval stage the genes are inhibited, and there is no further development. During the pupal stage, they are differentiated and developed rapidly to form adult organs. Adult wing buds located at the 2nd and 3rd larval segments are undifferentiated embryonic cell colonies. They are folded fleshy dermal cells. The Herold's glands are located at the 8th and 9th abdominal inter-segment membrane of male larvae. During the pupal stage, they are differentiated to spermatheca, ejaculatory vesicle, ejaculatory duct and accessary gland. The Ishiwada's fore glands are invaginations of the body wall, and consist of two parts. The Ishiwada's fore glands are located at two sides of the 8th medioventral line of female larvae. During the pupal stage, they are differentiated into copulatory pouch, spermatheca and the anterior portion of the ovipositor. The Ishiwada's hind glands are located at the two sides of the 9th medioventral line. Porcelain white spots can be observed on the external part of the 4th–5th instar larvae. During the pupal stage, they are differentiated into the posterior portion of the ovipositor and reproductive glands. The alluring glands, by which the female moths release sexual pheromone, are differentiated from the dermal cells between the hypospiracular line and the ventral line. Moreover, adult

organs such as compound eyes (from dermal cells adjacent to the larval ocelli), antennae (from the first segment of the larval thoracic legs), salivary glands (from the anterior cells of the larval salivary glands), penises or external reproductive organs of male moths (from the body wall adjacent to the supraventral line of the 11th segment) and papilla genitals of the female moths (from larval anal legs), etc. are all differentiated particularly from parts and tissues of the larvae during the pupal stage.

(5) Two-time metamorphosis: The muscles of the body wall and part of the tracheae undergo two rounds of histolysis during pupation and emergence. During metamorphosis, there are no great changes at the main tracheal trunk, but with the development of the organs, great changes take place at the tracheal termination distributed among different organs. The muscles around the exoskeleton and midgut are renewed twice during pupation and adult emergence. However, muscles that are distributed in the foregut and hindgut are degenerated during the pupal stage, and the flying muscles are differentiated into new tissues of adults during the pupal stage. Though no differentiations occur in the tracheae of the midgut and around it, the structure of epithelial cells of the pupal midgut differs from that of the adults. Therefore, it belongs to two-time metamorphosis at a cellular level.

2.1 Biochemistry of Metamorphosis

Histolysis and histogenesis during the pupal stage is not only the expression of profound morphological changes of tissues and organs, but also a reflection of physiological and biochemical changes.

2.1.1 Metamorphosis and Gas Metabolism

The pupal body is a closed system. The exchange of substance with the environment is mainly the gas metabolism. Figure 1 shows the changing curves of respiratory intensity of different voltine varieties of silkworms during the pupal stage. The pupal body of the multivoltine variety has the highest oxygen-consumption intensity; and that of the univoltine variety has the intermediate oxygen-consumption. The amount of oxygen consumed by the multivoltine variety per gram of pupal body weight during 1 hr is as high as 613 mm^3, the bivoltine variety 468 mm^3, and the univoltine variety only 389 mm^3 (Lv 2011).

From the physiological point of metamorphosis, we can see from Fig. 1 that, during the process of histolysis, the oxygen-consumption intensity of the pupal body decreases gradually. With the coming of histogenesis and differentiation, that is, the main period of histogenesis and differentiation, the oxygen-consumption intensity of the pupal body increases, and reaches its peak two days before emergence. On the eve of emergence, there are

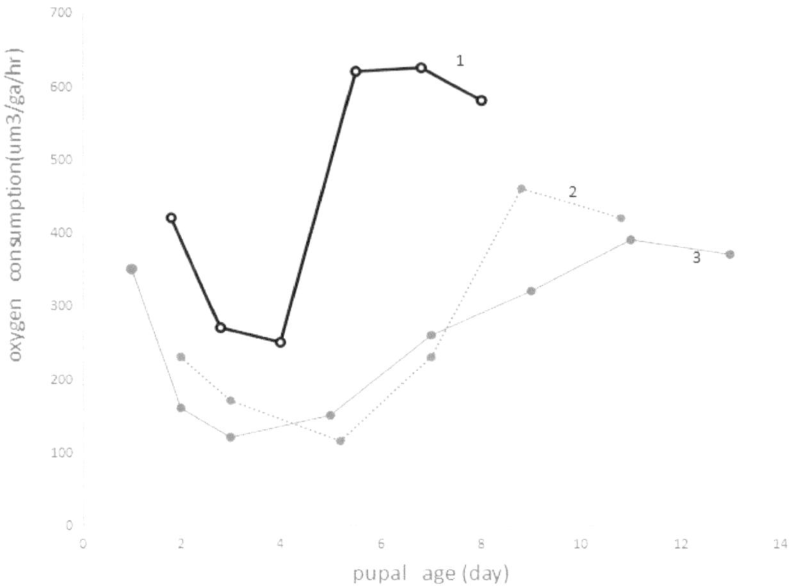

Fig. 1: The dynamic curves of oxygen consumption of the silkworm pupae. 1. multivoltine variety lan-xi 10, 2. biovoltine variety, 3. univoltine variety Baghdad (female).

some decreases. From this basis, we can roughly divide the histolytic stage and the histogenetic period according to the changing law of the intensity of the gas metabolism during the pupal stage. The histolytic stage ends with the decrease of the oxygen-consumption intensity to the lowest point; the increase of the oxygen-consumption intensity of the pupal body marks the beginning of the histogenetic stage. When the oxygen-consumption intensity of the pupal body is at its lowest point, it is the turning point of the histolytic stage and the histogenetic stage. The results of studies on the kinetics of hydrogen peroxidase activity of the pupal body fluid concur with the above law. That is, after pupation the activity of hydrogen peroxidase is low. With ongoing histolysis, the activity of this enzyme increases rapidly, and later stage histolysis reaches the highest point. With the beginning of the histogenetic stage, the activity of the hydrogen peroxidase enzyme reaches its lowest point.

2.1.2 Metamorphosis and Carbohydrate Metabolism

During the prepupal stage from spinning to pupation, the capacity of trehalase in the fat tissues decreases, and the synthesis of glycogen is strengthened. Hence, the accumulation of glycogen increases. After pupation, i.e., during the histolytic stage, the content of glycogen in the pupal body continues to increase gradually. During the histogenetic stage, the glycogen which was primarily synthesized in the fat body and stored as energy substance is gradually converted into trehalose which is released to the body fluid. One part of the trehalose is absorbed by the ovary and converted into ovary glycogen, the rest is consumed in the form of energy. Therefore, the total content of glycogen in the pupal body decreases gradually. Before emergence, there is an increase in the content of glycogen, which used as energy in the adult moths (Lv 2011).

2.1.3 Metamorphosis and Lipometabolism

The fat body is not only the site for the synthesis and storage of lipids, but also the main tissue of the synthesis, decomposition and storage of glycogen, trehalose, storage protein and vitellogenin. From the late stage of the 5th instar to the pupal stage, the fat body is well developed from the late stage of the 5th instar larval stage to the pupal stage, contributing 30% of the body weight, but it gradually decreases with the development of adult's body.

The activity of phosphatases of the microsomes in the fat cells of pupal body silkworm visibly changes according to the time of metamorphosis. At the beginning, the activity of this enzyme is very high; when histolysis is in full swing, it decreases. About 5 days after metamorphosis, the activity of this enzyme increases again (Fig. 2). This indicates that, at this moment,

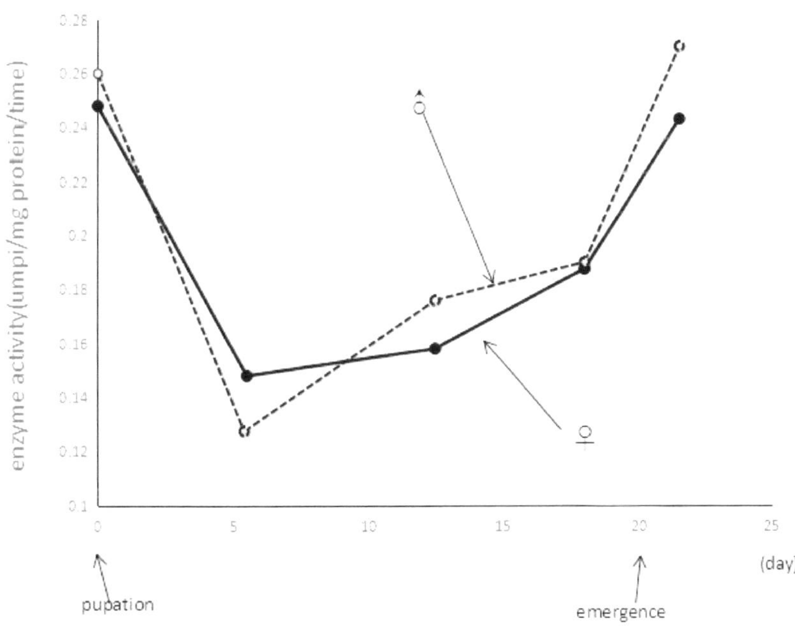

Fig. 2: Dynamics of activity of phosphatide phosphatase of *attacus atlas* pupae during metamorphosis.

apart from the synthesis of triphosphoglyceride or diphosphoglyceride which was used as energy, there is also synthesis of phospholipids necessary to the histogenesis of adults. During metamorphosis, the exchange of the function and direction of synthesis of the same tissue depend totally on the physiological requirement of the tissue.

2.1.4 Metamorphosis and Protein Metabolism

In silkworm, the fat body of the larvae absorbs the nutritional substance in the form of body fluids, upon which the storage proteins are synthesized (only the female synthesizes SP1, while both males and females can synthesize SP2) and secreted to the body fluids. Hence, before spinning, the content of storage proteins reaches its peak level. During the non-feeding stage, that is from spinning to pupation period, obvious changes take place in the morphology and function of the fat body. A large amount of storage proteins in the body fluid are absorbed by the fat body and temporarily stored (Fig. 3). After pupation, the storage proteins in the fat body are decomposed gradually and utilized during the synthesis of vitellognin. Four days after pupation, with the progress of histogenesis in the adults, vitellogenin, consisting of 50 ug proteins and 15 ug glycogen, is secreted to that of the silkworm. Twenty percent of the proteins are

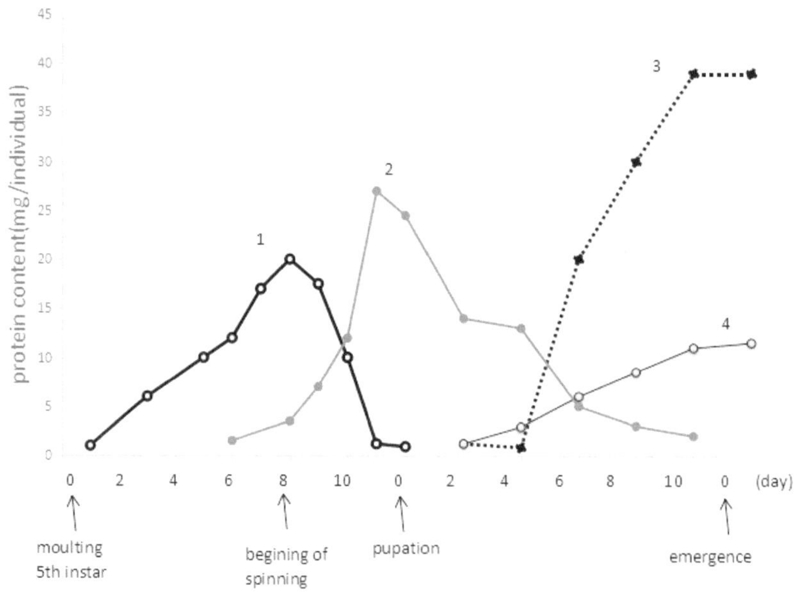

Fig. 3: The dynamics of contents of storage protein in the body fluid and fat body and that of vitellogenin and egg special protein in the ovary of the female silkworms. 1. storage protein in the haemolymph, 2. storage protein in the fat body, 3. vitellogenin in the ovary, 4. egg special protein.

vitellogenin, which is synthesized by the fat body and is released to the body fluid. It is absorbed by the ovary and enters the oocytes to form the yolk phosphoprotein. The other 20% forms special egg protein. It is self-synthesized during ovary development. It is a necessary protein for embryonic development. In the oocytes, there is another small molecular lipoprotein (molecular weight 27–29 kDa) which is also absorbed from the body. Fluid mRNA synthesizing this lipoprotein exists in the fat cells. However, during the prepupal stage, it decreases rapidly, and after pupation, the activity of this mRNA in the fat body disappears completely. The dynamic of contents to its products (small molecular lipoproteins) in the body fluid is as follows: On the 3rd day of the 5th instar, it increases rapidly. Between the spinning and pupation, it reaches the highest peak and then gradually decreases. Later, during emergence, it remains undetected.

2.1.5 Metamorphosis and Nucleic Acid Metabolism

During metamorphosis, the process of decomposition of larval tissues and generation of adult tissues is obviously reflected in the dynamic of the nucleic acid metabolism. During the histolysis, the dehradation of DNA occurs. Contrary to the gradual decrease of the radioactivity of DNA, the radioactivity of the acid soluble portion (products of DNA decomposition) increases significantly. Part of it is absorbed and utilized by the ovary at the end of pupal stage, another part is stored in the body fluid. Meanwhile, the silk gland degenerates after spinning, and two or three days after pupation it disappears. The DNA content in the silk gland of mature larva is more than 50% of the total DNA content in the larval body. During histolysis, the DNA in the cells of the silk gland is decomposed into several small fragments. Some of these fragments are absorbed and utilized by the fat tissues. The same phenomenon can also be observed in the midgut tissues. From prepupae to adults, the total DNA contents of the pupal body decreases, and RNA content decreases notably during histolysis. There is, however, some increase of RNA content with the generation and differentiation of adult tissues. Hence, during metamorphosis, the ratio of RNA/DNA is a U-shaped curve (Fig. 4). This reflects the timely superseding between histolysis and histogenesis and the changing of protein synthesis during metamorphosis (Chaudhuri and Medda 1985).

2.2 Regulation of Hormones in Growth and Development of the Silkworm

The growth and development of silkworm, including molting, metamorphosis and diapause, is all regulated and controlled by the endocrine system.

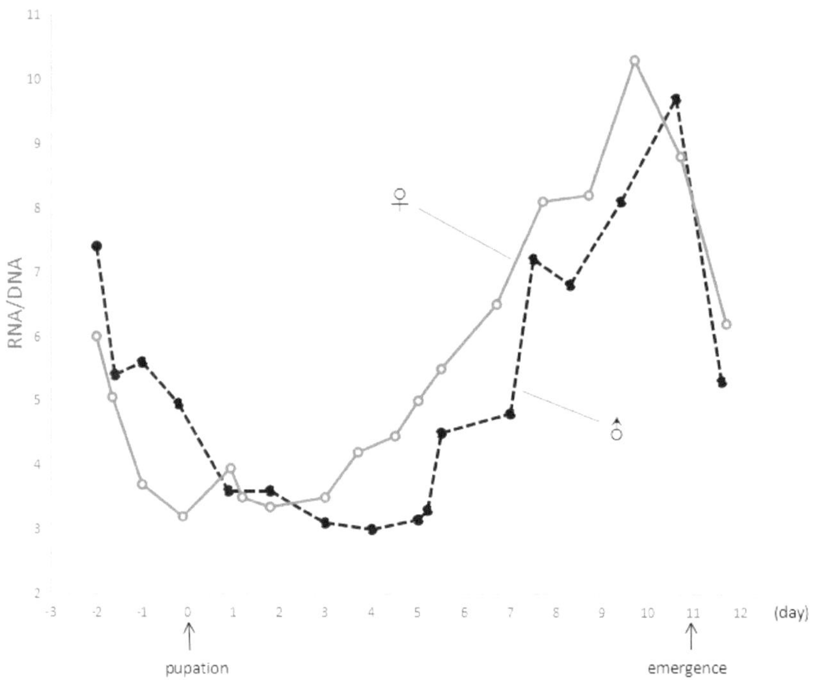

Fig. 4: The change of RNA/DNA ratio during metamorphosis of the silkworms.

The main hormones controlling growth and development of the silkworms are: brain hormone (BH) or prothoracotropic hormone (PTTH), juvenile hormone (JH), molting hormone (MH) and diapause hormone (DH). They are secreted by cerebral secretory cells, *corpus allatum,* prothoracic gland and subpharyngeal ganglion neurosecretory cells respectively (Morohoshi et al. 1977).

Neurosecretory cells are distributed in the whole nervous system and occupy a major position in the regulation and control of insect growth and development. They are connected to the central nervous system and monitor environmental information and epithelial endocrine glands controlling histo-differentiation. In silkworms, the most important neurosecretory cells are the cerebral neurosecretory cells. The studies of cobalt staining methods have indicated that, except for central neurosecretory cells and lateral neurosecretory cells, there is the frontal 2nd cell and the hind 3rd cell. It is also observed that the hormone secreted by these neurosecretory cells enters the *corpus allatum* through the *corpus cardiacum*. Apart from this, using histochemistry analysis, Chinese researchers have found that there is an additional hind group and a ventral lateral group of neurosecretory cells.

Central neurosecretory cells are located at the dorsal neuron of the inter cerebrum. They are the largest group of cerebral neurosecretory cells and are connected to the *corpus cardiacum* by a neural axon. The *corpus cardiacum* is a paired neuro-fluid organ located on two sides of dorsal vessel above the oesophagus, connected with the rear side of brain through the cardiac nerve. Brain hormone is stored in the *corpus cardiacum* and is released to the body fluid. The neurosecretory cells of the *corpus cardiacum* can also secrete hormone. The *corpus cardiacum* as a part of visceral nervous systems is differentiated from the dorsal wall of the stomodaem during embryonic development.

3. Molting Hormone

There is a united mechanism of the hormone regulation of molting and metamorphosis in insects. The main organs acting in this mechanism are the brain, *corpus allatum* and prothoracic gland.

In 1930s Bounhiol reported that, if the *corpus allatum* of the 3rd or 4th larvae of the silkworm is excised, the larvae could not molt and pupate. In 1940s, the endocrine function of the prothoracic gland was discovered and it was experimentally demonstrated that both *corpus allatum* and prothoracic gland are needed in order to induce larval molting, but pupation only needs the prothoracic gland. That is, the hormone secreted by the prothoracic gland can lead to larval molting and pupation, while the hormone secreted by the *corpus allatum* can inhibit pupation.

If the brain activity of the pupae is existed after pupation, dauer-pupae are obtained and the adult ecdysis is inhibited. If part of the brain tissues containing neurosecretory cells are transplanted into the dauer-pupae, the formation of adult body wall and emergence can be investigated. Other insects have also proved that the brain plays the role of central regulation and control.

3.1 Prothoracic Glands

Like the salivary gland, the prothoracic gland is an ectodermal invagination of the 2nd maxillary segment during the embryonic stage. It is a multi-cellular, branched and band-like gland, which tangling on the tracheal bush inside the first spiracle during larval stage, and can be divided into four parts (Fig. 5). The prothoracic gland periodically secretes molting hormone. After emergence, the gland deteriorates very quickly.

The prothoracic glands were wrapped in a thick layer of common membrane and crowded with polygonal or oval-shaped secretory cells of larger nuclei. In early silkworm instars, the nucleus is spherical in shape, while it is irregularly shaped dendrites in later developmental instars (Fig. 6). The secretion of prothoracic glands presented a periodic change during the growth of silkworm. In the early stage of the instar, the nucleus

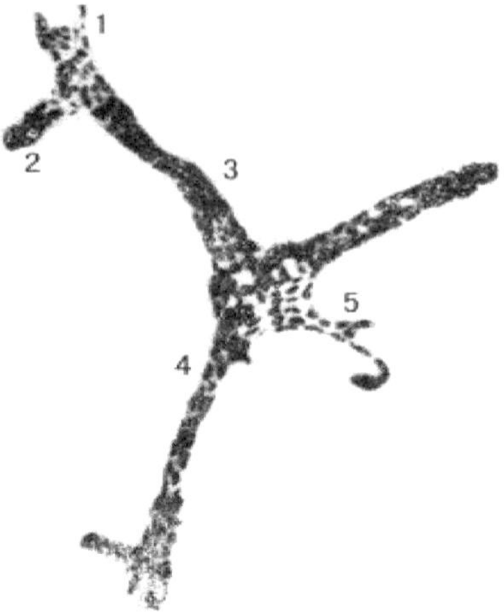

Fig. 5: Diagram of prothoracic glands of the silkworms. 1. fore-branch, 2. mid-branch, 3. trunk, 4. hind-branch, 5. back-prominent.

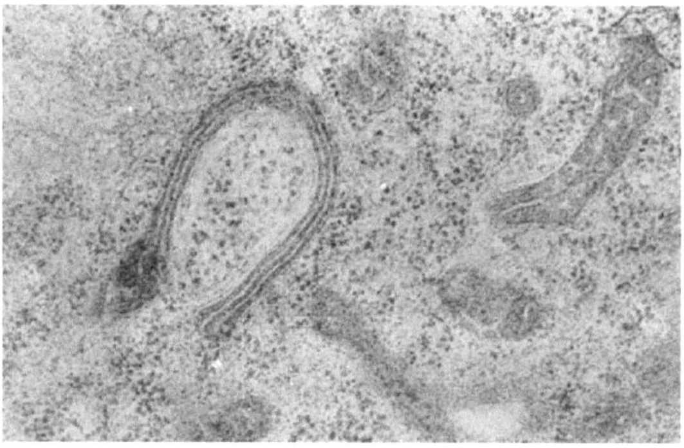

Fig. 6: Ultrastructure of prothoracic glands.

of prothoracic gland cells displayed a smooth wall, the chromosomes dispersed into tiny particles and the nucleus cytoplasm. By the end of the age, the nuclear wall becomes irregular, presenting cytoplasm protruding into the nucleus, with good nuclear staining, tiny granular secretions in the cytoplasm approaching the nucleus, and soon the particles become larger and move toward the edges of the cells' secretion layer.

3.2 Chemistry of Molting Hormone and its Action Way

3.2.1 Chemical Structure of Molting Hormone (MH)

Molting hormone is secreted by prothoracic gland. In 1954 Butenandt and Karlson separated crystallized MH from the pupae of silkworm.

Its chemical formula is C17H44o5, called α-ecdysone. After 11 years, the chemical structure was confirmed (Fig. 7). In the next year, a substance which has one additional hydroxyl group was separated from the pupae of silkworm, called 20-hydroxyedysone, ecdysterone or β-ecdysone. Its biological activity is three to five times higher than that of α-exdysone. Subsequently, the 3rd ecdysone of lower activity was separated from pupae of tobacco horny moths, called 20, 26-dihydroxyecdysone. At present, 17 steroids having molting hormone activity have been separated from insects and crustaceans (Karlson 1980).

The structure of molting hormones in insects differs from that of steroid hormones in vertebrates. The molting hormones in insects are all C_{27} steroids, while steroid hormones in vertebrates are almost always C_{18}, C_{19}, or C_{21} steroids. Both α-ecdysone and β-ecdysone have been separated from plants, and there are 58 other steroids having molting hormone activity existing in plants (Fig. 8). Besides the naturally occurring compounds, recently some α-ecdysone analogues have been artificially synthesized. For example, compound (I) in Fig. 9 is 22, 25-dideoxycdysterone, and its activity is even higher than α-ecdysone, β-ecdysone and most of the plant-originated ecdysone (Huber and Hoppe 1965).

All steroids promoting the physiological activity of molting, no matter whether they are artificially synthesized or naturally occurring, are called molting steroids. Ecdysone is one such example.

Exdysone is secreted by the prothoracic gland. It controls the apolysis and molting process. The silkworms cannot synthesize steroids. Therefore, it has to depend on the phytosterols which exist in the foods and can be converted into cholesterin. In the silkworms both α-ecdysone and β-ecdysone are found, though the latter is dominant. H-ecdysone injected

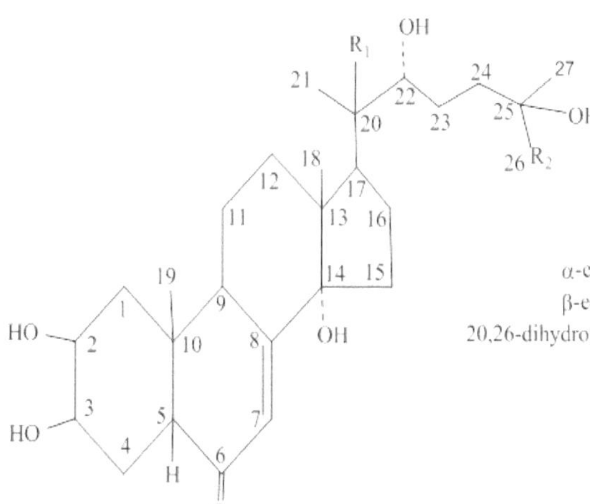

	R_1	R_2
α-ecdysone	H	CH_3
β-ecdysone	OH	CH_3
20,26-dihydroxyecdysone	OH	CH_2OH

Fig. 7: Chemical structures of molting hormones in insects.

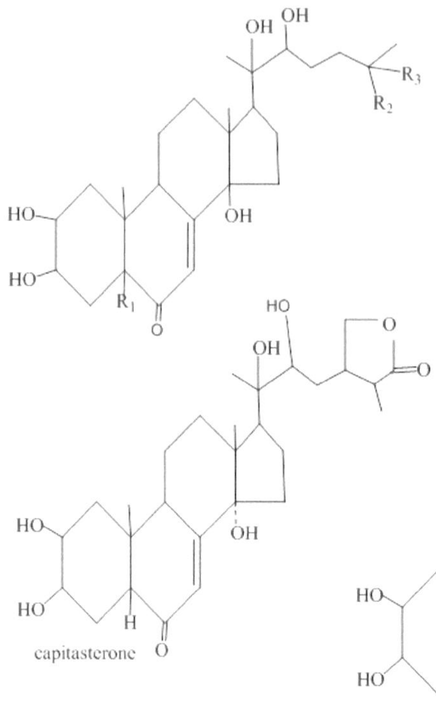

	R_1	R_2	R_3
polypodine B	OH	OH	CH_3
ponasterone A	H	H	CH_3
inokosterone	H	H	CH_2OH

Fig. 8: Molting steroids from plants.

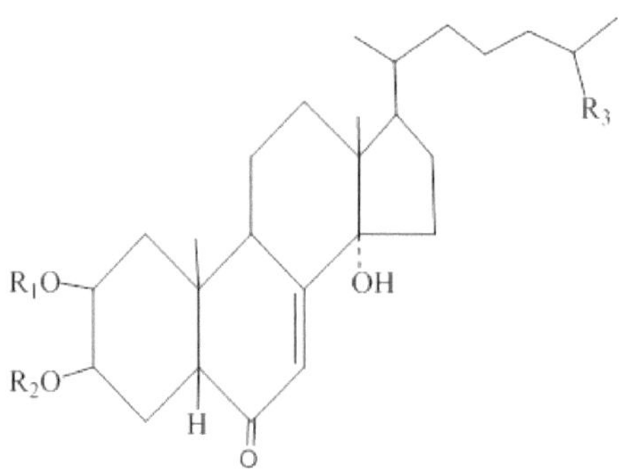

	R_1	R_2	R_3
(1)	H	H	H
(2)	H	H	CH_3
(3)	CH_3	H	CH_3
(4)	$(CH_3)_3Si$	$(CH_3)_3Si$	CH_3

Fig. 9: Synthesized molting steroids.

into the silkworm body is quickly hydroxylated to ecdysterone, which combines with proteins. Hence, most scholars believe that α-ecdysone is synthesized and secreted by the prothoracic gland. Whether α-ecdysone is combined with the targeted tissue protein or α-ecdysone is firstly converted into β-ecdysone and later acts on the targeted tissue, or whether both of them have their own function, remains elusive.

In recent years, a-deoxyecdysone, 2-deoxy-20 hydroxyedysone and 22-dideoxy-20-hydroxyedysone have been separated from the ovaries of the silkworms. Of these, a-deoxyecdysone has been separated from the eggs of the silkworms.

3.2.2 Role of Molting Hormone

Ecdysteroid acts directly on the tissues and organs of the ectodermal-origin (Niwa and Niwa 2016). Its main functions are:

(1) Stimulation and formation of the new skin and shedding of the old skin. Its role includes: (i) stimulation of ectodermal cells in a variety of enzyme activities, prompting cells division; (ii) Promotion of cell secretion activity, new upper epidermis formation, secretion of eczema, and activation of chitin lytic in the inner epidermis, and shedding of old skin; (iii) As skin cells continue to secrete epidermal secretions under the new epidermis, this prompts the formation of diphenol oxidase, increases enzyme activities and promotes tyrosine metabolism, resulting in tanning and coloring of the body.

(2) Stimulation of proteins and enzymes synthesis. Experiment of MH injection into blood stink nymphs like nymphs, have shown (within hours of injection) enlarged nucleolus and an increase of RNA content in the cytoplasm and fat body cells. Altogether, it indicates that MH played an important role on the nuclear, RNA and protein synthesis.

(3) Enhancement of cellular respiratory metabolism. Ecdysone promotes respiratory rate, number and size of mitochondria. The main difference between the energy supply system of diapause and non-diapause insects is the quantity of certain enzymes in the mitochondrial system. In other words, insects stop growing into the diapause state due to hormonal imbalances caused by physiological and biochemical reactions related to stagnation. The main mechanism of action is to reduce concentration of MH caused by cytochrome B and C deficiency, and replace with another special oxidase, that is cytochrome B5.

3.3 Metabolism of Molting Hormone

Ecdysone was periodically secreted in the larval stage, and basically, began to be secreted in the middle of all ages. Pupal stage also secretes until degeneration only in latent adults. Researchers have shown that the critical period of prothoracic gland secretion is about 2/5 of the stage in the last instar and about 3/5 of stage in other ages. In fact, the critical period of secretion depends on silkworm varieties, breeding methodology and nutrition, etc.

The secretion of the prothoracic gland is controlled by the brain hormones. The prothoracic gland is secreted before the brain hormones secretion, like secretion of thyroxine in human thyroid, which must be secreted by the pituitary gland to stimulate the thyroid gland. Likewise, the secretion of the prothoracic gland must be activated by the PTTH (prothoracicotropic hormone) hormone secreted in the brain.

3.3.1 Biosynthesis of Molting Hormone

Cholesterol is converted into the prohormone ecdysone via a series of reactions that occur in the ER, cytosol (suggested for other arthropod species) (Blais et al. 1996), and mitochondria, and are exported possibly by secretory vesicles into the haemolymph. The early gene broad (isoform BR-Z4) positively regulates the expression of Npc1a which encodes a cholesterol transporter, mediating the trafficking of cholesterol out of late endosomes into the ER, a critical step toward the synthesis of ecdysone (Niwa and Niwa 2016) (Fig. 10).

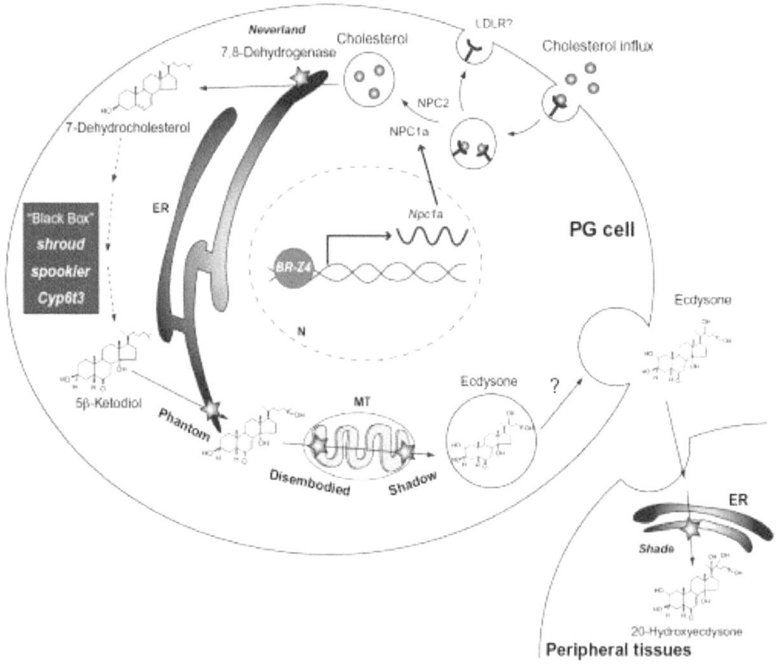

Fig. 10: Biosynthesis of molting hormone. LDLR, low-density lipoprotein receptor; MT, mitochondria; ER, endoplasmic reticulum; N, nucleus; PG, prothoracic gland.

3.3.2 Inactivity of Molting Hormone

Ecdysteroids such as ecdysone and 20-hydroxyecdysone may be excreted as such or as metabolites via the gut or Malpighian tubules. Ecdysteroid metabolizing enzymes, especially in fat body, midgut and Malpighian tubules, in addition to inactivating the hormone, may facilitate excretion. Variation in rate of excretion during development appears to be an important factor in determining the hormone titer. Hormone inactivation by the gut may be important in protecting the insect from any ingested phytoecdysteroids. Irreversible inactivation of ecdysteriods is also important in closed stages of insects where excretion is impossible (eggs and pupae) and metabolites accumulate in the gut (Koolman and Karlson 1985).

It is interesting that 3-dehydroecdysine produced by *Calliphora* was amongst the first ecdysteroid metabolites to be characterized, especially in view of the new significance. This compound has assumed to be a product of the prothoracic glands in Lepidoptera. Furthermore, perhaps 3-dehydro-20-hydroxyecdysone should be considered as an active hormone in some systems in view of its activity being higher than 20-hydroxyecdysome in inducing expression of the P1 gene in *Drosophila* fat body. Particularly puzzling is the occurrence of the enzymes for reversible interconversion of ecdysteroids and their 3-dehydro-derivatives in tissues of several species that are also incapable of producing 3-epi-ecdysteroids (Rees 1995) (Fig. 11).

Clearly, in the midgut cytosol of Lepidoptera, 3-dehydroecdysteroids serve as intermediates in the formation of 3-epi-ecdysteroids, which are further metabolised to phosphate conjugates. In *S. littoralis*, the conjugates have been identified as the 2- and 22-phosphates. Determination of the profiles of the three enzymatic activities involved in these transformations during the 6th instar of *S. littoralis* indicated that whilst all activities were detectable throughout the instar, ecdysone oxidase and the 2- and 22-phosphotransferases showed peaks early in the instar, whereas 3-dehydroecdysone 3α-reductase and 3β-reductase

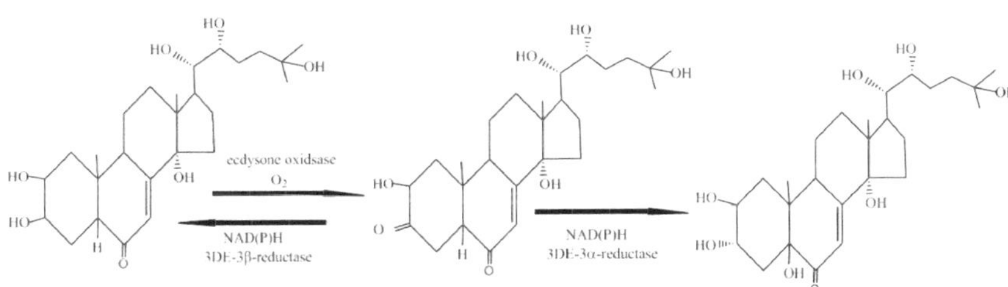

Fig. 11: Mutual transformation pathway of ecdysone.

showed maxima in mid and late instars, respectively. It is interesting that haemolymph 3-dehydroecdysone 3β-reductase, responsible for reduction of 3-dehydroecdysone produced by prothoracic glands also exhibited a maximum late in the instar, in agreement with that reported for *M. sexta* (Sakurai et al. 1989).

3.4 Action of Molting Steroids

Molting steroids can activate genes and induce chromosome puffing. Karlson (1980) suggested the assumed model of MH action (Fig. 12). After entering the cells, the molting hormone combines with the receipt protein of the cytoplasm and forms a compound. The structure of the receptor is changed. The compound enters the nucleus and combines with a particular position of DNA. As a result of mutual action, particular genes begin to transcript. The precursor of mRNA is converted into mRNA and is transported into the cytoplasm. mRNA combines with ribosome and begins to translate to direct protein synthesis.

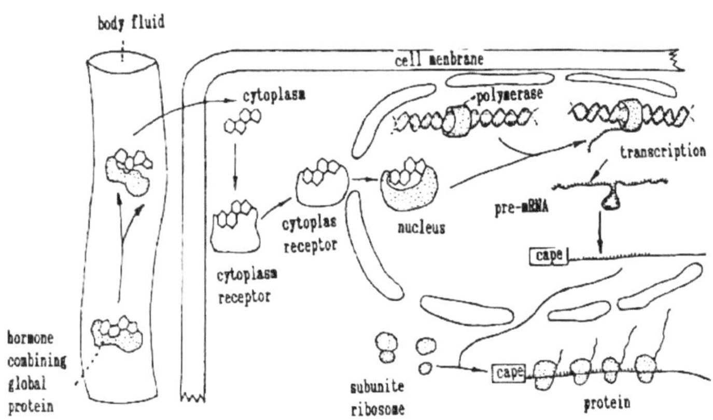

Fig. 12: Diagram explaining the molecular biology of the action mechanism of molting hormone.

4. Juvenile Hormone

4.1 Corpus allatum

Corpus allatum is the true epithelial endocrine gland. It is differentiated and developed from the buds between the mandibular segment and the maxillary segment during embryonic stage. *Corpus allatum* is located at the two lateral sides of the ventral alimentary canal, at the boundary between head and thorax. It is connected to *corpus cardiacum* by a sympathetic nerve (Fig. 13). *Corpus allatum* secretes juvenile hormone. As will be discussed later, different species of insects or even different development stages of the same insect can secrete juvenile hormones of different chemical structures. Recently it has been discovered that *corpus allatum* also has the function of storing and releasing brain hormone.

When observing the *corpus allatum* through an electron microscope, the ultrastructure of the neurosecretory granules can be seen. The entire *corpus allatum* is wrapped under a common membrane with axonal and trachea. The internal sides are secretory cells. Although these cells are morphologically different due to the period, they all contact each other in a loosely bound state. The axons seen in the common membrane have two secreted granules, one seen from the fine structure, which is considered to be neurosecretory granules migrated from the neurosecretory cells of the brain, the other lighter granules are of unknown origin. These two kinds of particles, through the branches of the axonal, are distributed in various parts of the *corpus allatum* cells (Fig. 14).

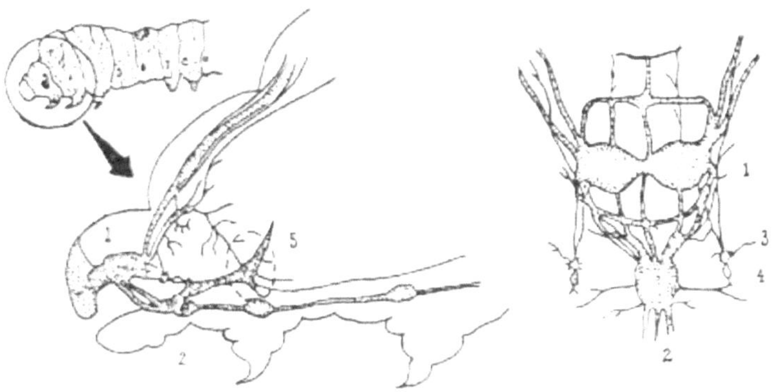

Fig. 13: Diagram of main nerve-endocrine organs of silkworms. 1. brain, 2. subpharyngeal ganglion, 3. *corpus cardiacum*, 4. *corpus allatum*, 5. prothoracic gland

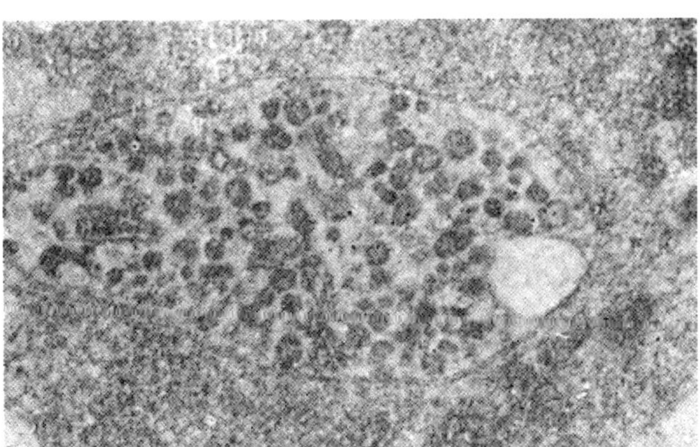

Fig. 14: Ultrastructure of *corpus allatum*.

4.2 Chemistry and Action Mode of Juvenile Hormone

4.2.1 Chemical Structure of Juvenile Hormone

Juvenile hormones secreted by the *corpus allatum* are terpenoids. There are four known juvenile hormones in insects: JH-O or c16JH, JH-I or C_{18}JH, JH-II or C_{17}JH and JH-III or C_{16}JH. Their skeleton structures are shown in Fig. 15.

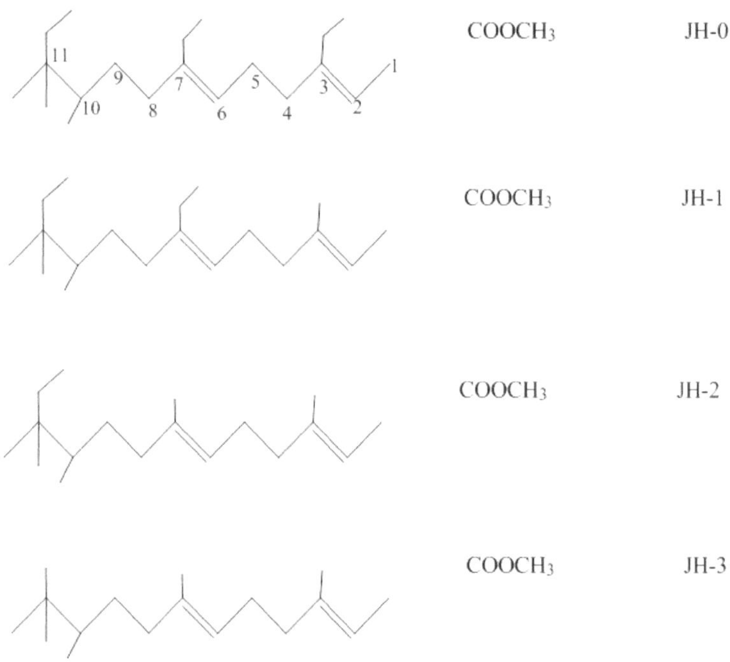

Fig. 15: Chemical structures of juvenile hormones.

In different insects, or at different developing stages of the same insect, there may be one or two juvenile hormones, or several juvenile hormones existing in different proportions. For example, in the larvae and adults of cockroach (Nauphoeta sp.) there are JH-I, JH-II, JH-III, but during the larval stage, 50% of them is JH-III, and during the adult stage 95% of them is JH-III. In the female adults of the tobacco horny moths, there exists only JH-III, but during the larval stage JH-I is dominant except for JH-III. JH-0 is discovered in the eggs of the tobacco horny moths. Hence, it is assumed that the kinds of JH are different between larval and adult stage, and their proportions are also different (Siddall et al. 1974).

4.2.2 Transportation and Metabolism of JH

JH is lipophilic, but in the body fluid it is transported by the protein carrier. In the body fluid two JH-protein complexes are formed. One is a complex formed from albumin or lipoprotein. It is of low affinity and low specialness, but it has high capacity of mutual action. The other is a complex formed from special protein carrier. It is of high affinity and high specialness, but it has low capacity of action. It has been found out that the special protein carrier in the tobacco horny moth is a single polypeptide. Its molecular weight is 28000 Da. The combining coefficient with JH-I is 3.3×10^{-6} M. During the 4th instar, the JH titer in the body fluid is 3.3×10^{-9} M and the dissolubility of JH-I in water substrate is 3.3×10^{-5} M. Hence, there are enough positions at the special protein carrier to combine all JH molecules. Besides the transportation of JH, the protein carrier has another function. That is, it can inhibit a non-special combination of JH with another protein or lipophilic surface. Thus, the protein carrier can reduce the combination of JH with non-target cells, and transport JH to a receptor whose affinity is higher than the protein carrier. This is the reason why JH can recognize the target cell.

JH is synthesized from acetic acid, propionic acid to mevalonic acid, but the exact pathway of biosynthesis is not clear. There are two main pathways of JH decomposition in insects. Carboxylesterase and epoxyhydrolase exist in many tissues, but there is no epoxyhydrolase in body fluid. The products of JH decomposition by these two enzymes are inactive. The decomposition of JH in tissues is important, but the activity of esterase and the existence of a certain critical factor in the body fluid have great influence on it. Their action can limit the amount of JH arriving at the target tissue. In the body fluid there are two common kinds of enzymes: one is general esterase, which can not affect the combination of JH with special protein carrier, the other is JH special esterase, which can act on the free JH and JH which can combine with special protein carrier to form a complex. Hence, the JH titer in the body fluid depends on: (i) The speeds of JH synthesis and release; (ii) The activity of the decomposition enzyme; (iii) The amount of protein carrier; (iv) The amount of JH absorbed and secreted by the tissues. The fluctuation of activity of JH special esterase plays an important role. Special protein carrier can protect JH from being decomposed by general esterase, but it is not resistant to the action of JH special protein carrier. The synthesis and release speed of *corpus allatum* plays a dominant role at the change of JH titer during different developing stages, and the protein carrier and decomposition mechanism can only reduce the signal of the hormone (Maxwell et al. 2002) (Fig. 16).

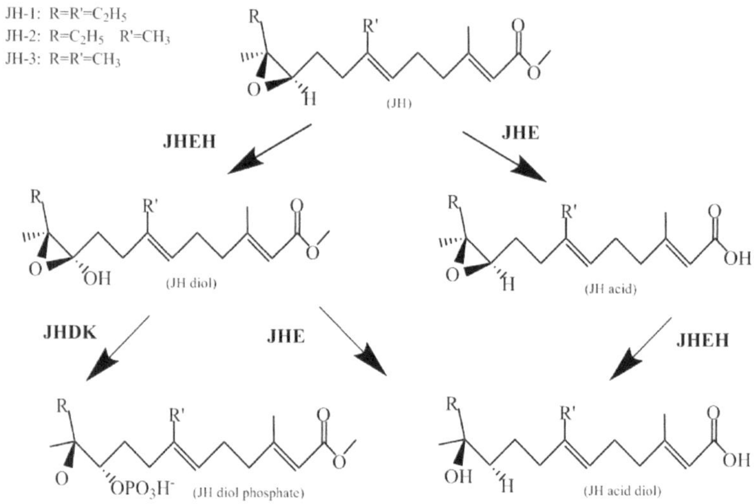

Fig. 16: The main dissimilation metabolism pathway of JH.

4.3 Role of Juvenile Hormone

Juvenile hormones do not have the specificity of "species" and "order". Any kind of *corpus allatum* can function in other insects. The main function of juvenile hormones is maintaining larval morphology, preventing the development of adult organ buds and preventing metamorphosis. At the same time, its function is also manifested in the following areas:

(1) The Juvenile hormone can promote DNA synthesis, inhibit RNA and protein synthesis. Only in the JH less secretion or stop secretion, we can find a large quantity of fat accumulating protein *in vivo* for rapid growth of the silk gland. The ovarian tissue can grow rapidly only in the absence of JH.

(2) JH up to a certain concentration can cause larval diapause of many insects, such as stem borers. This is because JH inhibits the activity of the brain and anterior thymus. Long light can reduce the activity of the pharyngeal side to remove the larvae diapause.

(3) JH can promote ovarian development, egg yolk accumulation and egg maturation, and maintain the function of the testis. For this reason, it is also known as "genitourin." The activity of the *corpus allatum* of the silkworm has nothing to do with the maturation of eggs, but it has a certain relationship with the number of eggs and their chemical composition.

(4) JH enhances the activity of alanine-glyoxylate aminotransferase, alanine-ketoglutarate transaminase, right-aminotransferase, and left aminotransferase in the posterior silk gland and fat body by 10–40%. This may be the mechanism of increased silk gain.

(5) JH can affect the transfer of the precursor of melanin–tyrosine—to ectoderm tissue or prevent the formation of melanin. Experiments with silkworm *corpus allatum* transplantation or injection of JH in 4th instar have found that low titer JH can increase the degree of darkness in the marker.

4.4 JH and Metamorphosis

The role of JH is to maintain the larval character. As discussed above, the excision of the *corpus allatum* of the 2nd and 3rd instar larvae can result in early maturation and pupation and lead to the production of very small moths. On the contrary, the transplantation of *corpus allatum* or the application of JH can prolong the duration of the instar and result in large moths and cocoons. General speaking, JH titer is high during the larval stage. It is low during the late stage of the 5th instar, particularly during pupation. This is the precondition of pupation. After the emergence of moths, the existence of JH is not needed. JH controls the extent and direction of histo-differentiation during every molting. The morphogenesis effects of JH depend on the time of its release. It must affect earlier than molting hormone, or it affects before cell division. The target tissue is sensitive to JH only at a limited stage. The sensitive stage is different at different instars or at different positions. Only with the existence of JH at early stages of every instar can the larval character be maintained.

JH and MH control the morphogenesis of the silkworms harmoniously, though their action times and models are different. MH regularly instructs whether all tissues and organs are developed or not. JH opens the genes controlling larval development and inhibits the genes of adult character and then decides to what degree the cells develop along a particular program of morphogenesis.

We still do not know enough about the mutual relation between JH and MH. If we spray excessive JH analogue (HA) to the 5th instar larvae of silkworm, pupation will be delayed and inhibited and cocooning will not take place, but after feeding of MH, most of the 5th instar larvae will cocoon and pupate. Under this circumstance, it looks like that excessive JH can lead to the lack of a certain endocrine secretion, but not because of the lack of response of the target tissue to MH.

5. The Control of Corpus Allatum and Prothoracic Gland by the Brain

5.1 Brain Hormone

The brain can produce various types of neuro-hormones, which are known by a joint name—brain hormone. It plays important roles during insect development. Of them, prothoracotropic hormone (PTTH) stimulates the prothoracic gland to secrete MH in order to control the molting and metamorphosis of the silkworms. Recently, studies by means of intro culture of organs and MH radio-immunoassay, etc., have demonstrated that PTTH in the tobacco horny moth is synthesized by a pair of cerebral neurosecretory cells and released by the *corpus allatum*. It can promote the speed of MH synthesis in prothoracic gland. Also, in silkworms, the situation is similar to that of the tobacco horny moth. In silkworms, the PTTH is secreted by the lateral large size neurosecretory cells. It is stored in the *corpus allatum* by way of internal lateral nerve-corpus *cardiacum* neurite and is released by the *corpus allatum*. It was reported that, in silkworms, the PTTH is a simple peptide. It contains tyrosine residue and its N-end is glycine. The amino acid analysis of PTTH hydrolysates indicates the integer proportion of amino acid residues is nearly as follows: serine 1, glutamic acid and glutamine 4, proline 1, glycine 4, alanine 3, aspartic acid and asparagine 4, cysteine 4, valine 4–5, leucine 5–6, tyrosine 1–2, phenylalanine 2, histidine 1, threonine 3, arginine3. According to composition of amino acids, the molecular weight of PTTH is 4330–4740 Da. This is in accord with the value measured by means of G-50 gel filtration of cross link glucosan (molecular weight 4400+/–400 Da).

Signals of outside environmental factors such as temperature and light cycle, and *in vivo* nutrition and physiology condition including hormone condition, are delivered to nerve center of the brains. After comprehensive analysis reaction is made, and PTTH or other various brain hormones are secreted. Except responding to the brain neurosecretory cells, the brain never center response to corpus allatum directly nervous stimulus.

5.2 Control of Prothoracic Gland by Brain

During the whole development stage, the secretion activity of ecdysone by the prothoracic gland changes regularly. The change of this secretion activity generally reflects the dynamic of ecdysterone content in body fluid. The secretion activity of the prothoracic gland is mainly controlled and regulated by PTTH secreted by brain neurosecretory cells and released by the *corpus allatum*. The action of PTTH on the prothoracic gland must last a long time, then the latter can be activated. The time that PTTH needs in order to activate the prothoracic gland is called the critical period. After the critical period, the function of secreting MH can independently be brought into play and even the brain and prothoracic gland are excised. Besides, the *in vitro* culture of prothoracic gland revealed that the existence of JH in the culture medium can inhibit the secretion activity of the cells, and the MH titer can also regulate the secretion activity of the gland cells. When the titer is high, it has feedback inhibition action on the prothoracic gland, while low titer can lead to positive feedback inhibition.

5.3 Control of Corpus Allatum by the Brain

In the silkworms, the secretion activity of the *corpus allatum* can be regulated in two ways: one is by the nervous stimulus of the brain, the other is by the regulation of brain neurosecretion. It was reported that, if single brain-*corpus allatum* is transplanted, JH titer is higher than that if single brain (or *corpus allatum*) is transplanted. This indicates that the brain can control JH secretion activity of the *corpus allatum* by stimuli of the nerve axon. About the control and regulation of *corpus allatum* by BH, some people assumed that, in the brain neurosecretion there exists a hormone which can promote and inhibit the activity of the *corpus allatum*.

5.4 Brain-corpusAllatum-prothoracic GlandSystemandMetamorphosis

To sum up, molting and metamorphosis in silkworms is a qualitative change that takes place when individual development reaches a certain stage. It is a complex physiological process, completed by the coordinative action of JH (secreted by the *corpus allatum*) and MH (secreted by the prothoracic gland) under the control of the central nerve center. In the silkworms, all the internal tissues and organs, the external morphological characters, and the behavior of the adult stage are decided by genetic genes, but the time and manner in which these genetic genes express themselves is influenced by the environmental conditions and regulated by the *in vivo* hormones. The physical conditions, like temperature, light cycle, etc., *in vivo* nutrition, can all be delivered to the central nerve center (brain) as signals. After comprehensive analysis, reaction is made: the brain neurosecretory cells secrete BH (or directly through nerve stimuli) in order to cause the *corpus allatum* to secrete JH or cause the prothoracic gland to secrete MH. The combined action of JH and MH leads to the molting of larvae. When a small amount of JH is secreted, MH leads to the pupation; when there is no JH, MH leads to the emergence of adults and the process of metamorphosis is complete.

6. Concluding Remark

The growth and development of silkworm are regulated by hormones, especially the juvenile hormone and the molting hormone. Studies on the regulation and mechanism of insect hormones are important in establishing a rational rearing method, to produce large quantities of high quality cocoons and silks.

The development of silkworm mainly includes the molting during larval stages and metamorphosis during the pupa stages. The silkworms belong to holometabolan. Silkworm metamorphosis is not only a huge change in the external form, the internal tissues and organs also undergo drastic changes. Histolysis and histogenesis during the pupal stage is not only the expression of profound morphological changes of tissues and organs, but also a reflection of physiological and biochemical changes such as gas exchange, carbohydrate, lipo and protein metabolism, etc. Studies have shown that the type and titer of hormones regulate the process of the metamorphosis.

The main hormones controlling growth and development of the silkworms are: brain hormone (BH) or prothoracotropic hormone (PTTH), juvenile hormone (JH), molting hormone (MH) and diapause hormone (DH). They are secreted by cerebral secretory cells, *corpus allatum*, prothoracic gland and subpharyngeal ganglion neurosecretory cells, respectively.

At present, 17 steroids that have molting hormone activity have been separated from insects and crustaceans. Ecdysteroid acts directly on the tissues and organs of the ectodermal-original, its main function is to excite and form the new skin and remove the old skin.

Juvenile hormones secreted by the *corpus allatum* are terpenoids. There are four known juvenile hormones in insects: JH-O, JH-I, JH-II and JH-III. The main functions of juvenile hormones are maintaining larval morphology, preventing the development of adult organ buds and preventing metamorphosis.

The brain can produce various types of neuro-hormones, known by the joint name "brain hormone", that play important roles during insect development. Of them, the prothoracotropic hormone (PTTH) stimulates the prothoracic gland to secrete MH in order to control the molting and metamorphosis of the silkworms.

References

Blais, C., C. Dauphin-Nivvlemant, N. Kovganko, J. P. Girault, C. Descouns and R. Lafont. 1996. Evidence for the involvement of 3-oxo-Δ^4 in intermediates in ecdysteroid biosynthesis. Biochem. J. 320: 413–419.

Chaudhuri, A. and A. K. Medda. 1985. Effect of thyroxine on the pattern of variation of protein, RNA and DNA contents of ovary of silkworm *Bombyx mori* during metamorphosis. Environ. Ecol. (Kalyani) 33: 418–423.

Huber, R. and W. Hoppe. 1965. On the chemistry of ecdysone. VII. Analysis of the crystal and molecular structure of the molting hormone in insects, ecdysone, using the atomized folding molecule method. Chem. Ber. 98(7): 2403–2424.

Karlson, P. 1980. Ecdysone in retrospect and prospect. *In*: Jules, A. Hoffmann (ed.). Progress in Ecdysone Research, Elsevier/North Holland, Amsterdam.

Koolman, J. and P. Karlson. 1985. Regulation of ecdysteroid titer: Degradation. pp. 343–361. *In*: G. A. Kerkut and L. I. Gilbert (eds.). Comprehensive Insect Physiology, Biochemistry and Pharmacology, Vol. 7. Pergamon Press, Oxford.

Lv, H. S. 2011. Principles of the Silkworm Grainage. Shanghai Science and Technology Press, Shanghai.

Morohoshi, S., T. Oshiki and I. Kikuch. 1977. The control of growth and development in *Bombyx mori*. Proc. Japan Academy 53(B). (5): 199–203.

Maxwell, R. A., W. H. Welch and D. A. Schooley. 2002. Juvenile hormone diol kinase. I. Purification, characterization, and substrate specificity of juvenile hormone-selective diol kinase from *Manduca sexta*. J. Biol. Chem. 277(24): 21874–21881.

Niwa, Y. S. and R. Niwa. 2016. Transcriptional regulation of insect steroid hormone biosynthesis and its role in controlling timing of molting and metamorphosis. Develop. Growth Differ. 58: 94–105.

Niwa, Y. S. and R. Niwa. 2016. Ouija board: A transcription factor evolved for only one target in steroid hormone biosynthesis in the fruit fly *Drosophila melanogaster*. Transcription. 7(5): 196–202.

Rees, H. 1995. Ecdysteroid biosynthesis and inactivation in relation to function. Eur. J. Entomol. 92: 9–39.

Sakurai, S., J. Warren and L. I. Gilbert. 1989. Mediation of ecdysone synthesis in *Manduca sexta* by a haemolymph enzyme. Arch. Insect Biochem. Physiol. 10: 179–197.

Siddall, J. B., K. J. Judy, D. A. Schooley, R. C. Jennings, B. J. Bergot and M. Sharon hall. 1974. Insect juvenile hormones and insect growth regulators. Biochem. Soc. Trans. 2(5): 1027.1–1027.

Wu, Z. D., J. L. Xu, W. D. Shen, M. Z. Zhou, F. L. Wu and S. L. Lv. 1981. Anatomic physiology of silkworm. China Agriculture Press, Beijing.

Zhou, W. S. and H. S. Lv. 1980. Regulation of the growth and silk production in silkworm *Bombyx mori* by phytogenous ecdysteroids. *In*: Jules, A. Hoffmann (ed.). Progress in Ecdysone Research, Elsevier/North Holland, Amsterdam.

CHAPTER 5

Post-translational Modification in Silkworm

Sungjo Park,[1,*] *Enoch Y. Park*[2] and *Andre Terzic*[1]

1. Introduction

Post-translational modification is a chemical adaptation of proteins, enabling to escape from nature's limited genetic confinement (Barber and Rinehart 2018, Prabakaran et al. 2012, Walsh et al. 2005). While the completion of the human genome project provided an estimated 25,000 genes, the size of the human proteome is expected to be at over 1 million proteins (Aebersold et al. 2018, Jensen 2004). Post-translational modification is an indispensable source of protein diversity, providing extended potential protein functions by attaching small chemical groups to the selected amino acid through a covalent bond. More than 200 different types of post-translational modifications (http://www.uniprot.org/docs/ptmlist) have been identified including phosphorylation, glycosylation, methylation and acetylation. Established databases such as UniProt, dbPTM (Lu et al. 2013), PTMcode (Minguez et al. 2013) and PTMCuration (Khoury et al. 2011) offer curated post-translational modification information. Statistical analysis of proteome-wide post-

[1] Department of Cardiovascular Diseases and Center for Regenerative Medicine, Mayo Clinic, Rochester, Minnesota, 55905, USA.

[2] Laboratory of Biotechnology, Research Institute of Green Science and Technology, Shizuoka University, Shizuoka, 422-8529, Japan.

E-mails: park.enoch@shizuoka.ac.jp; terzic.andre@mayo.edu

* Corresponding author: park.sungjo@mayo.edu

translational modification based on Swiss-Prot database reveals that there are about 90,000 experimentally identified post-translational modifications, and about 240,000 putative modifications on 530,000 proteins (https://selene.princeton.edu/PTMCuration/) (Khoury et al. 2011). Among them, phosphorylation dominates the frequency of experimental post-translational modifications, followed by acetylation and N-linked glycosylation (Barber and Rinehart 2018).

Post-translational modification significantly alters protein structure and function, including the regulation of protein activity, localization and interactions (Bah and Forman-Kay 2016, Barber and Rinehart 2018, Bode and Dong 2004, Jensen et al. 2002). Structural analysis of Protein Data Bank using root-mean-square deviation revealed that post-translational modifications induce conformational changes at both local and global levels, yet not extreme changes to protein structure. Only 7% glycosylated and 13% phosphorylated protein undergo more than 2 Å global changes (Xin and Radivojac 2012). Nevertheless, the conformational change induced by post-translational modification introduces a significant alteration of protein functions (Bah and Forman-Kay 2016, Barber and Rinehart 2018, Bode and Dong 2004, Jensen et al. 2002). More than 60% of post-translational modification occurs at functional domains of protein, creating binding sites for specific protein-protein interaction domains to regulate molecular and cellular function (Lu et al. 2013).

Due to the intricate molecular functions associated with post-translational modification, numerous approaches have been applied in order to produce authentic recombinant proteins for structural and functional studies as well as for theranostic applications (Midgett and Madden 2007, Walsh and Jefferis 2006). The widely used *Escherichia coli*-based expression approach provides the advantages of the relatively low production cost, a basic experimental setup and the abundant tools for gene manipulations, yet the limited capacity for post-translational modification in bacteria has confined the production of eukaryotic proteins to post-translational processes (Midgett and Madden 2007). The insect-based system offers a distinct advantage in this regard, since insects can execute post-translational modification in a similar fashion to eukaryotic cells (Aucoin et al. 2010, Jarvis 2009, Kost et al. 2005). Since the landmark production of α-interferon in silkworm larvae (Maeda et al. 1985), silkworm has been increasingly recognized as a reliable host for eukaryotic protein production platform, implementing a proper post-translational modification.

Recombinant protein production in silkworm is traditionally executed by co-transfecting the wild type *Bombyx mori* nucleopolyhedrovirus (BmNPV) genome and transfer plasmid DNA, following the isolation and amplification of recombinant baculoviruses (Kato et al. 2010). These amplified viruses are then infected in the silkworm larvae or pupae in order to express recombinant protein. The alternative approach developed a decade ago utilizes BmNPV bacmid, an *E. coli* and *B. mori* hybrid shuttle vector, to facilitate heterologous protein production without baculovirus construction and amplification of viruses in *B. mori* culture cells (Kato et al. 2010, Motohashi et al. 2005, Park et al. 2013, 2017). The recombinant BmNPV bacmid DNA is directly injected into silkworm larvae or pupae for foreign gene expression, dramatically expediting protein production time. Using this system, cellular, mitochondrial, secretory and membrane proteins were successfully generated (Dojima et al. 2010, Du et al. 2009, Hwang et al. 2014, 2015, Miyazaki et al. 2017, Park et al. 2013, 2017).

Herein, this chapter discusses three post-translational modifications such as phosphorylation, biotinylation and disulfide bond formation in heterologous proteins produced in a silkworm BmNPV bacmid-based expression system. These are human acetyl-CoA carboxylase 2 (ACC2), human malonyl-CoA decarboxylase (MCD) and porcine insulin-like peptide 3 (pINSL3), all of which are very difficult to produce with full functionality due to post-translational modifications. The rigorous evaluation of post-translational modifications highlights the silkworm *B. mori* as a fail-proof host for high fidelity recombinant eukaryotic protein production implementing proper post-translational modifications.

2. Post-translational Modifications

2.1 Phosphorylation

Phosphorylation is the covalent attachment of a phosphate group by protein kinase to the selected amino acid residues such as serine, threonine and tyrosine (Ubersax and Ferrell 2007). In addition to these canonical phosphorylation-prone residues, arginine, lysine, aspartic acid, glutamic acid and cysteine could be phosphorylated (Ciesla et al. 2011). Protein phosphorylation is an integral part of the cell, regulating many cellular processes in eukaryotes (Hunter 1995). Historically, the presence of phosphate in vitellin was reported in the early twentieth century (Levene and Alsberg 1906), yet the enzymatic reaction for phosphorylation was reported at a much later date (Burnett and Kennedy 1954). A PubMed search using the keyword "phosphorylation" currently returns more than 280,000 publication hits, and about 30% of all cellular proteins are phosphorylated on at least one residue (Ubersax and Ferrell 2007), underscoring the importance of prevalent protein phosphorylation in regulating many cellular processes in a biological system. Protein kinase catalyzes phosphorylation, whereas protein phosphatase processes the reverse dephosphorylation reaction (Fig. 1A).

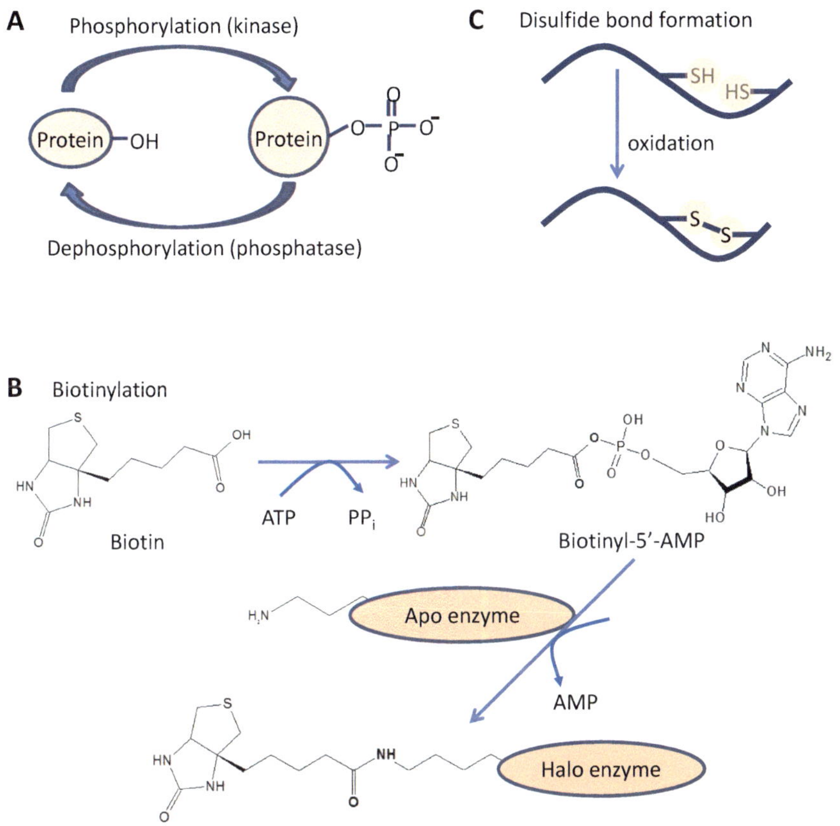

Fig. 1: Post-translational phosphorylation, biotinylation and disulfide bond formation. (A) Post-translational phosphorylation happens in hydroxyl group of amino acid residues by kinase, and de-phosphorylation by phosphatase. (B) Biotinylation happens in a two-step reaction mechanism by biotin protein ligase. First reaction is the formation of an intermediate biotinyl-5′-AMP through the condensation of biotin and ATP. The second step reaction is the covalent attachment of biotin to lysine residue in the biotin carboxyl carrier domain. (C) Disulfide bond formation between cysteine residues in proximity.

2.2 *Biotinylation*

Protein biotinylation is one of the post-translational modifications which attaches a biotin cofactor onto a specific metabolic enzyme. Unlike other cardinal post-translational modifications that often recognize a selected short amino acid sequence within the protein substrate for chemical modification, biotinylation requires a target lysine residue harbored within the highly conserved tertiary structure domain with ~ 70 amino acid residues (Beckett et al. 1999, Sternicki et al. 2017). Protein biotinylation plays a vital role in cellular metabolic pathways involving energy transduction, lipogenesis, gluconeogenesis and amino acid metabolism

(Chapman-Smith and Cronan 1999, Chaturvedi and Tyagi 2014, Sternicki et al. 2017). Biotinylation has also been increasingly recognized as a purification method due to its extremely high affinity and specificity in biotechnological applications (Leder 2015).

Biotin protein ligase, also known as holo-carboxylase synthetase, is accountable for all biotin-dependent enzymes, demonstrating high selectivity to produce no aberrant off-target biotinylation (Beckett et al. 1999, Chapman-Smith and Cronan 1999, Sternicki et al. 2017). Biotin protein ligase employs a two-step biotinylation reaction mechanism (Fig. 1B). The first step is a condensation reaction where biotin and ATP form the reaction intermediate of biotinyl-5′-AMP. The second reaction is the covalent attachment of biotin to a vital lysine residue located in the biotin carboxyl carrier protein (BCCP) domain with the release of AMP. There are five biotin-dependent enzymes in mammals, including acetyl-CoA carboxylase I and II, pyruvate carboxylase, propionyl-CoA carboxylase and 3-methylcrotonyl-CoA carboxylase. In humans, the congenital defect of biotin protein ligase and mishandling of biotin precipitate multiple carboxylase deficiencies, resulting in anything from minor symptoms such as a skin rash to more serious conditions such as developmental delay and even death (Online Mendelian Inheritance in Man (OMIM): 253270).

2.3 Disulfide Bond Formation

The disulfide bond plays an integral role, providing a stable three-dimensional structure as well as catalytic or regulatory functions (Hogg 2003, Sevier and Kaiser 2002). In a reaction between the sulfhydryl side chains of two cysteine residues in protein (Fig. 1C), the disulfide bond formation is affected by the spatial accessibility and proximity of the cysteine residues, the difference between the pKa of thiol groups and the pH of the local environment, and the redox environment. Disulfide bond formation occurs more commonly in the endoplasmic reticulum than in the cytosol of eukaryotes, because the endoplasmic reticulum intraluminal environment is more oxidizing than that of the cytosol (Bulleid 2012, Hogg 2003). In eukaryotes, protein disulfide isomerase, containing two thioredoxin-like active sites with two cysteine residues, is responsible for the disulfide bond formation of unfolded proteins (Wilkinson and Gilbert 2004). Based on the proteome analysis at protein data bank (PDB), eukaryotic proteins imprisoned to the secretory pathway have about three times higher frequency of disulfide bonds than bacterial proteins (Bosnjak et al. 2014). Furthermore, the average number of disulfide bonds is positively correlated with protein length for proteins of more than 200 amino acids, and negatively correlated for the shorter proteins (Bosnjak et al. 2014).

3. Post-translational Modifications in Acetyl-CoA Carboxylase, Malonyl-CoA Decarboxylase and Insulin-like Peptide 3

3.1 Acetyl-CoA Carboxylase 2 (ACC2)

The biotin-dependent enzyme, acetyl-CoA carboxylases (ACC), catalyzes the production of malonyl-CoA from acetyl-CoA, an integral metabolic modulator in anabolic and catabolic lipid metabolism (Fig. 2) (Brownsey et al. 2006, Tong 2013, Wakil and Abu-Elheiga 2009). *ACACA*-encoded ACC1 in the cytosol of lipogenic organs, such as the liver and adipose tissue, produces malonyl-CoA. This acts as a substrate for long-chain fatty acid synthesis, whereas *ACACB*-encoded ACC2 in the outer membrane of mitochondria of oxidative tissues, such as the heart, liver and skeletal muscle, delivers malonyl-CoA which is utilized as a negative regulator of lipid oxidation (Brownsey et al. 2006, Wakil and Abu-Elheiga 2009). The abundant malonyl-CoA restrains catabolic fatty acid β-oxidation by inhibiting carnitine palmitoyl transferase 1 (CPT-1), which catalyzes

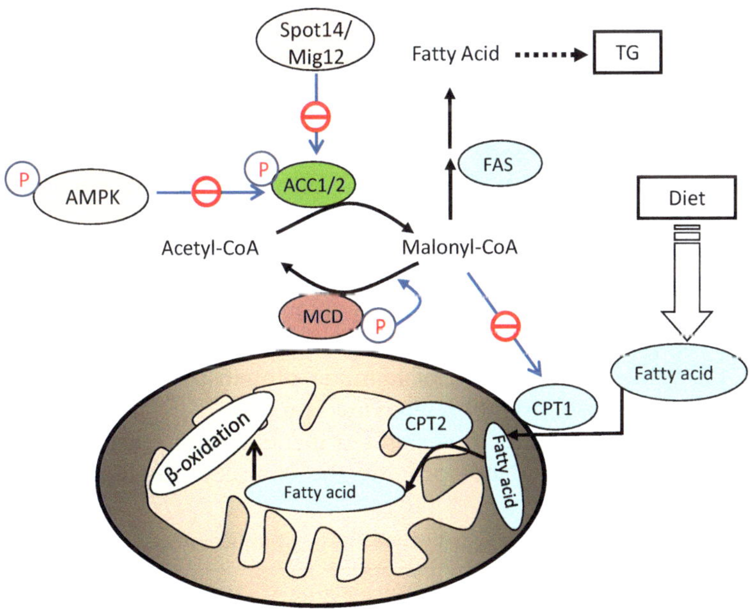

Fig. 2: Homeostatic regulation of fatty acid synthesis and β-oxidation through an acetyl-CoA/acetyl-CoA carboxylase/malonyl-CoA decarboxylase/malonyl-CoA axis. Phosphorylation of acetyl-CoA carboxylase (ACC) by AMP-activated kinase (AMPK) or inhibition of ACC by Spot14/Mig12 diminishes malonyl-CoA levels, subsequently promoting lipid oxidation. Phosphorylation of MCD enhances malonyl-CoA decarboxylation by reducing malonyl-CoA levels in the cytoplasm, which promotes stimulation of fatty acid oxidation by releasing the malonyl-CoA inhibition of carnitine palmitoyl transferase 1 (CPT1). FAS, fatty acid synthesis; TG, triglyceride.

the transfer of long-chain fatty acid into mitochondria. Because of dual functions, ACC as a bioenergetics controller promotes stem cell function and tissue regeneration (Folmes et al. 2013, Fullerton et al. 2013, Knobloch et al. 2013). Moreover, since anti-obesity effects and prevention of cardiac remodeling were observed in ACC2 knockout mouse, the activity regulation of ACC2 has been considered as an attractive target for deranged lipid metabolism cases, such as obesity, cardiovascular disease and diabetes (OMIM: 601557) (Abu-Elheiga et al. 2001, 2003, Kolwicz et al. 2012).

Eukaryotic ACCs carry out multiple distinct functions with a biotin carboxylase domain (BC), a biotin carboxyl carrier protein domain (BCCP) and a carboxyltransferase domain (CT) (Tong 2013). Allosteric regulation and post-translational biotinylation and phosphorylation in ACC are critical in implementing the full functionality of multi-step reactions (Lee et al. 2008, Tong 2005, 2013). Recently, the post-translational biotinylation and phosphorylation, and citrate-induced polymerization of human ACC2 were demonstrated utilizing the BmNPV bacmid-based silkworm expression system, providing a high yield of purified proteins (Table 1 and Fig. 3) (Hwang et al. 2014).

The biotinylation of recombinant ACC2 from silkworm was analyzed by Western blotting using an anti-biotin antibody, and further validated by streptavidin horseradish peroxidase conjugate because streptavidin is well-known to interact with biotin with an extremely high affinity (dissociation constant = 10^{-14} mol/L). This is one of the strongest non-covalent interactions (Green 1975). Purified ACC2 revealed the intense biotinylated bands in Western blotting compared with crude extracts of ACC2 from silkworm (Fig. 3B and 3C). Post-translational phosphorylation of ACC2 was also evaluated by immunoblotting analysis using a monoclonal anti-phosphoserine antibody, which demonstrated phosphorylated bands of ACC2. While the treatment of Lambda protein phosphatase, a Mn^{2+}-dependent dephosphorylation enzyme, progressively diminished the phosphorylation of ACC2, the extended incubation of phosphatase with ACC2 up to 6 hr could not provide the complete dephosphorylation

Table 1: Catalytic specific activities and yields of purified hACC2 and hMCD.

	Specific Activity (nmol/mg/min)	Yield	Reference
hACC2	0.79 ± 0.23	500 μg/pupa	Hwang et al. 2014
+ phosphatase	1.34 ± 0.44		
hMCD	48.2 ± 7.9	344 μg/pupa	Hwang et al. 2015
hMCD-S204G	24.4 ± 2.0		
hMCD-Y405F	31.2 ± 1.7		

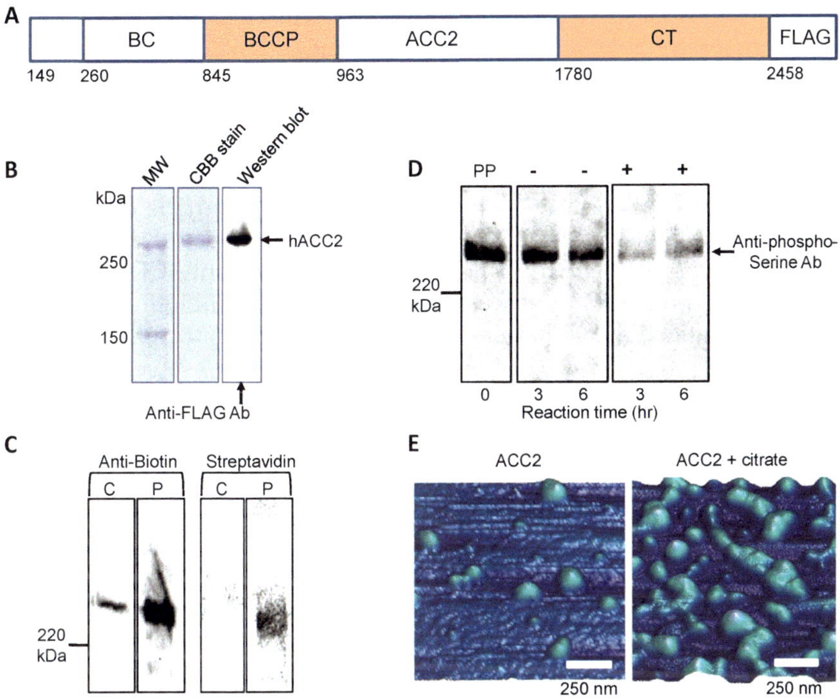

Fig. 3: Functional and structural properties of human acetyl-CoA carboxylase (hACC2). (A) Human ACC2 is a large polypeptide composed of an amino-terminal mitochondrial attachment signal domain, a biotin carboxylase domain (BC), a biotin carboxyl carrier protein (BCCP) domain and a carboxyltransferase domain (CT). In order to enhance the solubility of hACC2, the N-terminal mitochondrial target sequence was deleted for expression in silkworm. (B) The expression of hACC2 was confirmed with the analysis of SDS-PAGE Coomassie Brilliant Blue (CBB) staining and Western blot with anti-FLAG antibody. (C) The purified hACC2 from silkworm harbors post-translational biotinylation validated by Western blot using an anti-biotin antibody and horseradish peroxidase-conjugated streptavidin. C and P denote the crude extract and purified hACC2, respectively. (D) Dephosphorylation of hACC2 treated with the Lambda protein phosphatase was evaluated by Western blotting using an anti-phosphoserine antibody. The extended incubation with phosphatase could not provide the complete dephosphorylation of hACC2, probably due to the accessibility of every phosphorylated site. (E) Nanoscale imaging of ACC2 with high-resolution atomic force microscopy revealed the homogeneous particles of ACC2 and citrate-induced filament formation. (Modified from Hwang et al. 2014. Human acetyl-CoA carboxylase 2 expressed in silkworm *Bombyx mori* exhibits posttranslational biotinylation and phosphorylation. Appl. Microbiol. Biotechnol. 98: 8201–8209. https://creativecommons.org/licenses/by/4.0/.)

(Fig. 3D). This finding suggests that the phosphatase could not readily access every phosphorylation site in fully-assembled ACC2.

Dephosphorylation of ACC2 affects the catalytic function. Phosphorylated ACC2 purified from the silkworm provided a specific activity of 0.79 ± 0.23 nmol Pi/mg/min, whereas dephosphorylated

ACC2 provided a specific activity of 1.34 ± 0.44 nmol Pi/mg/min, about a 2-fold increase (Table 1). Similar enzymatic measurements were observed in knock-in mice model with the replacement of Ser212 with alanine, this to nullify the phosphorylation (Fullerton et al. 2013), suggesting the phosphorylation-associated modulation of enzymatic function. Nanoscale high resolution imaging using atomic force microscopy showed that ACC2 purified from silkworm revealed the homogeneous particle distribution as well as the filamentous polymeric forms in the presence of the allosteric regulator of citrate (Fig. 3E). Collectively, the recombinant ACC2 produced in silkworm possesses the necessary post-translational biotinylation and phosphorylation for maintaining structural and functional integrity.

3.2 Malonyl-CoA Decarboxylase

Malonyl-CoA decarboxylase (MCD) is a metabolic enzyme responsible for sustaining a homeostatic lipid metabolism along with ACC (Folmes et al. 2013). Encoded by *MLYCD*, MCD catalyzes the production of acetyl-CoA from malonyl-CoA, a metabolic intermediate in anabolic and catabolic lipid metabolism (Fig. 2). The acetyl-CoA/ACC/MCD/malonyl-CoA axis operates as a metabolic hub and integrates the output of multiple signaling pathways. Deficiency of MCD in humans triggers multiple disorders including cardiomyopathy, hypoglycemia, hypotonia, mild mental retardation, metabolic acidosis and malonic aciduria (OMIM: 248360) (Brown et al. 1984, Xue et al. 2012).

Catalytic function of MCD is modulated by post-translational modification. For example, acetylation of MCD enhances fatty acid β-oxidation, whereas deacetylation of MCD hampers decarboxylase activity, promoting *de novo* fatty acid biosynthesis (Laurent et al. 2013). The effects of post-translational phosphorylation on the catalytic function of MCD were identified utilizing the BmNPV bacmid-based silkworm expression system (Hwang et al. 2015). Silkworm-expressed human MCD (hMCD) demonstrated specificity and post-translational phosphorylation, validated by Western blotting analysis using anti-FLAG and anti-phosphoserine antibodies (Hwang et al. 2015). The specific activities of hMCD, measured in the coupled reactions by the production of reduced nicotinamide adenine dinucleotide, were 59.6 ± 7.7 nmol/mg/min from silkworm fat body and 48 ± 8 nmol/mg/min from silkworm pupae. These values are much higher than that of hMCD from *Escherichia coli*-based expression (Zhou et al. 2004), signifying that the post-translational modification executed in the silkworm expression system indeed modulates the catalytic activity of hMCD.

To elucidate hMCD catalytic activity associated with phosphorylation, potential phosphorylation sites were evaluated using the bioinformatics program NetPhos 2.0, neural network prediction for eukaryotic protein phosphorylation (Blom et al. 1999). Nine highly conserved residues, including Ser204, Ser275, Ser326, Ser380, Thr9, Thr60, Thr245, Thr396 and Tyr468, were putatively mapped as phosphorylation site candidates (Hwang et al. 2015). An additional proteomic search using mass spectrometry analysis identified exclusive phosphorylation residues of hMCD at Ser204 and Tyr405, eliminating other predicted phosphorylated residues. The hMCD Ser204Gly and Tyr405Phe point mutants displayed decreased intensities on immunoblotting (Hwang et al. 2015) as well as the specific decarboxylase activity (Table 1). Specific activities of hMCD Ser204Gly and Tyr405Phe mutants exhibited about 50% and 40% reduced activity, respectively, compared with wild type hMCD, implying that the hMCD catalytic function is indeed modulated by post-translational phosphorylation (Hwang et al. 2015). Collectively, the BmNPV bacmid-based silkworm expression system offers a fail-safe eukaryotic bioengineered protein production platform implementing proper phosphorylation.

3.3 Insulin-like Peptide 3

Insulin-like peptide 3 (INSL3), previously known as relaxin-like factor, is a member of the relaxin/insulin superfamily, having evolutionally conserved three intrinsic disulfide bonds (Fig. 4). Originally discovered in a boar testis cDNA library screen, INSL3 is essential for fetal testis decent as well as for sperm production and function (OMIM: 146738) (Anand-Ivell and Ivell 2014, Bullesbach and Schwabe 2002, Ferlin et al. 2017, Ivell et al. 2017, Minagawa et al. 2012). While the interest in INSL3 for research and clinical applications is proliferating, INSL3 isolation and purification methods that maintain the specific three disulfide bonds for active biological function are very limited (Bullesbach and Schwabe 2002, Gundlach et al. 2009, Luo et al. 2009, Minagawa et al. 2012). The silkworm-based expression system recently enabled the production of high-fidelity INSL3 fulfilling post-translational disulfide bond formation (Miyazaki et al. 2017).

Porcine INSL3 (pINSL3) with an amino-terminal FLAG tag was expressed in silkworm using the BmNPV bacmid and successfully secreted into larval hemolymph using its innate signal peptide composed of 20 amino acid residues (MDPHPLTWALVLLGPALALS) (Fig. 4). The collected hemolymph was pre-treated with HCl to aggregate silkworm proteins. After acidification and neutralization, recombinant pINSL3 was purified by FLAG-tag affinity chromatography, which migrated to a 14 kDa band resolved by SDS-PAGE and Western blot analysis, consistent with the expected molecular weight (Fig. 4). Typically, the 10 mL of larval hemolymph provides about 8 µg of purified recombinant pINSL3 (Miyazaki et al. 2017).

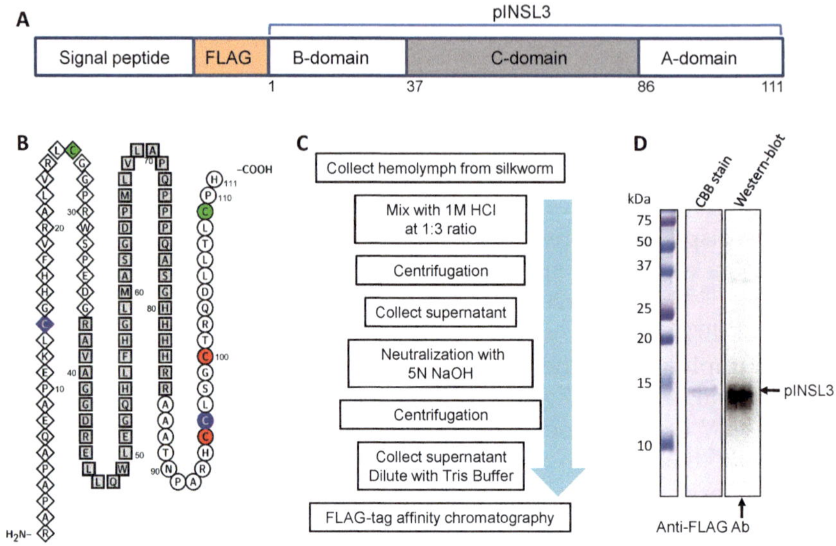

Fig. 4: A schematic presentation of recombinant porcine insulin-like peptide 3 (pINSL3) purification. (A) The pINSL3 construct was designed to include a FLAG-tag after intrinsic signal peptide sequence, which is expected to cleave off by signal peptidase in silkworm. (B) The amino acid residues of pINSL3 with B-, C- and A-domain are indicated by diamonds, squares and circle, respectively. Cysteine residues forming disulfide bonds in pINSL3 are colored with disulfide pairs in blue, green and red. Illustration of the pINSL3 primary sequence was generated using Protter (http://wlab.ethz.ch/protter/start/). (C) Purification strategy of pINSL3. (D) The molecular identity of purified pINSL3 from silkworm was confirmed by SDS-PAGE Coomassie Brilliant Blue (CBB) staining and Western blotting using an anti-FLAG antibody.

The structural integrity of disulfide bond formation in recombinant pINSL3 was evaluated by LC-MS/MS in the reduced and non-reduced conditions (Miyazaki et al. 2017). Mass spectrometry analysis revealed that the recombinant pINSL3 protein comprised the B-, C- and A-domains, including three disulfide bonds like native pINSL3. However, due to the nature of miscleavage from an endogenous signal peptidase, some of the recombinant pINSL3 protein included additional amino-terminal residues of Leu-Ser from the signal peptide. Additionally, recombinant pINSL3 had an internal cleavage between C- and A-domains when compared with the wild type protein (Minagawa et al. 2012).

To examine the evolutionally conserved three disulfide bond positions in recombinant pINSL3, the protein was digested and analyzed using higher-energy collisional dissociation (HCD)-MS/MS and electron transfer dissociation (ETD)-MS/MS (Miyazaki et al. 2017). Figure 5a illustrates the total ion chromatograms of the tryptic-digested recombinant pINSL3. For the non-reduced peptides, the two most intense peaks in the chromatogram (peaks I and II) were observed at 18.7 and

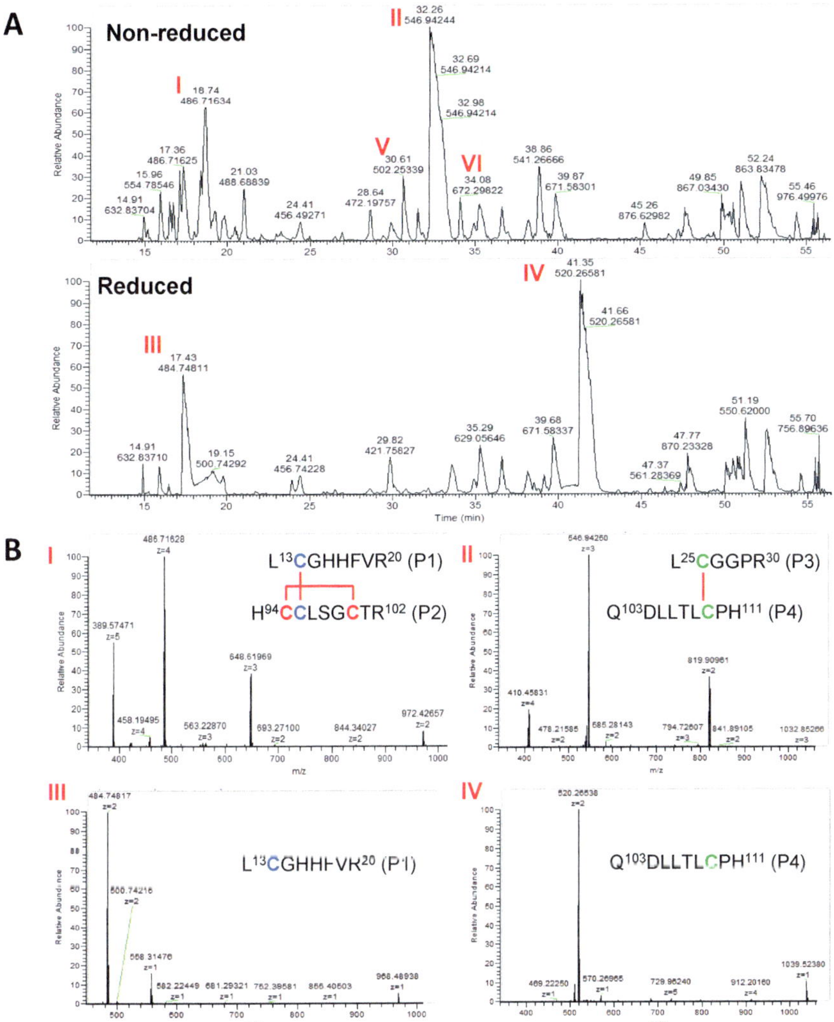

Fig. 5: Mass analysis of the tryptic-digested recombinant pINSL3. (A) Total ion chromatogram of pINSL3 digested with trypsin showed major peaks at 18.7 min (marked I) and 32.3 min (marked II) for the non-reduced form, and at 17.4 min (marked III) and 41.4 min (marked IV) for the reduced form. (B) MS1 survey spectra of these marked peaks (I–IV) revealed the precursor ions of cysteine peptide. Accurate masses of the observed peptide suggest that the peak I indicates that two peptides P1 (L^{13}CGHHFVR^{20}) and P2 (H^{94}CCLSGCTR^{102}) are connected by two disulfide bonds. The peak II indicates that two peptides P3 (L^{25}CGGPR^{30}) and P4 (Q^{103}DLLTLCPH^{111}) are linked by a disulfide bond. Peaks III and IV suggest peptide P1 and P4, respectively. Two unexpected peptide fragments were also eluted at 30.6 min (marked V) and 34.1 min (marked VI), representing that Cys109 in P4 (Q^{103}DLLTLCPH^{111}) are connected with Cys14 in P1 (L^{13}CGHHFVR^{20}) in peak V or one of the cysteines in P2 (H^{94}CCLSGCTR^{102}) in peak VI, respectively. (Modified from Miyazaki et al. 2017. Insulin-like peptide 3 expressed in the silkworm possesses intrinsic disulfide bonds and full biological activity. Sci. Rep. 7: 17339. https://creativecommons.org/licenses/by/4.0/.)

32.3 min, respectively, whereas for the reduced peptides, the two major peaks (peaks III and IV) were observed at 17.4 and 41.4 min, respectively. Figure 5b shows the most abundant precursor ions in the survey spectra of these peaks (peaks I–IV). The HCD-MS/MS analysis of the reduced digest identified cysteine-containing peptides P1 (residues 13–20, $L^{13}CGHHFVR^{20}$), P2 (94–102, $H^{94}CCLSGCTR^{102}$), P3 (25–30, $L^{25}CGGPR^{30}$) and P4 (103–111, $Q^{103}DLLTLCPH^{111}$), summarized in Table 2. Using the accurate molecular masses of the observed HCD-MS/MS matched peptides, peak I was expected to be peptides P1 and P2 with two inter- and intramolecular disulfide bonds. Peak II was also expected to be peptides P3 and P4 with one intermolecular disulfide bond. In addition, Peaks III and IV correspond to P1 and P4, which are located in the B- and A-domains, respectively.

The ETD-MS/MS analysis was utilized in order to confirm the predicted disulfide paired ions by showing MS/MS product ions that were predominantly intact cysteine peptides (Miyazaki et al. 2017). ETD fragmentation of the triple-charged product ion with an m/z value of 648.62 (from peak I) revealed two peptide fragments, P1 and P2, indicating that Cys14 in P1 produced a disulfide bond formation with cysteine residues in peptide fragment P2. However, identification of the exclusive cysteine residue among Cys95, Cys96 and Cys100 in P2 responsible for an intramolecular disulfide bond formation would remain problematic. The ETD fragmentation of the triple-charged ion with an m/z value of 546.9 (from peak II) produced two peptide fragments, P3 and P4, indicating that P3 and P4 formed an intermolecular disulfide bond between Cys26 and Cys109.

In addition to these predicted disulfide bond formations in pINSL3, the total ion chromatogram of the non-reduced form of pINSL3 shows minor peaks at 30.6 min (marked V) and 34.1 min (marked VI, Fig. 5A). The MS1 survey spectra of these peaks displayed two unexpected and misfolded peptide fragments, each containing 6% of the total cysteine peptides detected (Miyazaki et al. 2017). These unpredicted fragments indicate that Cys109 in P4 ($Q^{103}DLLTLCPH^{111}$) are connected with Cys14 in P1 ($L^{13}CGHHFVR^{20}$) or one of the cysteines in P2 ($H^{94}CCLSGCTR^{102}$) (Table 2). Collectively, these results suggest that about 90% of the recombinant pINSL3 expressed in silkworm possessed three correct disulfide bond pairs.

Functional activity of the purified pINSL3 was evaluated using a receptor-activating assay in relaxin family peptide receptor 2 (RXFP2)–expressing HEK293 cells (Miyazaki et al. 2017). This catalytic assay demonstrates similar Emax, a percentage of its maximum response, and EC_{50} values when compared with the synthetic human INSL3 (Miyazaki et al. 2017). However, about 30% less activity was detected

Table 2: Identified tryptic-digested peptides from recombinant pINSL3.*

Reduced/ Non-reduced	*m/z*	Peptide Fragments	Amino Acid Sequence	Percentage of Total Peak Areas
Reduced	968.5	P1	L^{13}CGHHFVR^{20}	–
	977.4	P2	H^{94}CCLSGCTR^{102}	–
	602.3	P3	L^{25}CGGPR^{30}	–
	1039.5	P4	Q^{103}DLLTLCPH^{111}	–
Non-reduced	648.6 (3+)	P1	L^{13}CGHHFVR^{20}	25%
		P2	H^{94}CCLSGCTR^{102}	
	546.9 (3+)	P3	L^{25}CGGPR^{30}	63%
		P4	Q^{103}DLLTLCPH^{111}	
	502.3 (4+) 669.3 (3+)	P1	L^{13}CGHHFVR^{20}	6%
		P4	Q^{103}DLLTLCPH^{111}	
	504.2 (4+) 672.3 (3+)	P2	H^{94}CCLSGCTR^{102}	6%
		P4	Q^{103}DLLTLCPH^{111}	

* Modified from Miyazaki et al. 2017. Insulin-like peptide 3 expressed in the silkworm possesses intrinsic disulfide bonds and full biological activity. Sci. Rep. 7: 17339. https://creativecommons.org/licenses/by/4.0/z.

in the recombinant pINSL3 from silkworm compared to the synthetic INSL3, which is probably due to the presence of isomers. Therefore, these findings indicate that the recombinant pINSL3 expressed in the silkworm BmNPV-based system harbors structural integrity through disulfide bond formation, demonstrating its functional ability to bind the receptor for cAMP production.

One should note that the silkworm produces an indigenous protein called bombyxin, a member of the relaxin/insulin superfamily. Similar to the functions of the relaxin/insulin superfamily that possesses diverse hormonal activities, the prothoracicotropic hormone (brain-secretory hormone) bombyxin reduces blood sugar concentration and stimulates the prothoracic gland to release the insect molting hormone (Mizoguchi and Okamoto 2013, Nagasawa et al. 1984). Although bombyxin is derived from invertebrates, the bombyxin signal peptide displays limited sequence conservation but shares a structural similarity with relaxin/insulin family peptides such as INSL3, insulin, insulin-like growth factor and human relaxin (Fig. 6). This exciting discovery suggests that the silkworm could be an ideal host for producing vertebrate-type peptide hormones, canonically characterized by three evolutionarily conserved disulfide bonds.

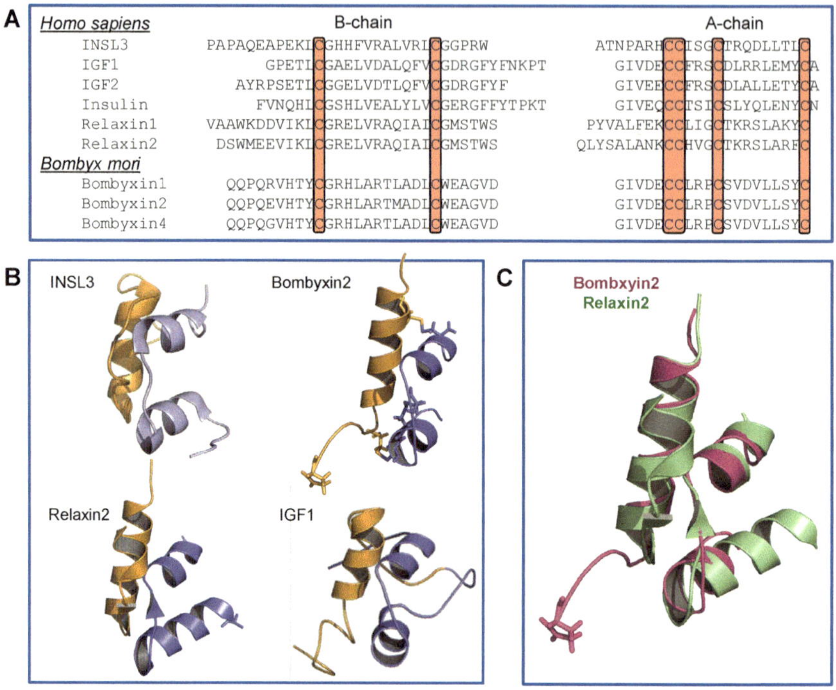

Fig. 6: Sequence and structural similarity of relaxin/insulin superfamily peptides. (A) Amino acid sequence alignment of relaxin/insulin superfamily peptides shows limited sequence conservation. Although bombyxin is derived from invertebrates, it is a member of the insulin family with sequence similarity. (B) Resolved crystal structures of relaxin/insulin superfamily, INSL3 (PDB ID: 2H8B), bombyxin 2 (PDB ID: 1BOM), relaxin 2 (PDB ID: 6RLX) and IGF1 (PDB ID: 3LRI). The common structural features of relaxin/insulin superfamily peptides include: (1) an A chain (light blue) with two α-helices joined by an extended loop, (2) a B chain (light orange) with a central α-helix, (3) three inter- and intra-chain disulfide bonds from 6 cysteine residues. (C) The root-mean-square deviation between bombyxin and relaxin is 1.45 Å within the common helical regions. Cartoon protein structures were generated using PyMol software.

4. Summary

Post-translational modification is a chemical adaptation of proteins, enabling to escape from nature's limited genetic imprisonment. As an integral source of protein diversity, post-translational modification plays a critical role in the functional and structural studies as well as having potential therapeutic applications. Numerous approaches have been applied in order to fabricate authentic heterologous proteins implementing proper post-translational modifications. The domesticated Silkworm *Bombyx mori* has been increasingly recognized as a reliable host for eukaryotic protein production with fail-safe post-translational modification. In this chapter, three post-translational modifications, including biotinylation, phosphorylation and disulfide bond formation in human acetyl-CoA carboxylase, malonyl-CoA carboxylase, and porcine insulin-like peptide 3 produced using a silkworm BmNPV bacmid-based expression system were discussed. The critical evaluation of post-translational modifications highlights that the silkworm BmNPV bacmid-based expression modality is a reliable platform for large-scale eukaryotic protein production implementing suitable post-translational modifications.

Acknowledgments

This work was supported by the National Institutes of Health, Marriott Foundation, and Mayo Clinic Center for Regenerative Medicine. A.T. holds the Marriott Family Professorship in Cardiovascular Diseases Research, and is the Michael S. and May Sue Shannon Family Director of the Mayo Clinic Center for Regenerative Medicine.

References

Abu-Elheiga, L., M. M. Matzuk, K. A. Abo-Hashema and S. J. Wakil. 2001. Continuous fatty acid oxidation and reduced fat storage in mice lacking acetyl-CoA carboxylase 2. Science. 291: 2613–2616.

Abu-Elheiga, L., W. Oh, P. Kordari and S. J. Wakil. 2003. Acetyl-CoA carboxylase 2 mutant mice are protected against obesity and diabetes induced by high-fat/high-carbohydrate diets. Proc. Natl. Acad. Sci. U.S.A. 100: 10207–10212.

Aebersold, R., J. N. Agar, I. J. Amster, M. S. Baker, C. R. Bertozzi, E. S. Boja, C. E. Costello, B. F. Cravatt, C. Fenselau, B. A. Garcia et al. 2018. How many human proteoforms are there? Nat. Chem. Biol. 14: 206–214.

Anand-Ivell, R. and R. Ivell. 2014. Insulin like factor 3 as a monitor of endocrine disruption. Reproduction (Cambridge, England) 147: R87–95.

Aucoin, M. G., J. A. Mena and A. A. Kamen. 2010. Bioprocessing of baculovirus vectors: A review. Curr. Gene Ther. 10: 174–186.

Bah, A. and J. D. Forman-Kay. 2016. Modulation of intrinsically disordered protein function by post-translational modifications. J. Biol. Chem. 291: 6696–6705.

Barber, K. W. and J. Rinehart. 2018. The ABCs of PTMs. Nat. Chem. Biol. 14: 188–192.

Beckett, D., E. Kovaleva and P. J. Schatz. 1999. A minimal peptide substrate in biotin holoenzyme synthetase-catalyzed biotinylation. Protein Sci. 8: 921–929.

Blom, N., S. Gammeltoft and S. Brunak. 1999. Sequence and structure-based prediction of eukaryotic protein phosphorylation sites. J. Mol. Biol. 294: 1351–1362.

Bode, A. M. and Z. Dong. 2004. Post-translational modification of p53 in tumorigenesis. Nat. Rev. Cancer. 4: 793–805.

Bosnjak, I., V. Bojovic, T. Segvic-Bubic and A. Bielen. 2014. Occurrence of protein disulfide bonds in different domains of life: A comparison of proteins from the Protein Data Bank. Protein Eng. Des. Sel. 27: 65–72.

Brown, G. K., R. D. Scholem, A. Bankier and D. M. Danks. 1984. Malonyl coenzyme A decarboxylase deficiency. J. Inherit. Metab. Dis. 7: 21–26.

Brownsey, R. W., A. N. Boone, J. E. Elliott, J. E. Kulpa and W. M. Lee. 2006. Regulation of acetyl-CoA carboxylase. Biochem. Soc. Trans. 34: 223–227.

Bulleid, N. J. 2012. Disulfide bond formation in the mammalian endoplasmic reticulum. Cold Spring Harb. Perspect. Biol. 4: a013219.

Bullesbach, E. E. and C. Schwabe. 2002. The primary structure and the disulfide links of the bovine relaxin-like factor (RLF). Biochemistry. 41: 274–281.

Burnett, G. and E. P. Kennedy. 1954. The enzymatic phosphorylation of proteins. J. Biol. Chem. 211: 969–980.

Chapman-Smith, A. and J. E. Cronan. Jr. 1999. The enzymatic biotinylation of proteins: A post-translational modification of exceptional specificity. Trends Biochem. Sci. 24: 359–363.

Chaturvedi, P. and S. C. Tyagi. 2014. Epigenetic mechanisms underlying cardiac degeneration and regeneration. Int. J. Cardiol. 173: 1–11.

Ciesla, J., T. Fraczyk and W. Rode. 2011. Phosphorylation of basic amino acid residues in proteins: Important but easily missed. Acta biochimica Polonica. 58: 137–148.

Dojima, T., T. Nishina, T. Kato, T. Uno, H. Yagi, K. Kato, H. Ueda and E. Y. Park. 2010. Improved secretion of molecular chaperone-assisted human IgG in silkworm, and no alterations in their N-linked glycan structures. Biotechnol. Prog. 26: 232–238.

Du, D., T. Kato, F. Suzuki and E. Y. Park. 2009. Expression of protein complex comprising the human prorenin and (pro)renin receptor in silkworm larvae using *Bombyx mori* nucleopolyhedrovirus (BmNPV) bacmids for improving biological function. Mol. Biotech. 43: 154–161.

Ferlin, A., L. De Toni, M. Sandri and C. Foresta. 2017. Relaxin and insulin-like peptide 3 in the musculoskeletal system: From bench to bedside. Br. J. Pharmacol. 174: 1015–1024.

Folmes, C. D., S. Park and A. Terzic. 2013. Lipid metabolism greases the stem cell engine. Cell Metab. 17: 153–155.

Fullerton, M. D., S. Galic, K. Marcinko, S. Sikkema, T. Pulinilkunnil, Z. P. Chen, H. M. O'Neill, R. J. Ford, R. Palanivel, M. O'Brien et al. 2013. Single phosphorylation sites in Acc1 and Acc2 regulate lipid homeostasis and the insulin-sensitizing effects of metformin. Nature Med. 19: 1649–1654.

Green, N. M. 1975. Avidin. Adv. Protein Chem. 29: 85–133.

Gundlach, A. L., S. Ma, Q. Sang, P. J. Shen, L. Piccenna, K. Sedaghat, C. M. Smith, R. A. Bathgate, A. J. Lawrence, G. W. Tregear et al. 2009. Relaxin family peptides and receptors in mammalian brain. Ann. N Y Acad. Sci. 1160: 226–235.

Hogg, P. J. 2003. Disulfide bonds as switches for protein function. Trends Biochem. Sci. 28: 210–214.

Hunter, T. 1995. Protein kinases and phosphatases: the yin and yang of protein phosphorylation and signaling. Cell. 80: 225–236.

Hwang, I. W., Y. Makishima, T. Kato, S. Park, A. Terzic and E. Y. Park. 2014. Human acetyl-CoA carboxylase 2 expressed in silkworm *Bombyx mori* exhibits post-translational biotinylation and phosphorylation. Appl. Microbiol. Biotechnol. 98: 8201–8209.

Hwang, I. W., Y. Makishima, T. Suzuki, T. Kato, S. Park, A. Terzic, S. K. Chung and E. Y. Park. 2015. Phosphorylation of Ser-204 and Tyr-405 in human malonyl-CoA decarboxylase expressed in silkworm *Bombyx mori* regulates catalytic decarboxylase activity. Appl. Microbiol. Biotechnol. 99: 8977–8986.

Ivell, R., A. I. Agoulnik and R. Anand-Ivell. 2017. Relaxin-like peptides in male reproduction—a human perspective. Br. J. Pharmacol. 174: 990–1001.

Jarvis, D. L. 2009. Baculovirus-insect cell expression systems. Methods Enzymol. 463: 191–222.

Jensen, L. J., R. Gupta, N. Blom, D. Devos, J. Tamames, C. Kesmir, H. Nielsen, H. H. Staerfeldt, K. Rapacki, C. Workman et al. 2002. Prediction of human protein function from post-translational modifications and localization features. J. Mol. Biol. 319: 1257–1265.

Jensen, O. N. 2004. Modification-specific proteomics: Characterization of post-translational modifications by mass spectrometry. Curr. Opin. Chem. Biol. 8: 33–41.

Kato, T., M. Kajikawa, K. Maenaka and E. Y. Park. 2010. Silkworm expression system as a platform technology in life science. Appl. Microbiol. Biotechnol. 85: 459–470.

Khoury, G. A., R. C. Baliban and C. A. Floudas. 2011. Proteome-wide post-translational modification statistics: Frequency analysis and curation of the swiss-prot database. Sci. Rep. 1: 90.

Knobloch, M., S. M. Braun, L. Zurkirchen, C. von Schoultz, N. Zamboni, M. J. Arauzo-Bravo, W. J. Kovacs, O. Karalay, U. Suter, R. A. Machado et al. 2013. Metabolic control of adult neural stem cell activity by Fasn-dependent lipogenesis. Nature. 493: 226–230.

Kolwicz, S. C., Jr., D. P. Olson, L. C. Marney, L. Garcia-Menendez, R. E. Synovec and R. Tian. 2012. Cardiac-specific deletion of acetyl CoA carboxylase 2 prevents metabolic remodeling during pressure-overload hypertrophy. Circ. Res. 111: 728–738.

Kost, T. A., J. P. Condreay and D. L. Jarvis. 2005. Baculovirus as versatile vectors for protein expression in insect and mammalian cells. Nat. Biotechnol. 23: 567–575.

Laurent, G., N. J. German, A. K. Saha, V. C. de Boer, M. Davies, T. R. Koves, N. Dephoure, F. Fischer, G. Boanca, B. Vaitheesvaran et al. 2013. SIRT4 coordinates the balance between lipid synthesis and catabolism by repressing malonyl CoA decarboxylase. Mol. Cell. 50: 686–698.

Leder, L. 2015. Site-specific protein labeling in the pharmaceutical industry: Experiences from novartis drug discovery. Methods Mol. Biol. 1266: 7–27.

Lee, C. K., H. K. Cheong, K. S. Ryu, J. I. Lee, W. Lee, Y. H. Jeon and C. Cheong. 2008. Biotinoyl domain of human acetyl-CoA carboxylase: Structural insights into the carboxyl transfer mechanism. Proteins. 72: 613–624.

Levene, P. A. and C. L. Alsberg. 1906. The cleavage productsof vitellin. J. Biol. Chem. 2: 127–133.

Lu, C. T., K. Y. Huang, M. G. Su, T. Y. Lee, N. A. Bretana, W. C. Chang, Yi. -Ju. Chen, Yu. -Ju. Chen and H. D. Huang. 2013. DbPTM 3.0: An informative resource for investigating substrate site specificity and functional association of protein post-translational modifications. Nucleic Acids R. 41: D295–305.

Luo, X., R. A. Bathgate, Y. L. Liu, X. X. Shao, J. D. Wade and Z. Y. Guo. 2009. Recombinant expression of an insulin-like peptide 3 (INSL3) precursor and its enzymatic conversion to mature human INSL3. FEBS J. 276: 5203–5211.

Maeda, S., T. Kawai, M. Obinata, H. Fujiwara, T. Horiuchi, Y. Saeki, Y. Sato and M. Furusawa. 1985. Production of human α-interferon in silkworm using a baculovirus vector. Nature. 315: 592–594.

Midgett, C. R. and D. R. Madden. 2007. Breaking the bottleneck: Eukaryotic membrane protein expression for high-resolution structural studies. J. Struct. Biol. 160: 265–274.

Minagawa, I., M. Fukuda, H. Ishige, H. Kohriki, M. Shibata, E. Y. Park, T. Kawarasaki and T. Kohsaka. 2012. Relaxin-like factor (RLF)/insulin-like peptide 3 (INSL3) is secreted from testicular Leydig cells as a monomeric protein comprising three domains B-C-A with full biological activity in boars. Biochem. J. 441: 265–273.

Minguez, P., I. Letunic, L. Parca and P. Bork. 2013. PTMcode: A database of known and predicted functional associations between post-translational modifications in proteins. Nucleic Acids Res. 41: D306–311.

Miyazaki, T., M. Ishizaki, H. Dohra, S. Park, A. Terzic, T. Kato, T. Kohsaka and E. Y. Park. 2017. Insulin-like peptide 3 expressed in the silkworm possesses intrinsic disulfide bonds and full biological activity. Sci. Rep. 7: 17339.

Mizoguchi, A. and N. Okamoto. 2013. Insulin-like and IGF-like peptides in the silkmoth *Bombyx mori*: discovery, structure, secretion and function. Front. Physiol. 4: 217.

Motohashi, T., T. Shimojima, T. Fukagawa, K. Maenaka and E. Y. Park. 2005. Efficient large-scale protein production of larvae and pupae of silkworm by *Bombyx mori* nuclear polyhedrosis virus bacmid system. Biochem. Biophys. Res. Commun. 326: 564–569.

Nagasawa, H., H. Kataoka, A. Isogai, S. Tamura, A. Suzuki, H. Ishizaki, A. Mizoguchi and Y. Fujiwara. 1984. Amino-terminal amino acid sequence of the silkworm prothoracicotropic hormone: Homology with insulin. Science. 226: 1344–1345.

Park, S., I. W. Hwang, Y. Makishima, E. Perales-Clemente, T. Kato, N. J. Niederlander, E. Y. Park and A. Terzic. 2013. Spot14/Mig12 heterocomplex sequesters polymerization

and restrains catalytic function of human acetyl-CoA carboxylase 2. J. Mol. Recognit. 26: 679–688.
Park, S., D. K. Arrell, S. Reyes, E. Y. Park and A. Terzic. 2017. Conventional and unconventional secretory proteins expressed with silkworm bombyxin signal peptide display functional fidelity. Sci. Rep. 7: 14499.
Prabakaran, S., G. Lippens, H. Steen and J. Gunawardena. 2012. Post-translational modification: nature's escape from genetic imprisonment and the basis for dynamic information encoding. WIREs Systems Biology and Medicine. 4: 565–583.
Sevier, C. S. and C. A. Kaiser. 2002. Formation and transfer of disulphide bonds in living cells. Nat. Rev. Mol. Cell. Biol. 3: 836–847.
Sternicki, L. M., K. L. Wegener, J. B. Bruning, G. W. Booker and S. W. Polyak. 2017. Mechanisms governing precise protein biotinylation. Trends Biochem. Sci. 42: 383–394.
Tong, L. 2005. Acetyl-coenzyme A carboxylase: Crucial metabolic enzyme and attractive target for drug discovery. Cell. Mol. Life Sci. 62: 1784–1803.
Tong, L. 2013. Structure and function of biotin-dependent carboxylases. Cell. Mol. Life Sci. 70: 863–891.
Ubersax, J. A. and J. E. Ferrell. Jr. 2007. Mechanisms of specificity in protein phosphorylation. Nat. Rev. Mol. Cell. Biol. 8: 530–541.
Wakil, S. J. and L. A. Abu-Elheiga. 2009. Fatty acid metabolism: Target for metabolic syndrome. J. Lipid Res. 50 Suppl: S138–143.
Walsh, C. T., S. Garneau-Tsodikova and G. J. Gatto. Jr. 2005. Protein post-translational modifications: The chemistry of proteome diversifications. Angew. Chem. Int. Ed. Engl. 44: 7342–7372.
Walsh, G. and R. Jefferis. 2006. Post-translational modifications in the context of therapeutic proteins. Nat. Biotechnol. 24: 1241–1252.
Wilkinson, B. and H. F. Gilbert. 2004. Protein disulfide isomerase. Biochim. Biophys. Acta. 1699: 35–44.
Xin, F. and P. Radivojac. 2012. Post-translational modifications induce significant yet not extreme changes to protein structure. Bioinformatics. 28: 2905–2913.
Xue, J., J. Peng, M. Zhou, L. Zhong, F. Yin, D. Liang and L. Wu. 2012. Novel compound heterozygous mutation of MLYCD in a Chinese patient with malonic aciduria. Mol. Genet. Metab. 105: 79–83.
Zhou, D., P. Yuen, D. Chu, V. Thon, S. McConnell, S. Brown, A. Tsang, M. Pena, A. Russell, J. F. Cheng et al. 2004. Expression, purification, and characterization of human malonyl-CoA decarboxylase. Protein Expr. Purif. 34: 261–269.

CHAPTER 6

Glycoprotein Biosynthesis of Silkworm

Kazuhito Fujiyama

1. Introduction

Silkworms (*Bombyx mori*) have traditionally been harnessed commercially in order to produce silk proteins. However, recently the potential for genetically engineered silkworms to work as "biofactories" for the production of recombinant proteins has gained a great deal of attention. Industrial-scale production of recombinant protein in transgenic silkworms could easily be controlled by changing the number of silkworms. This possibility increases the importance of understanding the silkworm production system and its mechanisms. Most pharmaceutical proteins are post-translationally modified in one way or another. However, post-translational modifications such as glycosylation are not identical in silkworms and humans. Thus, fundamental information on glycosylation in silkworms is critical in understanding the potential of insect production of pharmaceutical proteins. In this chapter, we shall describe in detail what has been learned about the mechanisms of glycosylation in silkworms and explain the potential to develop production of recombinant proteins with engineered N-glycans, which are N-linked oligosaccharides linked to Asn residues of peptides.

2. Glycan Structures

Structural analysis of proteins prepared from silkworms provides fundamental information on glycosylation of silkworms. Here the

International Center for Biotechnology, Osaka University, 2-1 Yamadaoka, Suita, Osaka 565-0871, Japan.
E-mail: fujiyama@icb.osaka-u.ac.jp

structures of N-glycans are focused on. The first research on this subject placed full-length mouse interferon-beta cDNA under the polyhedron promotor of *Bombyx mori* nucleopolyhedrovirus (BmNPV) or baculovirus. Recombinant mIFN-beta was produced in the haemolymph of infected silkworm larvae and purified for analysis of its glycan structures (Misaki et al. 2003). Nine N-glycan structures for recombinant mIFN-β were determined, as shown in Fig. 1. Most glycans were M3- or M2-based structures. Interestingly, difucosylated N-glycan M3F3F6 was found (12.0%). This structural analysis was the first to suggest the potential of glycosylation in silkworms. Among its findings:

(1) the predominant structure was M5A (28.5%);

	Name	Ratio (%)
Manα1-6 ; Fucα1-6 Manβ1-4GlcNAcβ1-4GlcNAc	M2BF6	6.4%
Fucα1-6 Manβ1-4GlcNAcβ1-4GlcNAc Manα1-2Manα1-3	M3CF6	16.3%
Fucα1-6 Manβ1-4GlcNAcβ1-4GlcNAc Manα1-3	M2AF6	3.7%
Manα1-6 Manβ1-4GlcNAcβ1-4GlcNAc Manα1-3	M3	11.8%
Manα1-6 ; Fucα1-6 Manβ1-4GlcNAcβ1-4GlcNAc Manα1-3	M3F6	1.1%
Manα1-6 ; Fucα1-6 Manβ1-4GlcNAcβ1-4GlcNAc Manα1-3 ; Fucα1-3	M3F3F6	12.0%
Manα1-6 Manα1-3 Manα1-6 Manβ1-4GlcNAcβ1-4GlcNAc Manα1-3	M5A	28.5%
Manα1-6 Manα1-2 Manα1-3 Manα1-6 Manβ1-4GlcNAcβ1-4GlcNAc Manα1-3	M6C	4.7%
Manα1-6 Manα1-2 Manα1-3 Manα1-6 Manβ1-4GlcNAcβ1-4GlcNAc Manα1-2 Manα1-3	M7D	15.5%

Fig. 1: N-glycans of mIFN-beta was produced in the haemolymph of silkworms infected by baculovirus.

(2) two fucosylation enzymes were involved, α1,3-fucosyltransferase (α1,3-FucT) and α1,6-fucosyltransferase (α1,6-FucT);

(3) N-acetylglucosamine (GlcNAc)-, galactose (Gal)- or N-acetylneuraminic acid (NeuAc)-extended glycans were not found.

As for the second example, the major structures of the N-glycans on membrane glycoproteins from Bm-N cells derived from *B. mori* were M3F6 (18%), M5A (13.6%), M6B (12.9%), and M7B (12.0%) (Fig. 2) (Kubelka et

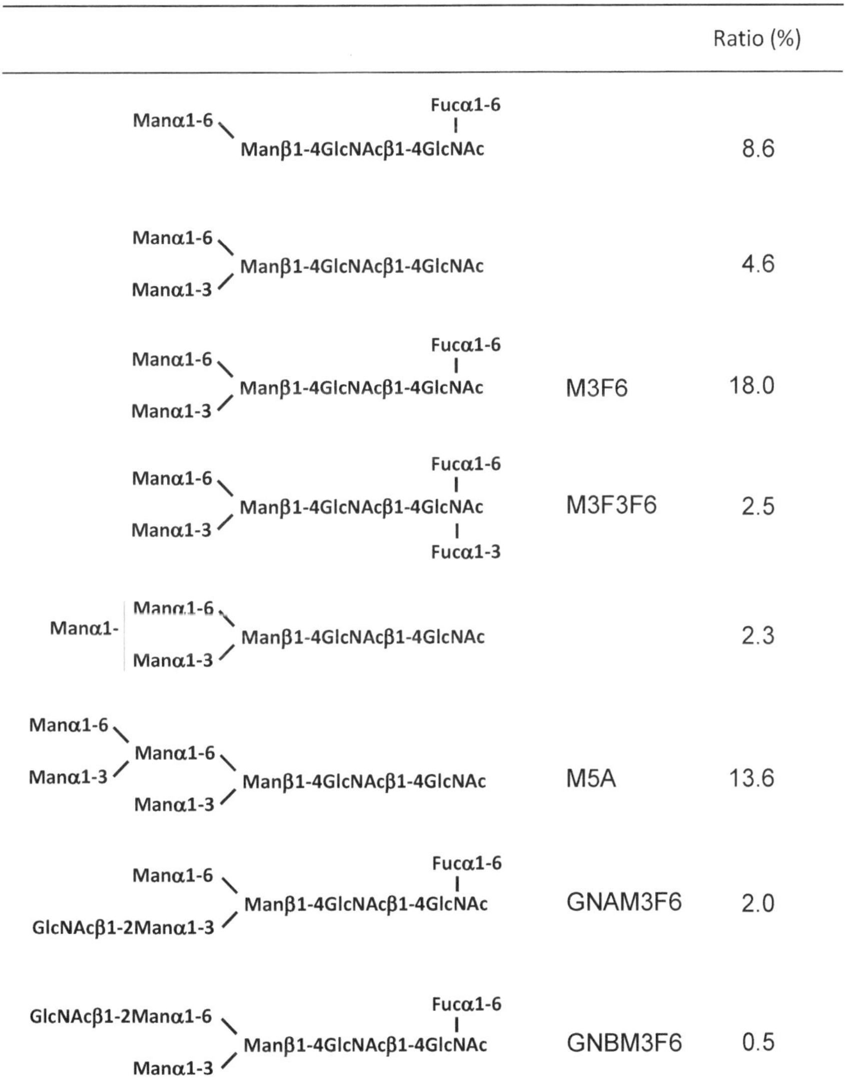

Fig. 2 contd....

...Fig. 2 contd.

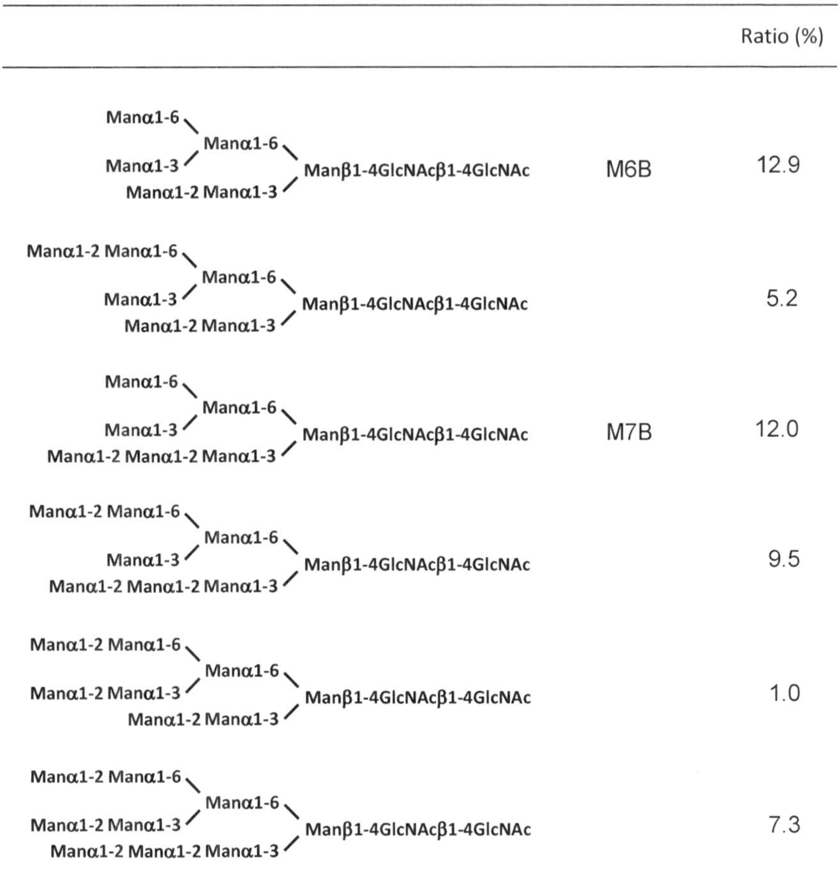

Fig. 2: N-glycan structures of endogenous membrane glycoproteins from Bm-N cells.

al. 1994). Like mIFNb produced by silkworm larvae, M3F3F6 was also detected (2.5%). Interestingly, two N-glycans with a GlcNAc residue, GNAM3F6 and GNAM3F6, were detected in the membrane glycoproteins from Bm-N cells. This suggested that silkworms have the intrinsic capability of transferring GlcNAc residues. However, GlcNAc residues of N-glycans are removed by N-acetylglucosaminidase (GlcNAcase) during intracellular trafficking. A membrane-bound GlcNAcase has been found in insect cells (Park et al. 2009).

In another experiment, human IgG was produced using baculovirus in silkworm (Park et al. 2009). Recombinant IgG (rIgG) was secreted into the haemolymph of 5th instar silkworm larvae. The N-glycans of the purified IgG were two major pauci-mannose-type oligosaccharides, M2BF6 (77.5%) and M3F6 (12.7%) (Fig. 3). Later, Dojima et al. (2010) co-expressed 5 molecular chaperones, such as calreticulin (CRT) and calnexin (CNX), to examine the effect of molecular chaperones on the N-glycosylation of the rIgG. Overexpression of molecular chaperones improved the production of IgG secretion in the larvae. The N-glycans of rIgG secreted in haemolymph were predominantly M2BF6 and M3F6 (Fig. 3); small amounts of the GlcNAc-extended structure GNAM3F6 was also detected.

Iizuka et al. (2009) also reported production of rIgG on cocoons of transgenic silkworms and performed structural analysis of rIgG N-glycans. Four major N-glycan structures in the rIgG were M6, M5A, GNAM3 and GN2M3 (Fig. 4). In addition, N-glycan structures of total proteins in cocoons, middle silk glands (MSGs) and fat bodies were also determined (Fig. 5) and revealed organ-specific N-glycan profiles. The N-glycan profile of MSGs were distinguishable from those of fat bodies. On the other hand, cocoons and MSGs shared similar N-glycan profiles, likely because cocoon proteins are produced in MSGs. In the profile of fat bodies, N-glycans were mannose-based structures, such as M2AF6 (37.4%) and M9 (17.3%). GlcNAc-extended glycans were not detected, while fucosylated glycan was predominant. Meanwhile, MSGs produced three

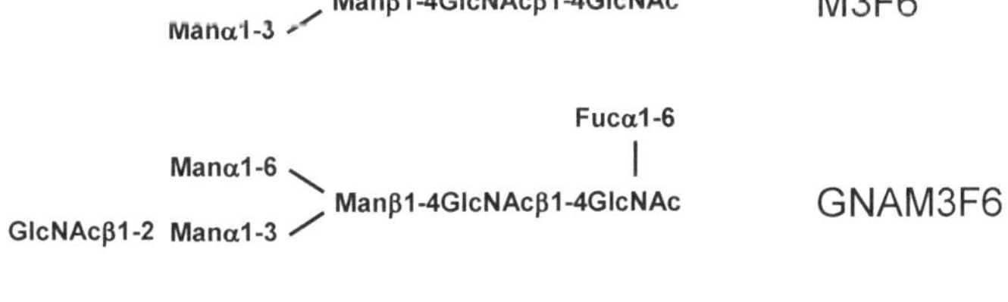

Structure	-	CRX	CRT	BiP	ERp57	Hsp70
M2BF6	47.4	49.1	46.1	45.7	44.4	46.0
M3F6	28.5	31.7	33.2	31.3	32.8	32.5
GnAM3F6	3.4	3.7	3.8	3.7	4.1	4.0

Fig. 3: N-glycan structures of rIgG co-expressed with molecular chaperons.

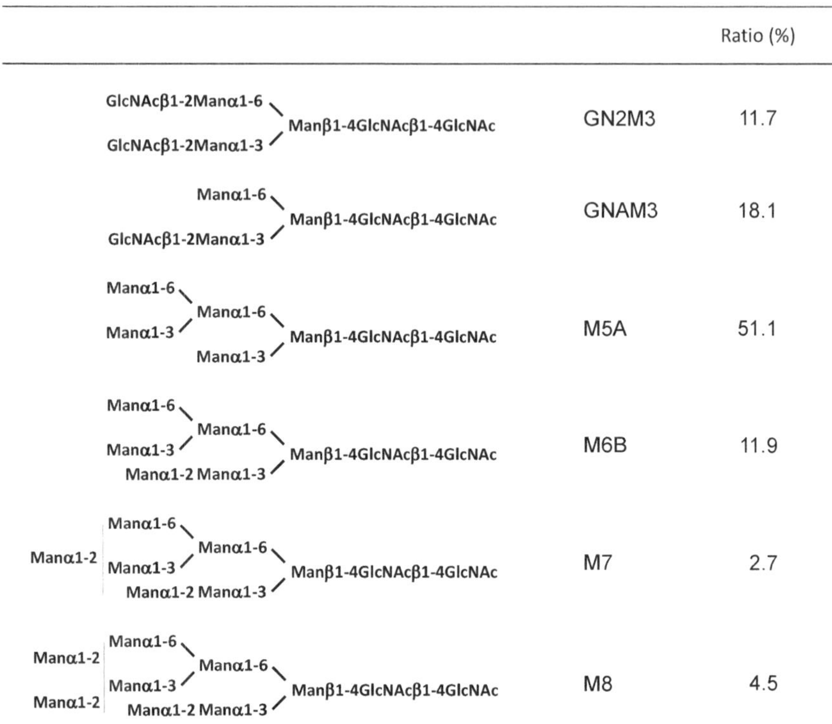

Fig. 4: N-glycan structures of rIgG on cocoons of transgenic silkworms.

GlcNAc-extended glycans, GNAM3 (10.9%), GNBM3 (7.2%) and GN2M3 (24.0%). Interestingly, fucosylated glycans were not detected. In profiles of both cocoons and MSGs, GNAM3 was more abundant than GNBM3. IgG purified from the haemolymph of silkworm larvae carried GNAM3 only. These data suggest the N-glycosylation potential in silkworms as well as organ-dependent glycosylation.

O-linked glycosylation occurs at Ser/Thr residues on the peptide. Structures in the *O*-glycan family are diverse. So far, no structural analysis of O-glycan has been reported.

3. Glycosylation of Proteins

Protein glycosylation is divided into two major types, N-linked and O-linked glycosylation. N-linked glycosylation occurs at Asn residues in the sequon Asn-X-Ser/Thr. Numerous enzymes are localized in the ER and Golgi and are involved in protein modification (Fig. 6). The core structure of N-linked glycan, GlcNAcβ1,2Manα1,6(GlcNAcβ1,2Manα 1,3)Manβ1,4GlcNAcβ1,4GlcNAc (GN2M3)-(Asn), is conserved among

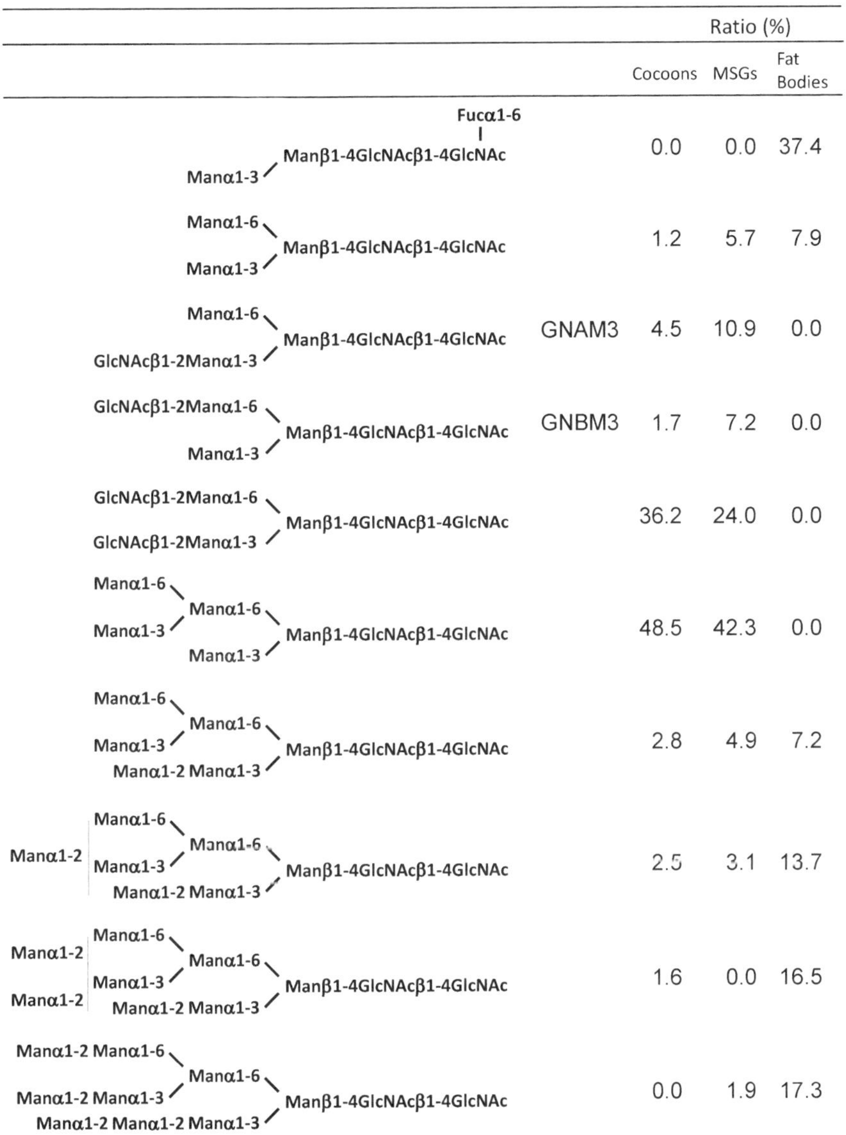

Fig. 5: N-glycan structures of total proteins in cocoons, MSGs and fat bodies.

mammals, plants and insects. However, extension of N-glycans toward the non-reducing end are diversified and dependent on the hosts and the organs of the host used for protein productions. The N-glycan in mammalian cells has β1,4-Gal, α2,3/α2,6-NeuAc and α1,6-Fuc. On the other hand, N-glycan in insect cells, as described above, shows the core

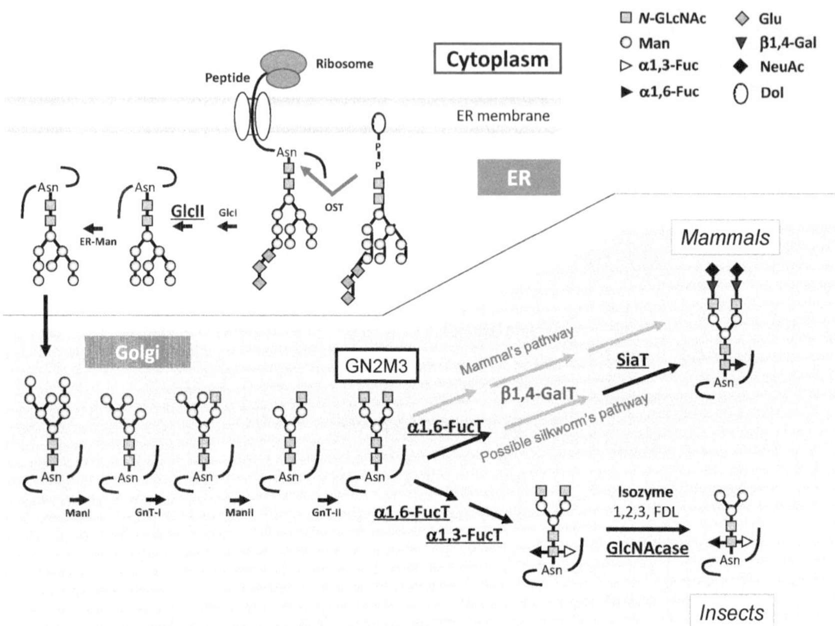

Fig. 6: A possible N-glycosylation pathway.

structure GN2M3 or a smaller glycan structure, pauci-mannose, such as M3, with α1,6-/α1,3-Fuc, but without β1,4-Gal or α2,3-/α2,6-NeuAc. This structural difference is intrinsic to the genome.

The silkworm genome structure was published in 2004 (Mita et al. 2004). We have intensively searched this library for candidate genes for glycosylation enzymes. Until now, cDNA of 8 glycosylation enzymes has been isolated (Fig. 6, underlined): Glucosidase II (Watanabe et al. 2013), α1,6-Fucosyltransferase (FucT) (Ihara et al. 2014), α1,3-FucT (Minagawa et al. 2015), four different GlcNAcases (Geisler and Jarvis 2010, Kokuho et al. 2010, Nomura et al. 2010, Okada et al. 2007) and sialyltransferase (Kajiura et al. 2015, Stanton et al. 2017). Following this, each corresponding recombinant enzyme was characterized. The following points have emerged from this work.

(1) In silkworms, glycoproteins carrying the core structure GN2M3 can be further modified by typical insect enzymes, such as α1,6-FucT and α1,3-FucT. FucT activities toward α1,6- and α1,3-linkages to the core GlcNAc of GN2M3 were found based on the glycan structures of silkworm glycoproteins. α1,6 FucT is generally present in mammals while α1,3 FucT is found in plants but not mammals. Silkworms, like *Drosophila*, have both types of FucTs, α1,6-FucT and α1,3-FucT.

The Fuc residue at α1,3-linked core GlcNAc of GN2M3 concerns allergenicity or immunogenicity. Though genes encoding FucT have been identified, enzyme properties such as substrate specificities of silkworm α1,6 FucT and α1,3 FucT have not been shown.

(2) GlcNAc residues at non-reducing ends of GN2M3 are mostly missing, because GlcNAc is hydrolyzed by hexosaminidases or GlcNAcase. In silkworms, 4 genes encoding GlcNAcase were identified (Fig. 7). Individual recombinant GlcNAcases were produced and used for characterization of enzyme properties. Okada et al. (Okada et al. 2007) examined the substrate specificities of BmGlcNAcases 1 and 2. Another GlcNAcase cDNA was isolated and named BmGlcNAcase 2 (Nomura et al. 2010); however, in this article it is called BmGlcNAcase 3 in order to avoid confusion. In 2010, Nomura et al. (Nomura et al. 2010) and Geisler and Jarvis (Geisler and Jarvis 2010) reported isolation of the 4th hexosaminidase, FDL (BmFDL). BmGlcNAcases 1, 2 and 3 showed specificity of hydrolytic activity first toward GlcNAcβ1-2Manα1-6 and then GlcNAcβ1-2Manα1-3, of N-glycan GN2M3, resulting in M3 (Fig. 7). However, BmFDL showed distinguishable specificity, because BmFDL hydrolyzes GlcNAcβ1-2Manα1-6 only of GN2M3. This strict specificity is conserved among FDLs from *Drosophila* (Leonard et al. 2006) and *Spodoptera* (Geisler et al. 2008).

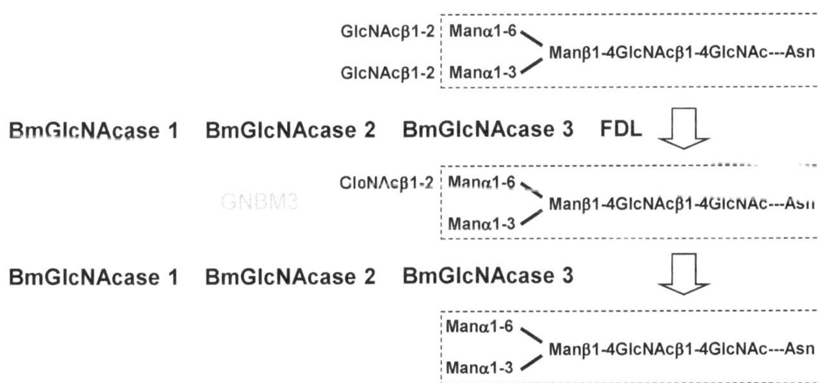

GlcNAcases	Reported by
BmGlcNAcase 1	Okada H et al. 2007
BmGlcNAcase 2	Okada H et al. 2007
BmGlcNAcase 3	Kokuho T et al. 2010
FDL	Nomura T et al. 2010; Geisler and Jarvis 2010

Fig. 7: A processing pathway by four GlcNAcases.

(3) Kajiura et al. (2015) isolated cDNA encoding sialyltransferase homolog (BmSiaT) and examined α2,6-sialyltransferase (SiaT) activity. The substrate specificity of BmSiaT was examined, showing that BmSiaT prefers GalNAcβ1,4-GlcNAc-linked N-glycan to Galβ1,4-GlcNAc-linked N-glycan as an acceptor substrate (Fig. 8). This preference is similar to that of *Drosophila* SiaT. These data suggest that silkworm carry the activity of β1,4-N-acetylgalctsaminyltransferase (β1,4-GalNAcT). The expression of BmST showed different levels in each organ and especially in stages. BmST was ubiquitously expressed among organs and stages. In *Drosophila*, SiaT is involved in nervous system function and development (Repnikova et al. 2010). Taking this into consideration, though SiaT is ubiquitously expressed, other sialylation-related enzymes, such as CMP-synthesis enzyme, might be regulated in order to develop the embryonic stage into the larval stage (Fig. 9). Hence, sialylation might not occur in 5th instar larvae or pupae. In other words, irregularly active sialylation beyond nervous system development in embryos might inhibit the normal development of larvae.

(4) The molecular biology of O-linked glycosylation enzymes in silkworms remains unclear.

4. Glyco-engineering for Production of Recombinant Protein in Silkworms

4.1 GlcNAc

The presence of pauci-mannose-type N-glycan in glycoproteins may not be suitable for expressing their biological activities for clinical purposes. Insect-specific β-N-acetylglucosaminidase (GlcNAcase) produces mannose-type N-glycan.

RNAi suppression of BmFDL in cultured silkworm cells resulted in production of a recombinant protein, puffer-fish saxitoxin- and

Acceptor substrate	Relative activity (%)
pNP-disaccharide	
Galβ1,4-GlcNAcβ-pNP	100
GalNAcβ1,4-GlcNAcβ-pNP	372
Asn-Fmoc-oligosaccharide	
$(Gal\beta1,4)_2(GlcNAc\beta1,2)_2(Man)_3(GlcNAc)_2$-Asn-Fmoc	35
$(GalNAc\beta1,4)_2(GlcNAc\beta1,2)_2(Man)_3(GlcNAc)_2$-Asn-Fmoc	140

Fig. 8: Acceptor substrate specificity of BmSiaT.

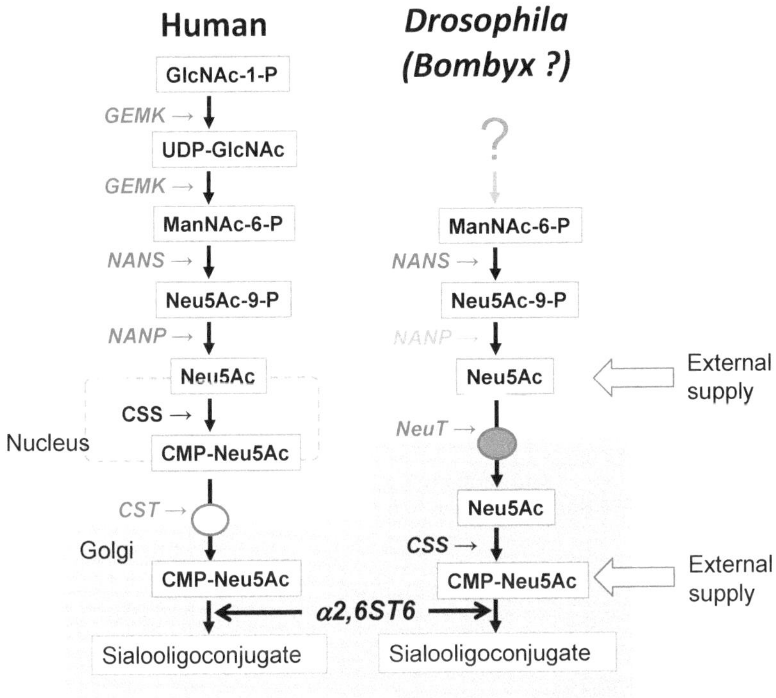

Fig. 9: A possible CMP-NeuAc synthetic pathway (Koles at al. (2009); modified). UDP-GlcNAc 2-epimerase/ManNAc kinase, GEMK; Neu5Ac-9-P synthase, NANS; Neu5Ac-9-P phosphatase, NANP; CMP-NeuAc synthase, CSS; CMP-NeuAc transporter, CST; α2,6-sialyltransferase, ST6.

tetrodotoxin-binding protein 3 (PSTBP3), using baculovirus expression systems (Nagata et al. 2013). The N-glycan analysis data showed the PSTBP3 carried complex-type N-glycans with sugar composition of GlcNAc2Man3GlcNAc2. The detailed structure was not elucidated.

Nomura et al. (2015) studied the expression of BmGlcNAcase1, BmGlcNAcase2, and BmFDL in fat bodies, which is the main tissue of recombinant protein expression by baculovirus and showed that BmFDL and BmGlcNAcase2 were mainly located in the fat bodies. A transgenic silkworm was constructed in order to reduce BmFDL transcripts by RNA interference (RNAi). This transgenic silkworm showed that the BmFDL transcript level and enzyme activity were vitiated to 25% and 50%, respectively, of that of the control silkworm, such that the composition of GlcNAc-linked N-glycan, GnAM3 and GnAM3F6, increased to 4.3% in the transgenic silkworm.

Dojima et al. (2009) expressed human α1,3N-acetylglucosaminyl transferase 2 (β3GnT2) fused to GFP in silkworm larvae using a baculovirus

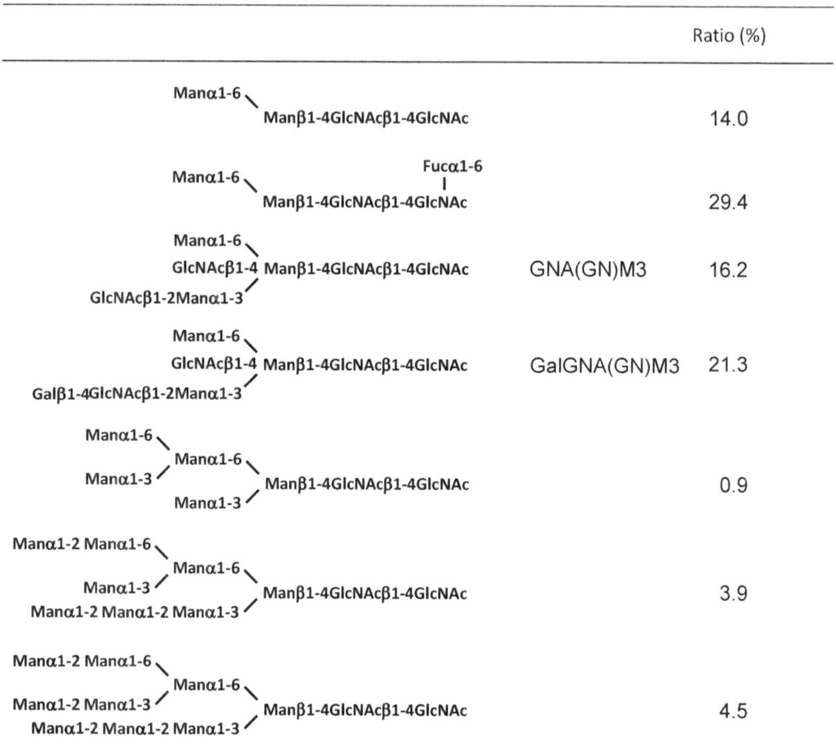

Fig. 10: N-glycan structures of human β3GnT2-fused to GFP produced in silkworm larvae using baculovirus system.

system. The recombinant β3GnT2 was analyzed for N-glycan structures (Fig. 10). Interestingly, N-glycan with Gal-extension at the Manα1,3 branch was detected (21.3%). In addition, N-glycans with bisecting GlcNAc were also detected at 37.5%, namely GNA(GN)M3 and GalGNA(GN)M3. This analysis suggests that silkworm might hold genes encoding GnTIII and GalT. The reason why expression of GnTIII and GalT was activated in silkworms is unclear.

4.2 Extension by GnTII

As shown by Iizuka et al. (2009), N-glycans carrying a single arm with extended GlcNAc, GNAM3 and GNBM3 were found on total proteins from MSGs (Fig. 5). GNAM3 seems to be a precursor of the N-glycosylation pathway (Fig. 6), while GNBM3 is derived from the enzyme reaction by BmGlcNAcases 1, 2 and 3 and FDL (Fig. 7). Introduction of the GnTII transgene can be used to increase the amount of GN2M3. Kato et al. (2017) examined the expression of hGnTII in silkworm pupae. GnTII co-expressed with IgG using baculovirus systems yielded GlcNAc-extended biantennary N-glycans, GN2M3 (4.5%) and GN2M3F6 (10.9%) with recombinant IgG (Fig. 11), showing that the expressed GnTII produced GlcNAc-extended N-glycans in silkworm pupae.

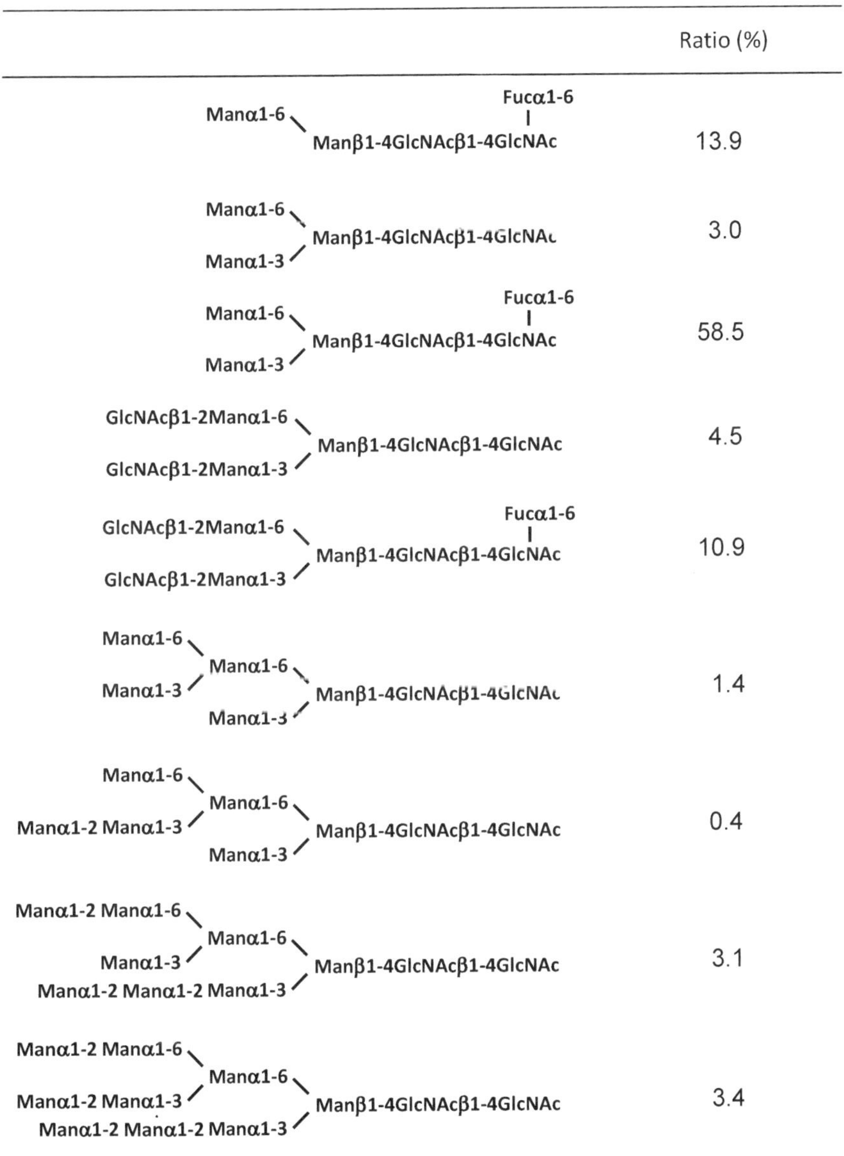

Fig. 11: N-glycan structures of rIgG with GnTII co-expression in silkworm pupae using baculovirus systems.

4.3 Extension by hGnTII and GalT

Mabashi-Asazuma et al. (2015) introduced and stably co-expressed hGnTII and bovine GalT in the posterior silk gland (PSG). Then, total proteins were prepared from PSG in transgenic silkworm larvae to determine N-glycan structures. Three Gal-extended N-glycans with compositions of Gal2GN2M3, GalGN2M3 and GalGNM3 were successfully produced. The total proportion of Gal-extended structures was about 15%. However, the predominant structure was M5A. Interestingly, fucosylated pauci-mannosidic N-glycans were found: M3F6 and Man2GlcNAc2Fuc. The N-glycosylation pathway in the silkworm PSG was successfully engineered.

As described above, Kato et al. (2017) co-expressed GnTII with IgG in silkworm pupae and examined the improvement of GlcNAc extension. Furthermore, using baculovirus systems they challenged co-expression in silkworm pupae of two glycosylation enzyme genes, hGnTII and hGalTI, with IgG. Gal-extended N-glycans were detected, with the three dominant structures being Gal2GN2M3F6 (11.4%), GalAGNM3 (7.0%) and GalBGN2M3 (3.9%). These Gal-extended N-glycans accounted for 27.8% of all N-glycans (Fig. 12). Biantennaries with GlcNAc-extension at each terminal, such as GN2M3F6, comprised 32.1%. However, N-glycans with GlcNAcβ1,2-Manα1,3-branching comprised 15.1%. Co-expression of hGnTII and hGalTI enables silkworms to produce recombinant proteins with biantennary, terminally galactosylated N-glycan.

4.4 Extension by hGnTII, GalT and SiaT

Glycoproteins produced in silkworms using a baculovirus system are secreted as pauci-mannose type N-glycans without NeuAc or Gal residues. Sialic acids such as NeuAc on N-glycans play important roles in protein functions. Sugamura et al. (2018) developed pathways for galactosylation and sialylation in silkworm using the baculovirus system.

Process 1: Sialylated N-glycans were successfully synthesized in silkworms by co-expressing GalT and SiaT when CMP-NeuAc was externally supplied (Fig. 13). However, without the supply of CMP-NeuAc, sialylated N-glycans were not produced, suggesting that the CMP-NeuAc synthesis pathway might not work in silkworm larvae.

Process 2: The external supply of CMP-NeuAc helped sialylation of N-glycans in silkworm larvae (Process 1). Co-expression of human CMP-NeuAc synthetase (CSS) with GalT and SiaT worked for sialylation

	Ratio (%)
Fucα1-6 Manα1-6 \ | Manβ1-4GlcNAcβ1-4GlcNAc	5.8
Manα1-6 \ Manβ1-4GlcNAcβ1-4GlcNAc Manα1-3 /	2.0
Fucα1-6 Manα1-6 \ | Manβ1-4GlcNAcβ1-4GlcNAc Manα1-3 /	26.3
Manα1-2 Manα1-6 \ Manα1-6 \ Manα1-2 Manα1-3 / Manβ1-4GlcNAcβ1-4GlcNAc Manα1-2 Manα1-2 Manα1-3 /	2.9
Manα1-2 Manα1-6 \ Manα1-6 \ Manα1-3 / Manβ1-4GlcNAcβ1-4GlcNAc Manα1-2 Manα1-2 Manα1-3 /	3.7
Manα1-6 \ Manα1-6 \ Manα1-2 Manα1-3 / Manβ1-4GlcNAcβ1-4GlcNAc Manα1-3 /	2.8
Manα1-6 \ Manα1-6 \ Manα1-3 / Manβ1-4GlcNAcβ1-4GlcNAc Manα1-3 /	4.9
Manα1-6 \ Manα1-6 \ Manα1-3 / Manβ1-4GlcNAcβ1-4GlcNAc Galβ1-4GlcNAcβ1-2 Manα1-3 /	2.0
Manα1-6 \ Manα1-3 / Manβ1-4GlcNAcβ1-4GlcNAc Galβ1-4GlcNAcβ1-2Manα1-3 /	0.8
Manα1-6 \ Manβ1-4GlcNAcβ1-4GlcNAc Galβ1-4GlcNAcβ1-2Manα1-3 /	0.5

Fig. 12 contd....

...Fig. 12 contd.

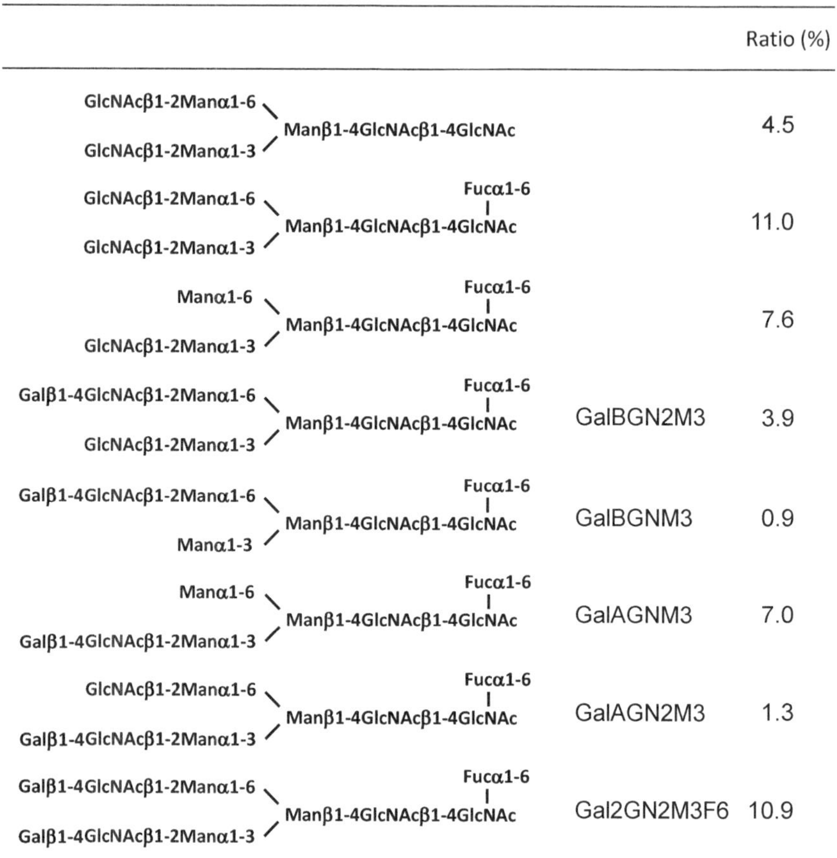

Fig. 12: N-glycan structures of rIgG with co-expression of GnTII and GalTI in silkworm pupae using baculovirus systems.

when NeuAc was supplied as a CSS substrate. These data suggest two possibilities for dysfunction of the CMP-NeuAc synthesis pathway in silkworm larvae: (i) CSS does not exist in silkworms, and (ii) NeuAc itself cannot be produced.

Process 3: For sialylation, co-expression of α2,3SiaT and α2,6SiaT with GalT resulted in α2,3-linked or α2,6-linked N-glycans, individually, in the supply of CMP-NeuAc.

Process 4: The co-expression of GnTII with GalT and SiaT in the supply of CMP-NeuAc improved the formation of di-sialylated N-glycan structures.

Results of co-expression of GalT and SiaT suggest that the CMP-NeuAc synthesis pathway does work in silkworms. The possible pathway

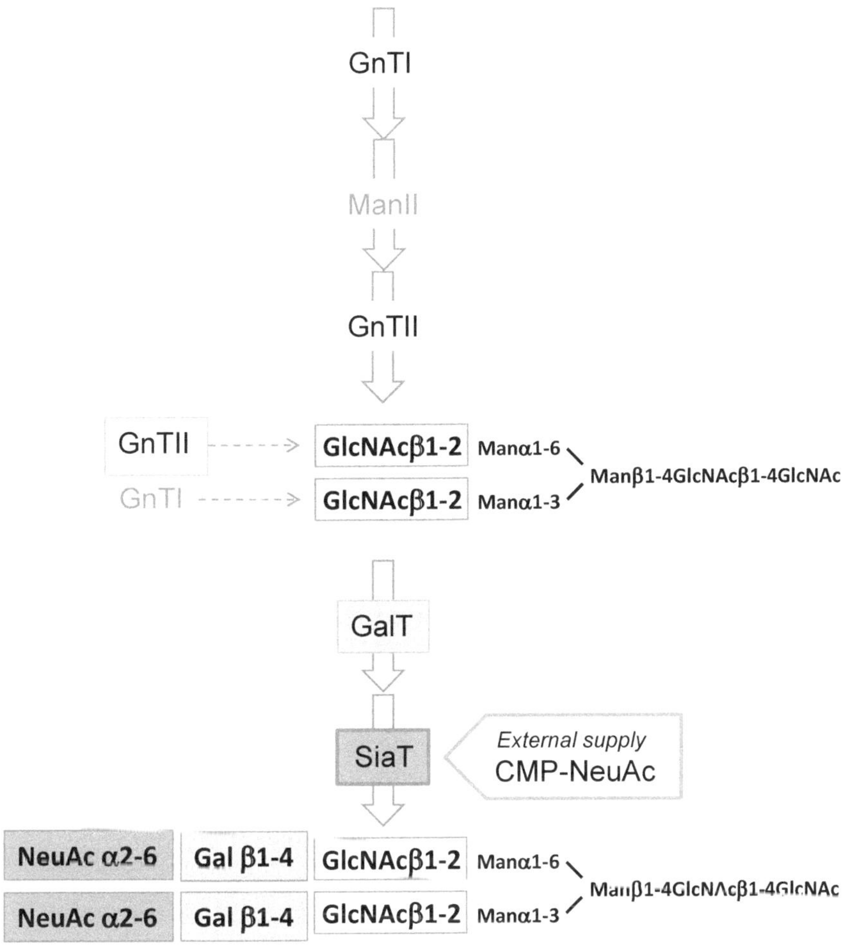

Fig. 13: Development of pathways for galactosylation and sialylation in silkworm using baculovirus system.

is shown in Fig. 9 (Koles et al. 2009). *Drosophila* does not have Neu5Ac-9-P synthase (NANS). If *B. mori* does not have NANS gene, supply of NeuAc or CMP-NeuAc for co-expression should be reasonable. However, *Drosophila* has a CMP-NeuAc synthase (CSS) gene and silkworm has a CSS homolog in its genome. Because co-expression of CSS with GalT and SiaT for sialylation requires an external supply of NeuAc, the CSS gene may not be expressed, particularly in silkworm larvae or pupae. This stage-specific expression is consistent with sialylation in *Drosophila* embryos. Expression of sialylation enzymes under organ-specific, particularly silk-gland specific, control would be feasible for formation of sialylated N-glycans.

5. Summary

The genome sequence of silkworms has been determined. However, the N-glycome of silkworms has not yet been revealed. First, we can learn the potential N-glycome from structural analysis of the N-glycans of total proteins prepared from larvae and pupae, and particular proteins, such as recombinant glycoproteins. N-Glycome data of *Anopheles gambiae, Aedes aegypti, Trichoplusia ni* and *Lymantria dispar* (Kurz et al. 2015, Kurz et al. 2016, Stanton et al. 2017) have provided new information on acidic N-glycosylation (sulphated, glucuronylated, etc.), Lewis-type antennal fucosylated structures and phosphorylcholine on terminal GlcNAc. However, these modifications were not found on N-glycans of silkworm glycoproteins. Intensive N-glycome study of the silkworm, *Bombyx mori*, will be necessary in order to better understand the potential of N-glycosylation.

Sialylation of N-glycans has been successfully achieved by transient co-expression of GalT and SiaT using baculovirus with an external supply of SiaT substrate, CMP-NeuAc.

Meanwhile, transgenic silkworms co-expressing human GnTII and bovine GalT in the PSG produced Gal-extended N-glycans and a small amount of fucosylated N-glycans. Transgenic silkworms carrying sialylation without an external supply of CMP-NeuAc would be expected to produce recombinant proteins with humanized N-glycans. Further understanding of organ- and stage-specific expression of glycosylation enzyme genes would make silkworms a more feasible biofactory for recombinant proteins with engineered N-glycans. Finally, silkworms may be alternative hosts for producing recombinant glycoproteins with sialylated complex N-glycans.

References

Dojima, T., T. Nishina, T. Kato, T. Uno, H. Yagi, K. Kato and E. Y. Park. 2009. Comparison of the N-linked glycosylation of human beta1,3-N-acetylglucosaminyltransferase 2 expressed in insect cells and silkworm larvae. J. Biotechnol. 143(1): 27–33. Doi: 10.1016/j.jbiotec.2009.06.013.

Dojima, T., T. Nishina, T. Kato, T. Uno, H. Yagi, K. Kato, H. Ueda and E. Y. Park. 2010. Improved secretion of molecular chaperone-assisted human IgG in silkworm, and no alterations in their N-linked glycan structures. Biotechnol. Prog. 26(1): 232–238. Doi: 10.1002/btpr.313.

Geisler, C., J. J. Aumiller and D. L. Jarvis. 2008. A fused lobes gene encodes the processing beta-N-acetylglucosaminidase in Sf9 cells. J. Biol. Chem. 283(17): 11330–11339. Doi: 10.1074/jbc.M710279200.

Geisler, C. and D. L. Jarvis. 2010. Identification of genes encoding N-glycan processing beta-N-acetylglucosaminidases in Trichoplusia ni and *Bombyx mori*: Implications for glycoengineering of baculovirus expression systems. Biotechnol. Prog. 26(1): 34–44. Doi: 10.1002/btpr.298.

Ignatyeva, D. O., G. A. Knyazev, P. O. Kapralov, G. Dietler, S. K. Sekatskii and V. I. Belotelov. 2016. Magneto-optical plasmonic heterostructure with ultranarrow resonance for sensing applications. Sci. Rep. 6: 28077. Doi: 10.1038/srep28077.

Ihara, H., T. Okada and Y. Ikeda. 2014. Cloning, expression and characterization of *Bombyx mori* alpha1,6-fucosyltransferase. Biochem. Biophys. Res. Commun. 450(2): 953–960. Doi: 10.1016/j.bbrc.2014.06.087.

Iizuka, M., S. Ogawa, A. Takeuchi, S. Nakakita, Y. Kubo, Y. Miyawaki, J. Hirabayashi and M. Tomita. 2009. Production of a recombinant mouse monoclonal antibody in transgenic silkworm cocoons. FEBS J. 276(20): 5806–5820. Doi: 10.1111/j.1742-4658.2009.07262.x.

Kajiura, H., Y. Hamaguchi, H. Mizushima, R. Misaki and K. Fujiyama. 2015. Sialylation potentials of the silkworm, *Bombyx mori*; *B. mori* possesses an active alpha2,6-sialyltransferase. Glycobiology. 25(12): 1441–1453. Doi: 10.1093/glycob/cwv060.

Kato, T., N. Kako, K. Kikuta, T. Miyazaki, S. Kondo, H. Yagi, K. Kato and E. Y. Park. 2017. N-Glycan modification of a recombinant protein via coexpression of human glycosyltransferases in silkworm pupae. Sci. Rep. 7(1): 1409. Doi: 10.1038/s41598-017-01630-6.

Kokuho, T., Y. Yasukochi, S. Watanabe and S. Inumaru. 2010. Molecular cloning and expression profile analysis of a novel beta-D-N-acetylhexosaminidase of domestic silkworm (*Bombyx mori*). Genes Cells. 15(5): 525–536. Doi: 10.1111/j.1365-2443.2010.01401.x.

Koles, K., E. Repnikova, G. Pavlova, L. I. Korochkin and V. M. Panin. 2009. Sialylation in protostomes: A perspective from *Drosophila* genetics and biochemistry. Glycoconj. J. 26(3): 313–324. Doi: 10.1007/s10719-008-9154-4.

Kubelka, V., F. Altmann, G. Kornfeld and L. Marz. 1994. Structures of the N-linked oligosaccharides of the membrane glycoproteins from three lepidopteran cell lines (Sf-21, IZD-Mb-0503, Bm-N). Arch. Biochem. Biophys. 308(1): 148–157. Doi: 10.1006/abbi.1994.1021.

Kurz, S., K. Aoki, C. Jin, N. G. Karlsson, M. Tiemeyer, I. B. Wilson and K. Paschinger. 2015. Targeted release and fractionation reveal glucuronylated and sulphated N- and O-glycans in larvae of dipteran insects. J. Proteomics. 126: 172–188. Doi: 10.1016/j.jprot.2015.05.030.

Kurz, S., J. G. King, R. R. Dinglasan, K. Paschinger and I. B. Wilson. 2016. The fucomic potential of mosquitoes: Fucosylated N-glycan epitopes and their cognate fucosyltransferases. Insect Biochem. Mol. Biol. 68: 52–63. Doi: 10.1016/j.ibmb.2015.11.001.

Leonard, R., D. Rendic, C. Rabouille, I. B. Wilson, T. Preat and F. Altmann. 2006. The Drosophila fused lobes gene encodes an N-acetylglucosaminidase involved in N-glycan processing. J. Biol. Chem. 281(8): 4867–4875. Doi: 10.1074/jbc.M511023200.

Mabashi-Asazuma, H., B. H. Sohn, Y. S. Kim, C. W. Kuo, K. H. Khoo, C. A. Kucharski and D. L. Jarvis. 2015. Targeted glycoengineering extends the protein N-glycosylation pathway in the silkworm silk gland. Insect Biochem. Mol. Biol. 65: 20–27. Doi: 10.1016/j.ibmb.2015.07.004.

Minagawa, S., S. Sekiguchi, Y. Nakaso, M. Tomita, M. Takahisa and H. Yasuda. 2015. Identification of core alpha 1,3-Fucosyltransferase gene from silkworm: An insect popularly used to express mammalian proteins. J. Insect Sci. 15: 110. Doi: 10.1093/jisesa/iev088.

Misaki, R., H. Nagaya, K. Fujiyama, I. Yanagihara, T. Honda and T. Seki. 2003. N-linked glycan structures of mouse interferon-beta produced by *Bombyx mori* larvae. Biochem. Biophys. Res. Commun. 311(4): 979–986.

Mita, K., M. Kasahara, S. Sasaki, Y. Nagayasu, T. Yamada, H. Kanamori, N. Namiki, M. Kitagawa, H. Yamashita, Y. Yasukochi, K. Kadono-Okuda, K. Yamamoto, M. Ajimura, G. Ravikumar, M. Shimomura, Y. Nagamura, T. Shin-I, H. Abe, T. Shimada, S. Morishita and T. Sasaki. 2004. The genome sequence of silkworm, *Bombyx mori*. DNA Res. 11(1): 27–35.

Nagata, Y., J. M. Lee, H. Mon, S. Imanishi, S. M. Hong, S. Komatsu, Y. Oshima and T. Kusakabe. 2013. RNAi suppression of beta-N-acetylglucosaminidase (BmFDL) for complex-type N-linked glycan synthesis in cultured silkworm cells. Biotechnol. Lett. 35(7): 1009–1016. Doi: 10.1007/s10529-013-1183-9.

Nomura, T., M. Ikeda, S. Ishiyama, K. Mita, T. Tamura, T. Okada, K. Fujiyama and A. Usami. 2010. Cloning and characterization of a beta-N-acetylglucosaminidase (BmFDL) from silkworm *Bombyx mori*. J. Biosci. Bioeng. 110(4): 386–391. Doi: 10.1016/j.jbiosc.2010.04.008.

Nomura, T., M. Suganuma, Y. Higa, Y. Kataoka, S. Funaguma, H. Okazaki, T. Suzuki, I. Kobayashi, H. Sezutsu and K. Fujiyama. 2015. Improvement of glycosylation structure by suppression of beta-N-acetylglucosaminidases in silkworm. J. Biosci. Bioeng. 119(2): 131–136. Doi: 10.1016/j.jbiosc.2014.07.012.

Okada, T., S. Ishiyama, H. Sezutsu, A. Usami, T. Tamura, K. Mita, K. Fujiyama and T. Seki. 2007. Molecular cloning and expression of two novel beta-N-acetylglucosaminidases from silkworm *Bombyx mori*. Biosci. Biotechnol. Biochem. 71(7): 1626–1635. Doi: 10.1271/bbb.60705.

Park, E. Y., M. Ishikiriyama, T. Nishina, T. Kato, H. Yagi, K. Kato and H. Ueda. 2009. Human IgG1 expression in silkworm larval hemolymph using BmNPV bacmids and its N-linked glycan structure. J. Biotechnol. 139(1): 108–114. Doi: 10.1016/j.jbiotec.2008.09.013.

Repnikova, E., K. Koles, M. Nakamura, J. Pitts, H. Li, A. Ambavane, M. J. Zeran and V. M. Panin. 2010. Sialyltransferase regulates nervous system function in *Drosophila*. J. Neurosci. 30(18): 6466–6476. Doi: 10.1523/JNEUROSCI.5253-09.2010.

Stanton, R., A. Hykollari, B. Eckmair, D. Malzl, M. Dragosits, D. Palmberger, P. Wang, I. B. Wilson and K. Paschinger. 2017. The underestimated N-glycomes of lepidopteran species. Biochim. Biophys. Acta. 1861(4): 699–714. Doi: 10.1016/j.bbagen.2017.01.009.

Suganuma, M., T. Nomura, Y. Higa, Y. Kataoka, S. Funaguma, H. Okazaki, T. Suzuki, K. Fujiyama, H. Sezutsu, K. I. Tatematsu and T. Tamura. 2018. N-glycan sialylation in a silkworm-*baculovirus* expression system. J. Biosci. Bioeng. 126(1): 9–14. Doi: 10.1016/j.jbiosc.2018.01.007.

Watanabe, S., A. Kakudo, M. Ohta, K. Mita, K. Fujiyama and S. Inumaru. 2013. Molecular cloning and characterization of the alpha-glucosidase II from *Bombyx mori* and *Spodoptera frugiperda*. Insect Biochem. Mol. Biol. 43(4): 319–327. Doi: 10.1016/j.ibmb.2013.01.005.

Part II
Utilization of Silkworm as a Biofactory

CHAPTER 7

BacMam System in Silkworm

Takashi Tadokoro

1. Introduction

The BacMam system (baculovirus-mediated gene transduction of mammalian cells) has been established as a relatively new technique, utilizing a modified baculovirus as a gene delivery vehicle to mammalian cells (Goehring et al. 2014, Kost et al. 2005, Kost et al. 2010). In recent years, the technique has been widely utilized in various fields, including pharmacological research and industry, due to the high gene delivery efficiency, its large genome (~ 130 kb) allowing it to harbor multiple genes or large inserts, its non-replicative feature and low cytotoxicity inside the mammalian host cells, and the ease of recombinant virus construction and production. Compared to the other gene transfer systems for mammalian cells (e.g., plasmid transfection or other viral vector systems), baculoviruses offer a rapid, safe and convenient alternative for gene transfer and protein expression in mammalian cells. Thus, the BacMam system has been exploited for a growing number of *in vitro* applications, such as high-level production of proteins, virus-like particles and virus vectors (Kost et al. 2005). At present, several assay systems utilizing a BacMam system are commercially available, such as BacMam 2.0 (Thermo Fisher Scientific) and cADDis cAMP BacMam Assay (Montana Molecular) for live cell labeling and cell-based assay.

Center for Research and Education on Drug Discovery, Faculty of Pharmaceutical Sciences, Hokkaido University, Kita-12, Nishi-6, Kita-ku, Sapporo 060-0812, Japan.
E-mail: tadokorot@pharm.hokudai.ac.jp

About a quarter of a century ago, Hofmann et al. (Hofmann et al. 1995) and Boyce and Bucher (Boyce and Bucher 1996) first reported that the recombinant baculoviruses harboring a mammalian promoter-driven gene expression cassette could transduce mammalian cells effectively. Their findings are shown with the *Autographa californica* multiple nucleopolyhedrovirus (AcMNPV)-based gene transfer system. Later, this AcMNPV-mediated BacMam (AcBacMam) system became widely accepted due to its aptitude for effective transduction in mammalian host cells and advantages in terms of efficacy, cost, safety and host ranges (Dukkipati et al. 2008, Goehring et al. 2014). For example, BacMam virus transduction is generally well tolerated without apparent cytotoxicity to the host cells and can do so in a biosafety level 1 facility. BacMam stocks can be stored for a relatively long period without significant loss of transduction ability as long as they are shielded from light and not submitted to repeated freeze/thaw cycles. One of the reasons for the wide use of the AcBacMam system is probably because of the availability of various options of vectors and systems for AcMNPV generation (Barford et al. 2013, Berger et al. 2004, Bieniossek et al. 2012). On the other hand, *Bombyx mori* nucleopolyhedrovirus (BmNPV) is another type of baculovirus that is next most well-studied. This chapter will describe the use of the BmNPV for the BacMam system and the possibility of wider applications. The BmNPV-mediated BacMam strategy and its possible applications, discussed later (see Subheading 5), are illustrated in Fig. 1.

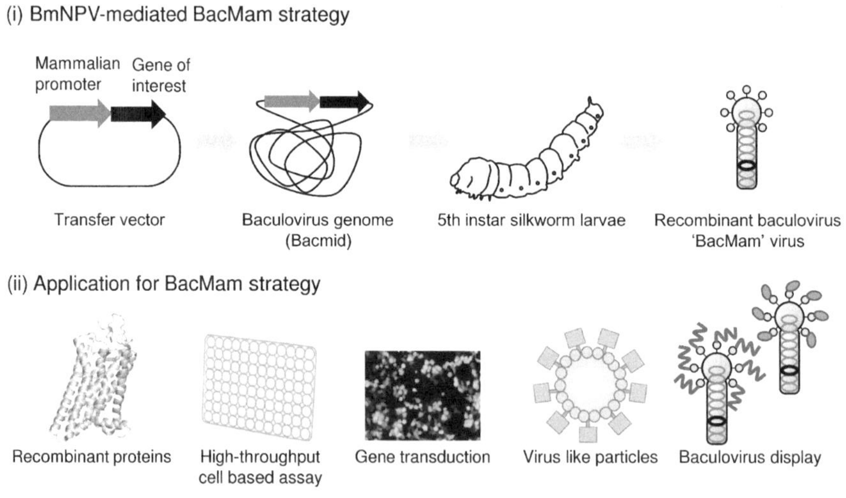

Fig. 1: BmNPV-mediated BacMam system and its possible application.

2. Silkworm Baculovirus and the Preparation of Recombinant Virus

In this chapter, the BacMam system is described mainly in terms of the modified Bac-to-Bac system derivatives for baculovirus generation. Motohashi et al. (2005) have established an effective baculovirus expression vector system (BEVS) using BmNPV bacmid DNA, which is directly applicable to silkworm expression. With this system, *E. coli* BmDH10Bac, which contains BmNPV bacmid and helper plasmid (pMON7124) (Motohashi et al. 2005), was used to construct the recombinant baculoviral genome via transfer vector, as used in the conventional Bac-to-Bac system (Thermo Fisher Scientific). Subsequently, the viral DNA is extracted and used to transfect silkworm larvae, instead of insect cells, by injecting the mixture solution of viral DNA and transfection reagent in order to generate the recombinant virus. As described above, the insect expression cassette in the transfer vector should be replaced by a mammalian expression cassette. All other features of the Bac-to-Bac system are retained. Hence, the entire procedure is simple and easy to conduct and recombinant BacMams can be usually generated in one week with this method, once the appropriate transfer vector is constructed.

The use of modified BmNPV for BacMam has several advantages over modified AcMNPV and other mammalian cell transduction systems, in terms of cost, ease of use and safety. One of the most significant points is that the use of BmNPV and silkworm, rather than AcMNPV and insect cells, can save time when preparing high-titer baculoviruses. For example, only 5–6 days are required in order to obtain BmNPV virus stock with a high titer (~ 10^9 plaque forming unit (pfu)/mL), whereas it usually takes at least several weeks or a month to obtain a comparable titer of AcMNPV virus. Another point is its cost-effectiveness. To maintain silkworms, only a basic incubator and food are required, while a bio-safety cabinet and expensive culture medium with serum, which is required for AcMNPV generation, are not necessary.

3. BmNPV-mediated BacMam

Although there has been a remarkable increase in published reports of the use of the BacMam system, most of them are about AcMNPV-mediated BacMam (AcBacMam), and the number of reports for BmNPV-mediated BacMam (BmBacMam) is still limited. To our knowledge, Kenoutis et al. (2006) have demonstrated the efficient transduction of mammalian cells *in vitro* for the first time. They showed that BmNPV-based vectors constitute an efficient alternative to AcMNPV-based vectors for gene transduction into mammalian cells. They showed that significantly more substantial levels of green fluorescence protein (GFP) expression were achieved upon addition of 1 μM of the histone deacetylase inhibitor trichostatin A (TSA) (Yoshida et al. 1990) when the HEK293 cells were infected with BmNPV/CMV-GFP, which contains GFP gene under control of human cytomegalovirus (CMV) promoter (Fig. 2). They also confirmed that the efficiency of transduction became maximised when using phosphate-buffered saline (PBS) as the transduction medium, as well as incubating at 28°C upon transduction, as has been reported for AcMNPV-based vectors (Hsu et al. 2004). The achieved transduction rate for HEK293 cells (greater

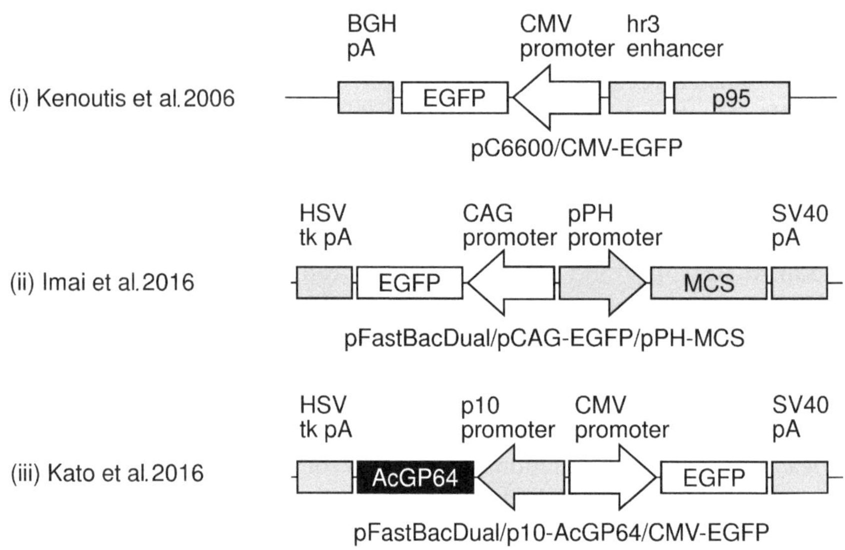

Fig. 2: Schematic presentation of the transfer vectors for BmBacMam.

Table 1: Comparison of the condition of BmNPV-mediated BacMam transduction.

MOI	Medium	Additives	Temperature	Exposure Time	Reference
500	PBS	1 μM TSA	28°C	24 hr	Kenoutis et al. 2006
500	40% DMEM, 60% PBS	4 mM sodium butyrate	30°C	6 d	Imai et al. 2016
300	Complete medium	None	room temperature, then 37°C	10 min, then 48 hr	Kato et al. 2016*

* The optimum transduction condition was not described in detail, and thus the condition described in the reference was listed.

than 90%) is comparable to the one achieved with mammalian virus-based vectors (Hofmann et al. 1995, Shoji et al. 1997), as well as that achieved with AcMNPV-based vectors (Hsu et al. 2004, Spenger et al. 2004).

More recently, Imai et al. (2016) and Kato et al. (2016) demonstrated the BacMam system using a silkworm baculovirus. The two groups showed the different optimum conditions for the gene transduction with different approaches.

Imai et al. (2016) reported that the BmBacMam gene transductions into mammalian cells were improved by using PBS and sodium butyrate during the gene transduction process. For this, they constructed the modified silkworm baculovirus, encoding EGFP under the CAG promoter control (Fig. 2). The recombinant baculovirus was amplified in the haemolymph of the 5th instar silkworm larvae and the transduction condition for the virus was examined. As shown in Fig. 3, the more the PBS proportion increased, the higher the observed EGFP expression. Around 3 to 6 days after virus infection under 60% and 80% PBS conditions, the percentage of EGFP positive cells reached a saturated level (Fig. 3). They tested the sodium butyrate, another histone deacetylase inhibitor, combined with PBS, and revealed that the EGFP expression was remarkably improved when more than 2 mM sodium butyrate was added together with 60% PBS, though only minor improvement was observed when sodium butyrate alone was added (Fig. 3). Since BacMam transduction was reportedly improved by lowering the temperature (Goehring et al. 2014), the temperature for transduction was also examined. After the addition of recombinant BmNPV, the incubation temperature was shifted to 30°C. The EGFP expression level was higher at 30°C than at 37°C. These features are similar to the case for AcBacMam, as previously reported (Dukkipati et al. 2008, Goehring et al. 2014, Hsu et al. 2004, Mahonen et al. 2010).

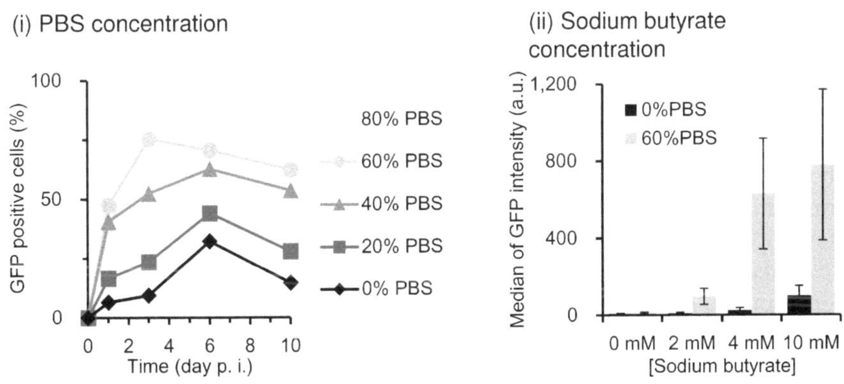

Fig. 3: Effect of the PBS and histone deacetylase inhibitor on BmBacMam transduction.

Kato et al. (2016) demonstrated that the gene transduction efficiency was improved by displaying Glycoprotein 64 (GP64) of AcMNPV (AcGP64) on the surface of BmNPV. It is known that GP64, which exists on the surface (envelope) of AcMNPV, plays an important role in the entry of AcMNPV into mammalian cells through dynein- and clathrin-dependent endocytosis and micropinocytosis (Kataoka et al. 2012). In their study, the BmGP64 was disrupted in order to construct the BmNPVΔbgp bacmid, and the AcGP64 gene under the control of the p10 promoter and the EGFP gene under the control of the CMV promoter (Fig. 2) were introduced using a modified pFastBacDual transfer vector, designated BmNPVΔbgp/AcGP64/EGFP. The GFP expression was observed in the cells transduced with BmNPVΔbgp/AcGP64/EGFP. However, the transduction efficiency of recombinant BmNPVΔbgp/AcGP64/EGFP was lower than that of the BacMam 2.0 (Thermo Fisher Scientific), which is based on AcMNPV. One possibility is the promoter for GP64. The GP64 promoter is known to work immediately upon infection before budding to the cells, whereas the p10 promoter functions at a very late stage of infection (Whitford et al. 1989, Zhou et al. 2003), which indicates that GP64 expression was achieved at a very late stage of infection, leading to inefficient GP64 expression in cells and inefficient display of GP64 on the baculovirus particles. Another reason why BmNPV inefficiently transduces foreign genes into mammalian cells as compared to AcMNPV could be the difference between AcGP64 and BmGP64 in terms of their capacity to fuse with the plasma membrane at different pH conditions. It is reported that AcGP64 can fuse with the plasma membrane at a lower pH than BmGP64 (Katou et al. 2010). This evidence indicates that AcGP64 facilitates its fusion with the plasma membrane at a higher pH than BmGP64.

Kato et al. (2017) further investigated the production of antibodies against *Neospora caninum antigen* (NcSRS2) using AcGP64-displaying BmNPVs containing the gene encoding NcSRS2 under the control of the CMV promoter, designated BmNPVΔbgp/AcGP64/SRS2. NcSRS2 antigen was similarly expressed in HEK293T cells as mentioned above for modified BmNPV encoding EGFP. Moreover, BmNPVΔbgp/AcGP64/SRS2 transduced mice produce NcSRS2-specific antibodies, suggesting that BmBacMam is capable of producing antigen-specific antibodies in immunized mice and can be used for antibody production and vaccine development.

Taken together, optimizing the culture conditions for infection and replacing an envelope protein on the surface of BmNPV enable this virus to transduce foreign genes into mammalian cells efficiently. Schematic structures of the transfer vectors (Fig. 2) and transduction condition (Table 1) in the studies described above are summarized. The culture optimization method (i.e., the addition of PBS and histone deacetylase inhibitor) is likely easier and can be applied to the *in vitro* expression of foreign genes in cultured cells. However, the application of this method is difficult to apply *in vivo*, compared to the AcGP64-displaying BmNPV method mentioned by Kato et al. (Kato et al. 2016, 2017). On the other hand, the replacement of the AcGP64 from BmGP64 is not always desirable since AcMNPV viruses exhibit relevant cell type preference, as described below. Therefore, optimizing the culture condition must be progressed depending on the cell types.

As mentioned above, BmNPV is better than AcMNPV for the large-scale preparation of baculovirus particles because BmNPV can be prepared from silkworm haemolymph, which contains high-titer BmNPV particles, without the use of specialized equipment (Kato et al. 2009). However, the purification strategy of BmNPV from haemolymph is limited by factors such as size exclusion chromatography and sucrose density gradient centrifugation (Kato et al. 2011, 2016). The proteomic study on the silkworm haemolymph revealed the existence of abundant 30K proteins and lipoproteins (Li et al. 2012). Thus, those components must be considered, especially when planning to use BmBacMam for *in vivo* study. The components may induce unexpected immune responses in mice. Therefore, the need to investigate how to separate recombinant BmNPV from the silkworm haemolymph and which components in the haemolymph contribute to inducing the unfavorable immune responses must be addressed.

Combination of the target cell type and mammalian expression element is also of concern for the successful gene transduction. In earlier research, it was shown that cells that are not transduced by a baculovirus expressing β-galactosidase under the control of a CMV promoter can be transduced efficiently by a baculovirus expressing the same reporter protein under the control of the stronger CAG promoter, consisting of the β-actin promoter and CMV immediate-early enhancer (Shoji et al. 1997). In addition to the described promoters, several mammalian promoters are available, such as SV40 (simian virus 40) and human EF-1α (elongation factor-1 α), thus, the different promoters need to be examined to various type of cell lines.

4. Preference of Cell Types

One of the unique features of AcBacMam is the ability to transduce a variety of mammalian cell lines, HEK293, HeLa, monkey kidney (COS), and Chinese Hamster ovary (CHO) cells (Airenne et al. 2000, Kost and Condreay 2002, Ping et al. 2006, Tani et al. 2001). Here the cells which reportedly succeeded in transducing foreign genes by BacMam are summarized (Table 2) and highlighted for the cells transduced by BmNPV-

Table 2: List of mammalian cell types successfully transduced by recombinant baculoviruses.

Human cells	*Rabbit cells*
HeLa	Primary hepatocytes
Huh7	
HEK 293*	*Non-human primate cells*
HEK 293T*	COS-7
HepG2	CV-1
KATO-III	Vero
IMR32	
MT-2	*Rodent cells*
Pancreatic β-cells	CHO
Keratinocytes	BHK
Bone marrow fibroblasts	RGM I
CHP212	PC12
Primary neural cells	Mouse pancreatic β-cells
W12	N2a
SK-N-MC	Primary rat hepatocytes
Saos-2	L929
WI38	**Mouse embryo fibroblasts***
Primary hepatocytes	**Rat Schwann cells***
FLC4	
143TK-	*Porcine cells*
DLD-1	CPK
Embryonic lung fibroblasts	FS-L3
Primary foreskin fibroblasts*	PK-15
MRC5	
MG63	*Bovine cells*
	MDB
	BT
	Ovine cells
	FLL-YFT

* The cells transduced by the BmNPV-mediated BacMam were highlighted with bold.

mediated BacMam. In the case of BmBacMam, many cell types were not tested, besides human foreskin fibroblasts and mouse embryo fibroblasts (Fig. 4; Tadokoro, unpublished data). The trial of various cell types is

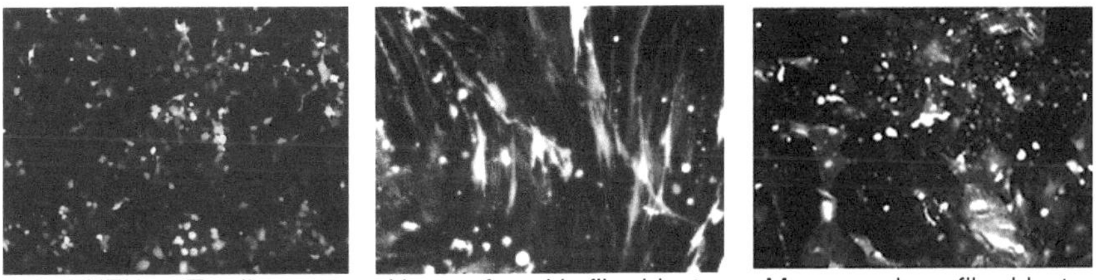

Fig. 4: Representative images of mammalian cells transduced by BmBacMam (EGFP).

required in order to expand the applicable field of use of BmBacMam. Indeed, quite low levels of transduction by AcBacMam were reported for several cell lines of hematopoietic origin, such as THP-1, U937 and K562 cells (Kost and Condreay 2002). AcBacMam transduction to monocyte-like cell lines is likely difficult, thus, cell type preference should be examined for BmBacMam in terms of whether it has an advantage over AcBacMam.

Primary neurons still tend to be refractory to plasmid transfection in cell culture. Some studies reported significant transduction rates for embryonic or postnatal cerebellar and hippocampal neurons, but gene transfer into adult neurons, non-dividing neurons, can be considerably less efficient (at best 5–10% by combination of calcium phosphate precipitation and lipofection methods or recombinant lentivirus delivery system) (Karra and Dahm 2010). The methods have problems, such as limited applicability for neuron cell cultures, cytotoxicity, risk of insertional mutations, late onset of transgene expression, expense for transient transfection and/or requirement for biosafety level 2 (Cockrell and Kafri 2007, Karra and Dahm 2010, Russell 2000, Zeitelhofer et al. 2007). Levin et al. (2016) provide the first evidence that a BacMam virus can efficiently transduce genes of interest into primary neurons of the adult central nervous system, as well as the peripheral nervous system. Importantly, transduction rates remarkably exceeded previously-reported results of other gene transfer methods without obvious neuro-cytopathic effects. Thus, the BacMam system would be quite useful in this field.

5. Possible Application for BmBacMam

5.1 Structural Study

One of the main concerns for biological research and medical use of recombinant protein is appropriate biological activities to the proteins of interest, which is usually achieved by appropriate post-translational modifications, such as glycosylation, disulfide formation, phosphorylation and acetylation. However, glycosylation in insect cells differs in many aspects from that in mammalian cells. Thus, the AcBacMam system is sometimes employed for the recombinant protein production and structural studies due to its aptitude for large-scale production. Milligram quantities of a GPCR–protein ligand complex, suitable for crystallization trials using a combination of the BacMam and HEK293 GnTI−cell line, were successfully produced by (Dukkipati et al. 2008). Morales-Perez et al. (2016) also succeeded in expressing, purifying and growing diffraction-quality crystals of the nicotinic acetylcholine receptor using the BacMam system. Those studies indicated the potential usefulness of BmBacMam in this field.

5.2 High-throughput Cell Based Assay

The AcBacMam is utilized for the expression of certain functional proteins such as transporters (Hassan et al. 2006), ion channels (Pfohl et al. 2002), and nuclear receptors (Clay et al. 2003), which lead to the development of various cell-based assays. Mazina et al. (2012) have developed the high-throughput cell based assay system of a Förster resonance energy transfer (FRET) biosensor for cAMP in combination with a BacMam transduction system. In order to identify the biological activity of various GPCR-specific agents, they used the BacMam transduction system to introduce a biosensor protein that is directly activated upon cAMP binding, altering the distance between the two fluorophores. Thus, monitoring the second-messenger cAMP level allowed them to characterize the different GPCR activation activities. The BacMam system is suitable for this assay due to its relatively low toxicity to host cells and applicability for a high-throughput assay in comparison to the plasmid transfection methods.

5.3 Virus Like Particles (VLPs)

The AcBacMam has been proved as an alternative strategy for VLP production since the system is easy to scale up and purify. Further details about VLPs will not be given here (see Chapters 12 and 16). Tang et al. (2011) demonstrated that the influenza VLPs can be generated from cells transduced by the BacMam which contains the hemagglutinin, neuraminidase and matrix genes of influenza virus. Moreover, they succeeded in vaccinating mice with VLPs generated with BacMam, as well as plasmid transfection-produced VLPs.

5.4 Baculovirus Display

Baculovirus display is another strategy for displaying heterologous peptides or proteins on the surface of baculovirus particles by fusing the peptide or protein to the baculovirus surface glycoprotein, gp64. Zhang et al. (2014) generated an improved BacMam virus which displayed the S1 glycoprotein (contains the neutralizing epitopes for infectious bronchitis virus) on the baculovirus envelope and was capable of expressing it in mammalian cells (Zhang et al. 2014). The BacMam virus elicited strong humoral and cell-mediated immune responses and induced greater cytotoxic T lymphocyte responses. Thus, the results indicated that BacMam virus-based surface display strategies could serve as powerful tools in designing vaccines for infectious diseases.

5.5 Large Gene Delivery

Takata et al. (2011) successfully generated the iPS cells from mouse embryonic fibroblasts (MEF) using the BacMam transduction system. These contain a polycistronic plasmid which encodes four defined, tandemly fused reprogramming factors, Oct4, Klf4, Sox2 and c-Myc, via self-cleaving 2A peptides under the control of CAG promoter. They employed the capability carrying a sizable foreign DNA fragment (> 15 kb), one of the features of baculovirus (Motohashi et al. 2005). Viral gene delivery systems utilized for the iPS generation have a limited insert size (< 2.5 kb for adeno-associated virus, 2.5–5 kb for lentivirus) and a requirement for biosafety level 2 for lentivirus (Karra and Dahm 2010). Their work shows the potential to overcome these problems.

6. Conclusions and Future Prospective

This chapter described the possibility that the BacMam system will ultimately become as commonplace as the use of recombinant baculoviruses for gene delivery systems. This relatively new technique offers several advantages, including the inability of the virus to replicate in mammalian cells, the absence or virtual absence of cytotoxicity, technical simplicity and a superior biosafety profile as compared to mammalian cell-derived viral vectors. Moreover, the BmBacMam system has several advantages over not only the conventional viral and non-viral transgene methods, but also the AcBacMam, which is often employed as a rapid and cost-effective approach for recombinant baculovirus generation. Nevertheless, accumulating evidence about the baculovirus-cell interactions that result in efficient gene transfer will further guide our efforts in investigating superior vector design and/or engineering cells in order to increase transduction efficiency. Finally, the efforts for possible applications described in this chapter, achieved with AcBacMam system, will provide an alternative approach for addressing and developing pharmacological and structural studies and provide the potential for broad applications to research in the life sciences, as well as the industrial and medical fields.

References

Airenne, K. J., M. Hiltunen, M. Turunen, A. Turunen, O. Laitinen, M. S. Kulomas and S. Ylä-Herttuala. 2000. Baculovirus-mediated periadventitial gene transfer to rabbit carotid artery. Gene Ther. 7: 1499–1504.

Barford, D., Y. Takagi, P. Schultz and I. Berger. 2013. Baculovirus expression: Tackling the complexity challenge. Curr. Opin. Struct. Biol. 23: 357–364.

Berger, I., D. Fitzgerald and T. Richmond. 2004. Baculovirus expression system for heterologous multiprotein complexes. Nat. Biotechnol. 22: 1583–1587.

Bieniossek, C., T. Imasaki, Y. Takagi and I. Berger. 2012. MultiBac: Expanding the research toolbox for multiprotein complexes. Trends Biochem. Sci. 37: 49–57.

Boyce, F. and N. Bucher. 1996. Baculovirus-mediated gene transfer into mammalian cells. Proc. Natl. Acad. Sci. U S A. 93: 2348–2352.

Clay, W. C., J. P. Condreay, L. B. Moore, S. L. Weaver, M. A. Watson, T. A. Kost and J. J. Lorenz. 2003. Recombinant baculoviruses used to study estrogen receptor function in human osteosarcoma cells. Assay Drug Dev. Technol. 1: 801–810.

Cockrell, A. S. and T. Kafri. 2007. Gene delivery by lentivirus vectors. Mol. Biotechnol. 36: 184–204.

Dukkipati, A., H. H. Park, D. Waghray, S. Fischer and K. C. Garcia. 2008. BacMam system for high-level expression of recombinant soluble and membrane glycoproteins for structural studies. Protein Expr. Purif. 62: 160–170.

Goehring, A., C. H. Lee, K. H. Wang, J. C. Michel, D. P. Claxton, I. Baconguis, T. Althoff, S. Fischer, K. C. Garcia and E. Gouaux. 2014. Screening and large-scale expression of membrane proteins in mammalian cells for structural studies. Nat. Protoc. 9: 2574–2585.

Hassan, N. J., D. J. Pountney, C. Ellis and D. E. Mossakowska. 2006. BacMam recombinant baculovirus in transporter expression: A study of BCRP and OATP1B1. Protein Expr. Purif. 47: 591–598.

Hofmann, C., V. Sandig, G. Jennings, M. Rudolph, P. Schlag and M. Strauss. 1995. Efficient gene-transfer into human hepatocytes by baculovirus vectors. Proc. Natl. Acad. Sci. U S A. 92: 10099–10103.

Hsu, C. S., Y. C. Ho, K. C. Wang and Y. C. Hu. 2004. Investigation of optimal transduction conditions for baculovirus-mediated gene delivery into mammalian cells. Biotechnol. Bioeng. 88: 42–51.

Imai, A., T. Tadokoro, S. Kita, M. Horiuchi, H. Fukuhara and K. Maenaka. 2016. Establishment of the BacMam system using silkworm baculovirus. Biochem. Biophys. Res. Commun. 478: 580–585.

Karra, D. and R. Dahm. 2010. Transfection techniques for neuronal cells. J. Neurosci. 30: 6171–6177.

Kataoka, C., Y. Kaname, S. Taguwa, T. Abe, T. Fukuhara, H. Tani, K. Moriishi and Y. Matsuura. 2012. Baculovirus GP64-mediated entry into mammalian cells. J. Virol. 86: 2610–2620.

Kato, T., S. Manoha, S. Tanaka and E. Y. Park. 2009. High-titer preparation of *Bombyx mori* nucleopolyhedrovirus (BmNPV) displaying recombinant protein in silkworm larvae by size exclusion chromatography and its characterization. BMC Biotechnol. 9.

Kato, T., F. Suzuki and E. Y. Park. 2011. Purification of functional baculovirus particles from silkworm larval haemolymph and their use as nanoparticles for the detection of human prorenin receptor (PRR) binding. BMC Biotechnol. 11.

Kato, T., S. Sugioka, K. Itagaki and E. Y. Park. 2016. Gene transduction in mammalian cells using *Bombyx mori* nucleopolyhedrovirus assisted by glycoprotein 64 of *Autographa californica* multiple nucleopolyhedrovirus. Sci. Rep. 6.

Kato, T., K. Itagaki, M. Yoshimoto, R. Hiramatsu, H. Suhaimi, T. Kohsaka and E. Y. Park. 2017. Transduction of a *Neospora caninum* antigen gene into mammalian cells using a modified *Bombyx mori* nucleopolyhedrovirus for antibody production. J. Biosci. Bioeng. 124: 606–610.

Katou, Y., H. Yamada, M. Ikeda and M. Kobayashi. 2010. A single amino acid substitution modulates low-pH-triggered membrane fusion of GP64 protein in *Autographa californica* and *Bombyx mori* nucleopolyhedroviruses. Virology. 404: 204–214.

Kenoutis, C., R. C. Efrose, L. Swevers, A. A. Lavdas, M. Gaitanou, R. Matsas, K. Iatrou. 2006. Baculovirus-mediated gene delivery into mammalian cells does not alter their transcriptional and differentiating potential but is accompanied by early viral gene expression. J. Virol. 80: 4135–4146.

Kost, T. A. and J. P. Condreay. 2002. Recombinant baculoviruses as mammalian cell gene-delivery vectors. Trends Biotechnol. 20: 173–180.

Kost, T. A., J. P. Condreay and D. Jarvis. 2005. Baculovirus as versatile vectors for protein expression in insect and mammalian cells. Nat. Biotechnol. 23: 567–575.

Kost, T. A., J. P. Condreay and R. Ames. 2010. Baculovirus gene delivery: A flexible assay development tool. Curr. Gene Ther. 10: 168–173.

Levin, E., H. Diekmann and D. Fischer. 2016. Highly efficient transduction of primary adult CNS and PNS neurons. Sci. Rep. 6.

Li, J., J. Li and B. Zhong. 2012. Proteomic profiling of the hemolymph at the 5th instar of the silkworm *Bombyx mori*. Insect Sci. 19: 441–454.

Mahönen, A., K. E. Makkonen, J. P. Laakkonen, T. O. Ihalainen, S. P. Kukkonen, M. U. Kaikkonen, M. Vihinen-Ranta, S. Ylä-Herttuala and K. J. Airenne. 2010. Culture medium induced vimentin reorganization associates with enhanced baculovirus-mediated gene delivery. J. Biotechnol. 145: 111–119.

Mazina, O., R. Reinart-Okugbeni, S. Kopanchuk and A. Rinken. 2012. BacMam system for FRET-based cAMP sensor expression in studies of melanocortin MC1 receptor activation. J. Biomol. Screen. 17: 1096–1101.

Morales-Perez, C. L., C. M. Noviello and R. E. Hibbs. 2016. Manipulation of subunit stoichiometry in heteromeric membrane proteins. Structure. 24: 797–805.

Motohashi, T., T. Shimojima, T. Fukagawa, K. Maenaka and E. Y. Park. 2005. Efficient large-scale protein production of larvae and pupae of silkworm by *Bombyx mori* nuclear polyhedrosis virus bacmid system. Biochem. Biophys. Res. Commun. 326: 564–569.

Pfohl, J. L., J. F. 3rd. Worley, J. P. Condreay, G. An, C. J. Apolito, T. A. Kost and J. F. Truax. 2002. Titration of K-ATP channel expression in mammalian cells utilizing recombinant baculovirus transduction. Receptors Channels. 8: 99–111.

Ping, W., J. Ge, S. Li, H. Zhou, K. Wang, Y. Feng and Z. Lou. 2006. Baculovirus-mediated gene expression in chicken primary cells. Avian Dis. 50: 59–63.

Russell, W. 2000. Update an adenovirus and its vectors. J. Gen. Virol. 81: 2573–2604.

Shoji, I., H. Aizaki, H. Tani, K. Ishii, T. Chiba, I. Saito, T. Miyamura and Y. Matsuura. 1997. Efficient gene transfer into various mammalian cells, including non-hepatic cells, by baculovirus vectors. J. Gen. Virol. 78: 2657–2664.

Spenger, A., W. Ernst, J. P. Condreay, T. A. Kost and R. Grabherr. 2004. Influence of promoter choice and trichostatin A treatment on expression of baculovirus delivered genes in mammalian cells. Protein Expr. Purif. 38: 17–23.

Takata, Y., H. Kishine, T. Sone, T. Andoh, M. Nozaki, M. Poderycki, J. D. Chesnut and F. Imamoto. 2011. Generation of iPS cells using a BacMam multigene expression system. Cell Struct. Funct. 36: 209–222.

Tang, X. C., H. R. Lu and T. M. Ross. 2011. Baculovirus-produced influenza virus-like particles in mammalian cells protect mice from lethal influenza challenge. Viral Immunol. 24: 311–319.

Tani, H., M. Nishijima, H. Ushijima, T. Miyamura and Y. Matsuura. 2001. Characterization of cell-surface determinants important for baculovirus infection. Virology. 279: 343–353.

Whitford, M., S. Stewart, J. Kuzio and P. Faulkner. 1989. Identification and sequence-analysis of a gene encoding-gp67, an abundant envelope glycoprotein of the baculovirus *Autographa californica* nuclear polyhedrosis-virus. J. Virol. 63: 1393–1399.

Yoshida, M., M. Kijima, M. Akita and T. Beppu. 1990. Potent and specific-inhibition of mammalian histone deacetylase both *in vivo* and *in vitro* by trichostatin-A. J. Biol. Chem. 265: 17174–17179.

Zeitelhofer, M., J. P. Vessey, Y. Xie, F. Tübing, S. Thomas, M. Kiebler and R. Dahm. 2007. High-efficiency transfection of mammalian neurons via nucleofection. Nat. Protoc. 2: 1692–1704.

Zhang, J., X. W. Chen, T. Z. Tong, Y. Ye, M. Liao and H. Y. Fan. 2014. BacMam virus-based surface display of the infectious bronchitis virus (IBV) S1 glycoprotein confers strong protection against virulent IBV challenge in chickens. Vaccine. 32: 664–670.

Zhou, Y., Y. Yi, Z. Zhang, J. He and Y. Zhang. 2003. Cetyltriethylammonium bromide stimulating transcription of *Bombyx mori* nucleopolyhedrovirus gp64 gene promoter mediated by viral factors. Cytotechnology. 41: 37–44.

CHAPTER 8

Transgenic Silkworm

Hideki Sezutsu

1. Introduction

The domesticated silkworm, *Bombyx mori*, has been used in the production of silk for 5,000 years, sericulture itself having been an important industry for thousands of years. In the 1980s, an epoch-defining method of making useful proteins in silkworm body fluids by infecting the silkworms with a recombinant baculovirus was established (Maeda 1989). TORAY and ZENOAQ in Japan have already sold medications for dogs and cats that were produced using this method (see Chapter 17 of this book). Furthermore, Tamura et al. (2000) established another innovative method for generating transgenic silkworms. The transgenic silkworm is a genetically modified organism that has a foreign gene (transgene) in its genome. Transgenic silkworms are established by germline transformation and the transgene is inherited by the next generation, although baculovirus-infected silkworm is usually incapable of reproduction.

The transgenic silkworm method has greatly expanded the possible utilities of silkworms and silk. Various approaches involving transgenic silkworms have been used in order to clarify the biological functions of silkworms as well as to develop industrial applications. In the industrial sphere, transgenic silkworms have mostly been used for producing new high-performance silks or useful recombinant proteins in the silk glands, the organ that produces high amounts (0.2–0.5 g) of silk protein. In addition, researchers have been developing a silkworm as a new model

Transgenic Silkworm Research Unit, National Agriculture and Food Research Organization, 1-2 Owashi, Tsukuba, Ibaraki, 305-8634, Japan.
E-mail: hsezutsu@affrc.go.jp

animal for testing medicines or as a new biosensor. These applications illustrate the suitability and potential of transgenic silkworms for medical and other uses. Some of these recombinant proteins have already seen commercial use, and a new apparatus plus a large-scale plant for mass rearing of silkworms has been constructed. Finally, farmers have started to rear the transgenic silkworms that produce modified silk.

This chapter focuses on the development of transgenic silkworms and their applications. The rest of this chapter is organized as follows: Section 2 introduces the methods for constructing transgenic silkworms by germline transformation, using a transposon vector or genome editing. Section 3 is the main section of this chapter and describes the applications of transgenic silkworm technology. This section introduces the gene expression systems used in transgenic silkworms, in addition to their applications in analyses of gene functions, modification of gene functions, production of recombinant proteins and production of modified silks. Section 4 introduces the trials currently being undertaken for large-scale rearing of transgenic silkworms in new industrial sericulture. Section 5 concludes this chapter.

2. Construction of Transgenic Silkworm

This section presents the methods for constructing transgenic silkworms. Transgenic silkworm lines are established by incorporating a foreign gene (a transgene), usually originating from another species, into silkworm's genome. It is important that the transgene can be inherited by the next generation and is suitable for large-scale rearing and industrial use. Initially, transgenic silkworms were constructed by germline transformation using a transposon as a gene transfer vector. Recently, a new technology called "genome editing" has been rapidly developed because it enables easy modification of any target gene.

2.1 Germline Transformation using Transposon Vector

Tamura et al. (2000) established a method of constructing transgenic silkworm by injecting the transgene DNA into the egg. Figure 1 summarizes the method. DNA of a transgenic vector is microinjected into the eggs by glass capillary. Because the egg shell of the silkworm is too hard to inject through directly by glass capillary, a tungsten needle is used in order to make a small hole prior to the microinjection by glass capillary. In the injected generation (G_0), the transgene will be incorporated into a part of the cells of the silkworm, forming a mosaic. If the transgene is integrated into the genome of the germline, transgenic silkworms will be obtained in the next generation (G_1) and can be identified by screening for the transgenic marker.

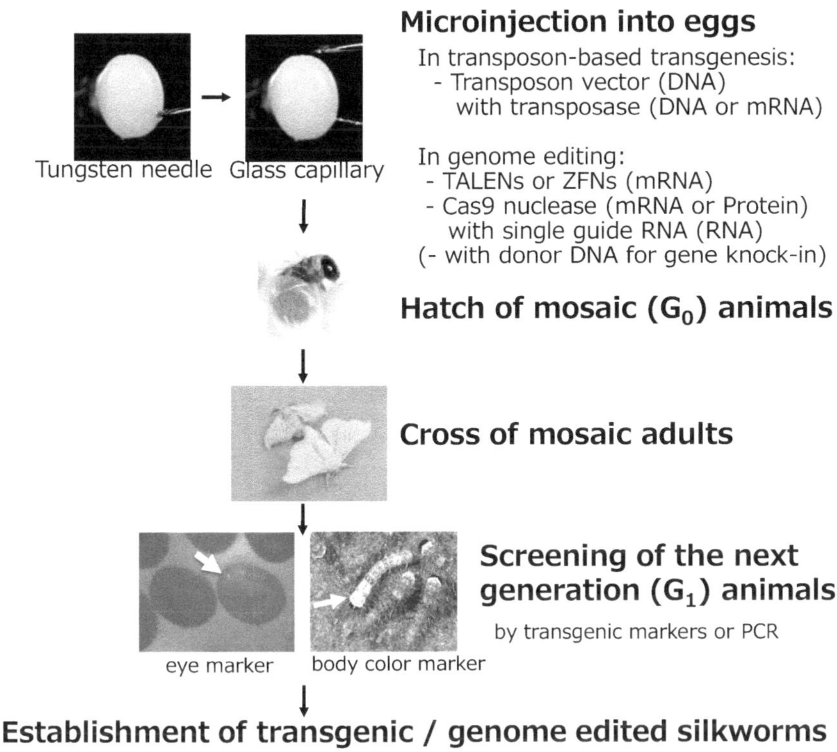

Fig. 1: Process for generating transgenic or genome-edited silkworms.

Transposons have been widely used for constructing transgene vectors in insects. DNA transposons are mobile DNA elements that are capable of moving within the genome. One kind of DNA transposon, *piggyBac*, has been predominantly used in manipulating silkworms (Tamura et al. 2000). As shown in Fig. 2a, *piggyBac* randomly inserts into TTAA sequences in the genome using its transposase. To construct the *piggyBac* vector plasmid (Fig. 2b), transgene and marker gene with promoters for tissue-specific expressions are integrated between the inverted terminal repeats (ITR). The DNA of the transposon vector, with the helper plasmid providing transposase and/or mRNA of transposase, is co-injected into silkworm eggs (Fig. 1). As markers of the success of genetic modification, genes of jellyfish and coral that encode fluorescent proteins are expressed in the eyes or body of the transgenic silkworms (Horn et al. 2000). Body color marker genes can also be used (Osanai-Futahashi et al. 2012). Table 1 summarizes the efficiency of producing transgenic silkworms using the *piggyBac*-based vector. In the first report of Tamura et al. (2000), the efficiency was 0.7% or 3.9%, but it has been significantly improved.

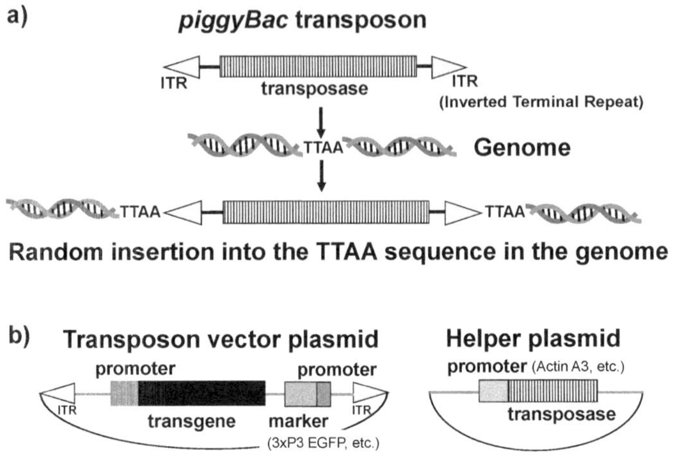

Fig. 2: *piggyBac* transposon and the transposon-based vector for the transgenesis: (a) Integration mechanism of *piggyBac* transposon into the genome; (b) Transposon vector plasmid and its helper plasmid for the transposition.

Table 1: Efficiency of producing transgenic silkworms using transposon-based vector.

Number of Embryos Injected	Hatched %	G_0 Yielder %: % of Adults producing Transgenic Silkworm	References
1058	66	0.7	(Tamura et al. 2000)
1440	33	3.9	(Tamura et al. 2000)
671	41	14	(Tamura et al. 2007)
706	62	12	(Tamura et al. 2007)
1092	28	26	(Tamura et al. unpublished)
466	54	51	(Tamura et al. unpublished)

2.2 Genome Editing

Genome editing is a recent innovative technology that has been rapidly developed for modification of the genome in many organisms. Genome editing has also been available in the silkworm (Tsubota and Sezutsu 2017). Genome editing tools such as artificial nuclease or RNA-guided nuclease can be engineered to induce double strand break in the target sequence of DNA (Fig. 3). An artificial nuclease, the zinc finger nuclease (ZFN), is a chimeric enzyme composed of the zinc finger domain for DNA recognition and the nuclease domain of FokI restriction enzyme. Another artificial nuclease, TALEN (Transcription Activator-Like Effector Nuclease), features the TALE domain for DNA recognition and the FokI nuclease domain. The DNA binding domains of ZFNs and TALENs can

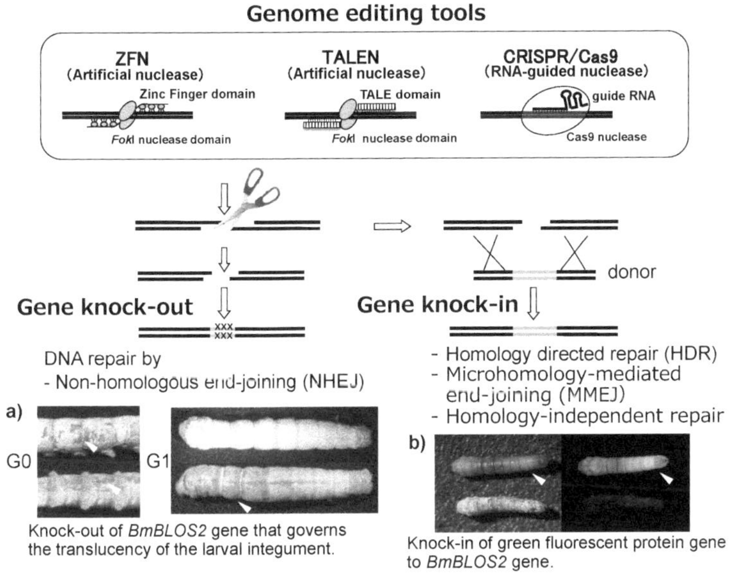

Fig. 3: Genome editing in the silkworm: (a) knock-out of *BmBLOS2* gene shows the translucency of the larval integument (modified from Takasu et al. 2013); (b) knock-in of green fluorescent protein (GFP) gene into *BmBLOS2* showed translucent integument and GFP expression (modified from Nakade et al. 2014).

be customized. CRISPR/Cas9 (Clustered Regularly Interspaced Short Palindromic Repeats/CRISPR associated protein 9) is a system using an RNA-guided nuclease. Guide RNA can be designed for a target DNA sequence, which the Cas9 nuclease will cut. Because CRISPR/Cas9 is especially easy to use, the CRISPR/Cas9 system has been widely used in many organisms. In the genome editing process, the RNAs or proteins of genome editing tools are microinjected into the egg (Fig. 1).

After the double strand breaks are induced, DNA will be repaired by DNA repair mechanisms (Fig. 3). Non-homologous end-joining (NHEJ) ligates the broken DNA but is also likely to induce small insertion or deletion mutations. When the function of the target endogenous gene is disrupted, "gene knock-out" individuals will be obtained. The broken DNA can be repaired using donor sequence by homology directed repair (HDR), microhomology-mediated end-joining (MMEJ) or homology-independent repair. In all of these cases, donor DNA can be integrated into the target gene. This establishes "gene knock-in" individuals.

Figure 3 shows the "knock-out" or "knock-in" silkworm in the *BmBLOS2* gene that governs the translucency of the larval integument. The efficiencies of gene knock-out and knock-in in the silkworms are summarized in Tables 2 and 3. At first, gene knock-out efficiencies using

Table 2: Efficiency of gene knock-out using genome editing tools.

Genome Editing Tools	Target Gene	Target	Number of Embryos Injected	Hatched %	G0 Mosaics%	G0 Yielder %: % of producing Knock-out G1	G1 Knock-out % in all G1 Individuals	References
ZFN	BmBLOS2	BL-1	480	18	72	10	0.3	(Takasu et al. 2010)
TALEN	BmBLOS2	BLT-1	480	37	15	11	0.7	(Sajwan et al. 2013)
TALEN	BmBLOS2	BLTS-4	191	32	82	100	10.4	(Takasu et al. 2013)
TALEN	BmBLOS2	BLTS-6	191	51	100	100	49.9	(Takasu et al. 2013)
TALEN	re	RE	192	46	72	100	77.3	(Takasu et al. 2013)
CRISPR/Cas9	BmBLOS2	od2	192	18	100	25	0.5	(Daimon et al. 2014)
CRISPR/Cas9	BmBLOS2	S1+S2	480	9	100	100	NA	(Wang et al. 2013)

NA: Not analyzed.

Table 3: Efficiency of gene knock-in using genome editing tools.

Genome Editing Tools	Target Gene	Marker Gene	Method	Number of Embryos Injected	Hatched %	G0 Mosaics %	G0 Yielder %	G1 Knock-in %	References
ZFN	BmBLOS2	BmA3-EGFP	HDR[1)]	264	NA	9	NA	0.008	(Daimon et al. 2014)
TALEN	BmBLOS2	hsp90-EGFP	PITCh[2]	264	34	NA	23	NA	(Nakade et al. 2014)
TALEN	ku80	hsp90-EGFP	PITCh[2)]	181	48	83	4	NA	(Tsubota and Sezutsu 2017)
TALEN	BmBLOS2	–	ssODN[3)]	88	64	NA	79	NA	(Takasu et al. 2016)

[1)] HDR (homology directed recombination),
[2)] PITCh (Precise Integration into Target Chromosome) technique,
[3)] ssODNs (short single-stranded oligodeoxynucleotides) was used as donors. NA: Not analyzed.

ZFNs, TALENs and CRISPR/Cas9 were relatively low (Daimon et al. 2014, Sajwan et al. 2013, Takasu et al. 2010), but they have been improved through the optimization of TALENs and CRISPR/Cas9 (Takasu et al. 2013, Wang et al. 2013). Gene knock-in is still relatively difficult in the silkworm. However, the efficiencies have been improved by the PITCh (Precise Integration into Target Chromosome) technique (Nakade et al. 2014, Tsubota et al. 2017), using the microhomology of donor and target sequences, or by using ssODNs (short single-stranded oligodeoxynucleotides) (Takasu et al. 2016). The disruption and/or replacement of endogenous genes by genome editing should become increasingly important in the future.

3. Application of Transgenic Silkworm

This section introduces the great potential of applications that use transgenic silkworms. Development of gene expression systems enabled a tissue-specific expression of the transgene in the silkworm. The gene expression systems are useful for the functional analysis of genes in numerous fields, such as genetics, physiology and developmental biology. Based on the fundamental research, transgenic silkworms show a very high potential for producing new materials, as well as applications in a new field. Researchers and companies have been especially focused on the production of useful proteins for pharmaceutical application and the development of novel silk materials.

3.1 Gene Expression Systems using the Transgenic Silkworms

Various transgene expression systems have been developed in the silkworm. A promoter is a DNA sequence that initiates the transcription of a particular gene in a tissue-specific manner. By ligating a promoter sequence and a foreign gene, the transgene can be expressed under the direct control of the promoter. However, the direct expression system shows a relatively low level of transgene expression because of the lack of a genomic enhancer sequence. Furthermore, if the transgene itself is deleterious, transgenic silkworm cannot be obtained.

Imamura et al. (2003) developed a useful binary expression system for the silkworm using the GAL4/UAS system. Figure 4a explains the GAL4/UAS binary system. GAL4 is a yeast transcription activator protein that activates the upstream activation sequence (UAS). UAS lines that have a target gene and a reporter gene, such as a fluorescent protein gene or other effector gene, such as an apoptosis-induction gene, can be used for forced expression, visualization, apoptosis induction or other functions (Fig. 4a). Importantly, UAS-transgene is only expressed alongside GAL4 protein, allowing transgenic silkworms carrying a toxic gene to be obtained. GAL4

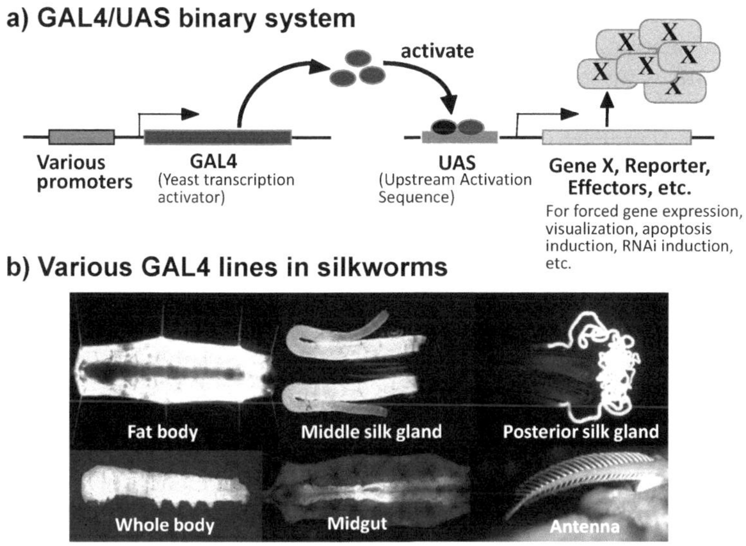

Fig. 4: GAL4/UAS binary system for transgene expression and various GAL4 lines in the silkworm: (a) GAL4/UAS system for various purposes; (b) GAL4 transgenic lines which express GFP reporter protein in various organs/tissue.

expression under the various promoters enhances the expression of the following gene to UAS. Many GAL4 lines with various promoters have been already established (Fig. 4b). Another binary system, the IE1/HR3 system, which uses the IE1 transactivator and the HR3 enhancer of a baculovirus (*Bombyx mori* nucleopolyhedrovirus—BmNPV), has also been established in the silkworm (Tomita et al. 2007). These binary systems using transactivators show higher transgene expression level than can be achieved under the direct expression system.

3.2 Analyses of Gene Functions

After a long history of sericulture, many natural mutants or regional varieties of silkworms, in addition to mutants artificially induced by exposure to radiation or chemical agents, have been established and conserved (Banno et al. 2010). A nearly complete genome sequence of *B. mori* was released in 2008 (International Silkworm Genome 2008). Post-genomic studies have actively sought to identify the genes responsible for these mutants. In these studies, forced expression and disruption of the genes that are putatively responsible for the mutations are used in order to understand their functions. The GAL4/UAS system was effectively used for mutant rescue experiments through forced expression of the normal types of genes in the mutant strains (Sakudoh et al. 2007, Tan et al. 2005).

Embryonic RNAi in the silkworm was also used in many experiments for the disruption of the putative genes (Quan et al. 2002). Recently, gene knock-out using genome editing has been used in the analyses of gene functions (Daimon et al. 2014). Thus, transgenesis, gene knock-out and RNAi are all powerful tools for the functional genomics of silkworms.

3.3 Modification of Gene Functions

The development of new uses for silkworms by modifying gene functions is ongoing. Some researchers have been trying to alter the glycosylation modification in the silkworm. This is very useful for the production of recombinant protein, both in transgenic silkworm and baculovirus-infected silkworm. Others have been developing a silkworm that is resistant to viral infection, a new human disease model, new biosensors and other important applications. Subbaiah et al. (2013) constructed a transgenic silkworm carrying the inverted repeat sequences of the essential gene of a baculovirus, *Bombyx mori* nucleopolyhedrovirus (BmNPV). The inverted repeats induced RNA interference (RNAi) of the BmNPV gene, after which, the transgenic silkworm exhibited stable protection against baculovirus infection.

As a new human disease model, transgenic silkworms with introduced mammal genes have been developed in order to promote new drug discoveries. Matsumoto et al. (2014) established transgenic silkworms expressing the human insulin receptor for the evaluation of insulin receptor agonists. Nikaido et al. (2015) evaluated novel μ-opioid receptor agonist compounds using a transgenic silkworm with a G protein-coupled receptor.

Using the pheromone-detecting capability of the male adult silkworms (silkmoths), a research group has been developing silkmoths that can replace police dogs as biosensors. Sakurai et al. (2011) successfully modified the selectivity of the sex pheromone receptor using transgenic silkmoths expressing the sex pheromone receptor PxOR1 of the diamondback moth *Plutella xylostella*. The transgenic silkmoths sensed the sex pheromone of diamondback moth and displayed sexual behavior. This suggested that transgenic silkmoths can be used as biosensors by introducing odorant receptor genes.

3.4 Productions of Recombinant Proteins

Because a silkworm larva has the capacity to synthesize 0.2 to 0.5 g of silk proteins, transgenic silkworms are expected to have great potential for producing recombinant proteins for pharmaceutical or cosmetic use. Cocoon silk fiber consists of core fibroin filaments coated with sericin glue (Fig. 5a). Fibroin fiber is difficult to dissolve and requires the use of strong protein-denaturing agents, whereas sericin glue is easily dissolved in a neutral buffer or alkaline solution. Fibroin constitutes 70–75% of cocoon silk and sericin contributes the remaining 25–30%. Fibroin consists of Fibroin light and heavy chain proteins. *B. mori* produces three types of sericin proteins: Sericin 1, Sericin 2 and Sericin 3. The silk proteins are produced in a pair of silk glands (Fig. 5b). Fibroin proteins are expressed in the posterior silk gland and secreted in the lumen. Sericin proteins are expressed and secreted in the middle silk gland. Because sericins are soluble, it is easy to recover and purify the recombinant proteins from the sericin layer. Therefore, systems for producing recombinant proteins in the middle silk gland have been developed.

Figure 6 shows the production systems of recombinant proteins using the GAL4/UAS or IE1/HR3 systems with the promoter of the *sericin 1* gene (Sericin 1 promoter). The GAL4 line, in which the GAL4 transactivator is expressed in the middle silk gland under the control of the Sericin 1 promoter, is crossed to the UAS line that has a target protein gene (Tatematsu et al. 2010). In the IE1/HR3 system, the target gene expresses in the middle silk gland under the control of the HR3-linked Sericin 1 promoter and IE1 gene (Tomita 2011). In the hybrid silkworm, the target gene expresses in the middle silk gland and the recombinant protein is co-secreted with the native sericin protein. The recombinant proteins are easily recovered and purified from the silk gland or cocoon using a neutral buffer.

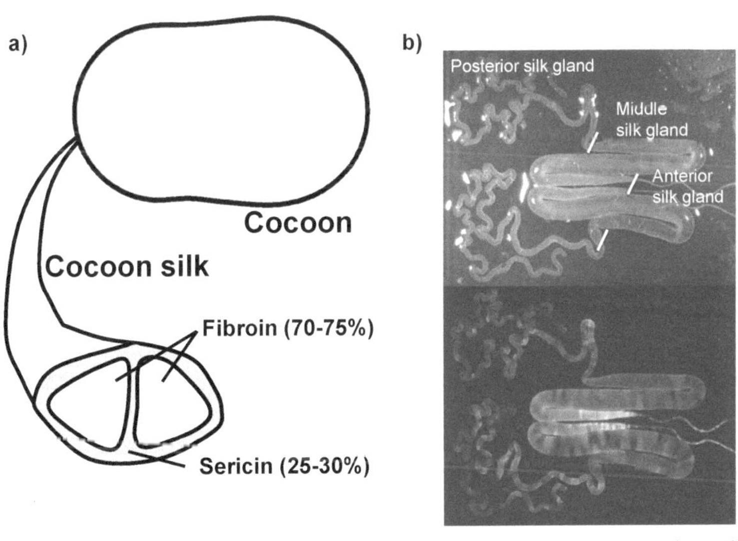

Fig. 5: Structure of cocoon silk and a pair of silk glands: (a) Silk fiber consists of core fibroin filaments coated with sericin glue; (b) A pair of silk glands of the transgenic silkworm that is expressing green fluorescent protein in a mosaic pattern. Each cell of the silk gland can be observed.

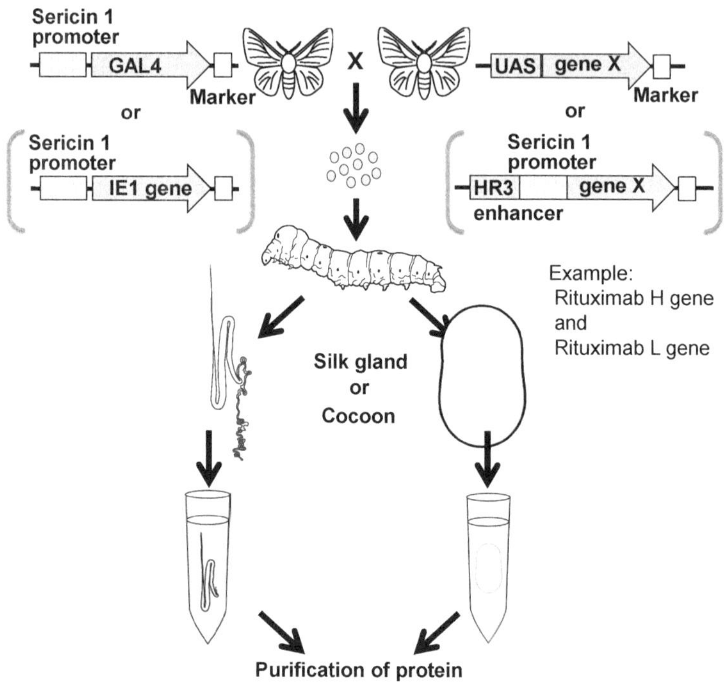

Fig. 6: The systems for the production of recombinant proteins using transgenic silkworm.

Several test reagents and cosmetics produced using the above method have already been commercialized by Japanese companies since 2011. The production of monoclonal antibody drugs for cancer (Iizuka et al. 2009, Tada et al. 2015), human drugs for orphan disease (Itoh et al. 2016), animal drugs and others is underway. The protein production system is currently capable of producing from 0.1 mg to a maximum of 20 mg recombinant protein per larva/cocoon. The amounts differ among target proteins and vector constructs. Transgenic silkworms can produce many kinds of recombinant proteins (Tomita 2011). Even proteins with complicated structures, such as antibodies, can be stably produced by these systems (Iizuka et al. 2009, Tada et al. 2015).

Figure 7 shows an example of the production of recombinant monoclonal antibodies (mAbs). Tada et al. (2015) constructed a transgenic silkworm stably expressing a human-mouse chimeric anti-CD20 mAb (rituximab). Rituximab heavy chain (H) and light chain (L) genes were introduced in the transgenic silkworms as shown in Fig. 6. As a result, rituximab H and L proteins were successfully produced in the silk gland and cocoon (Fig. 7a). When rituximab H and L proteins were expressed in one larva, the molecules of the H and L chain were assembled to H_2L_2-

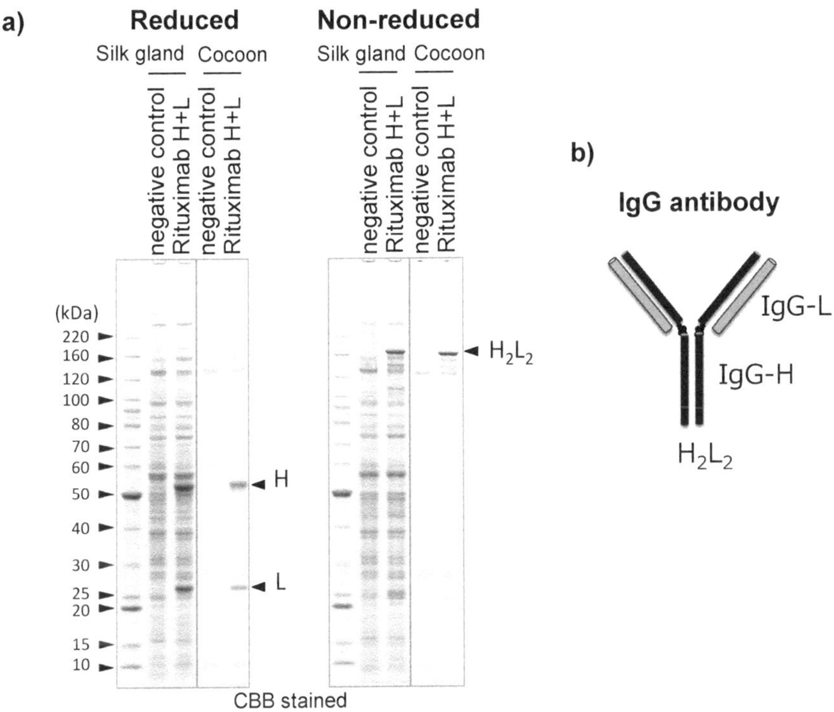

Fig. 7: Production of anti-CD20 monoclonal antibody (rituximab) in the silk gland or cocoon using transgenic silkworm: (a) SDS-page of the crude proteins from the silk gland or cocoon in a reduced or non-reduced condition. H and L chain of rituximab are shown as arrowheads (modified from Tada et al. 2015); (b) Possible model of the assembled H_2L_2-subunit structures of IgG antibody.

sub-unit structures (Fig. 7b) and secreted into the cocoons (Fig. 7a). The anti-CD20 mAb produced by the transgenic silkworm showed stronger antibody-dependent cellular cytotoxicity (ADCC) than the CHO-derived mAbs (MabThera), possibly due to the lack of the core-fucose in the N-glycan in the silkworm-derived mAbs (Tada et al. 2015). This suggests that the transgenic silkworm can serve as a suitable production system for the tumor-targeting mAbs with higher ADCC activity.

3.5 Productions of Modified Silks

It is possible to create new types of silk by incorporating a synthetic fibroin gene into silkworms. The methods for fibroin modification in the posterior silk gland were developed using vectors containing the fibroin heavy or light chain genes. Figure 8 shows the transgenic vectors for the fibroin modifications. In these vectors, the target genes are fused with the fibroin light chain gene (Tomita et al. 2003) or the fibroin heavy chain gene

(Kojima et al. 2007). The modified fibroin light or heavy chain proteins will be secreted with endogenous fibroin proteins into the lumen of the silk glands.

Iizuka et al. (2013) developed the fluorescent silks by incorporating the fibroin heavy chain vectors with various kinds of fluorescent protein genes. As shown in Fig. 9, the silks fluoresce with various colors under the excitation light. Even under natural light, the silks display colors and are useful as colored fibroins (Fig. 9a, c). Using the fluorescent silks, some trial products, such as a silk dress, have been made (Fig. 10).

Another modified silk, spider silk, has been developed by incorporating a gene from a spider species encoding a dragline protein of *Nephila clavipes* or *Araneus ventricosus* (Kuwana et al. 2014, Teule et al. 2012). A silk for improving the adhesive abilities or the growth of cells was also constructed

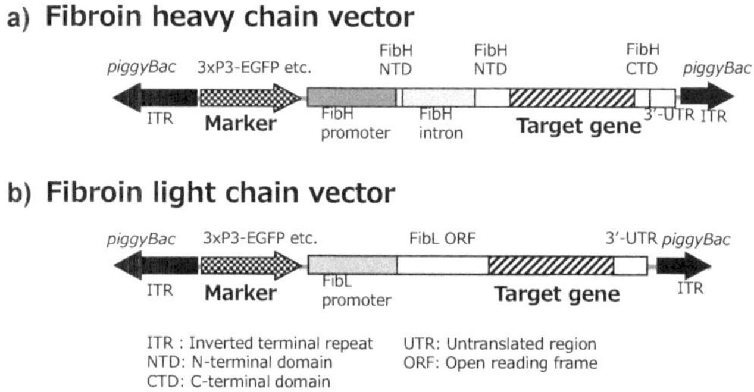

Fig. 8: Transgenic vectors for modifying the silk: (a) fibroin heavy chain vector; (b) fibroin light chain vector.

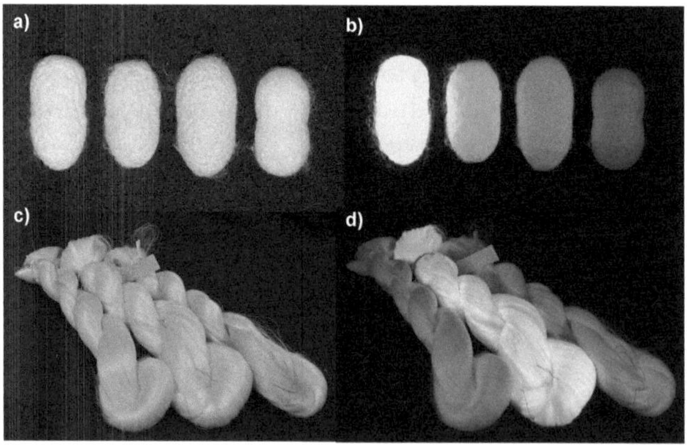

Fig. 9: The cocoons and the raw silks made of the fluorescent silks spun by transgenic silkworms: (a), (c) the green, orange and red fluorescent silks under natural light; (b), (d) under a blue light source with a yellow filter.

by introducing RGD motifs of fibronectin or vascular endothelial growth factor (Saotome et al. 2015). Using the silks, an artificial blood vessel and the sponge sheet, such as that shown in Fig. 11, has been developed. The silk-based materials improved the revascularization properties and can be used in regenerative medicine. Sato et al. (2017) developed "affinity" silk containing the fibroin L-chain fused with the single-chain variable fragment (scFv). The silk thin film, as shown in Fig. 11a, can be used in enzyme-linked immunosorbent assay (ELISA) for the detection of a target protein.

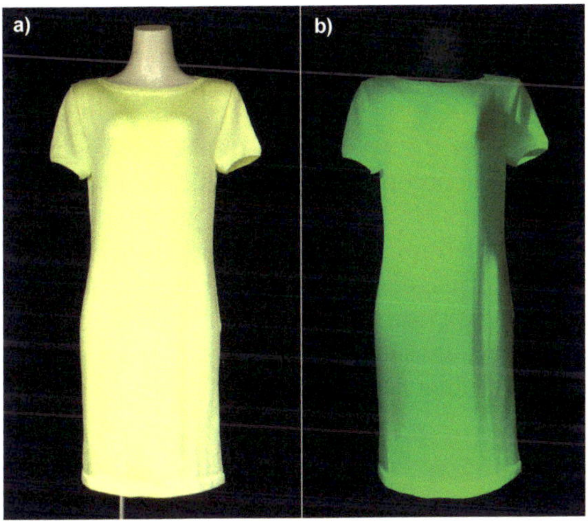

Fig. 10: Knit dress made of the green fluorescent silk: (a) under natural light; (b) under a blue light source with a yellow filter.

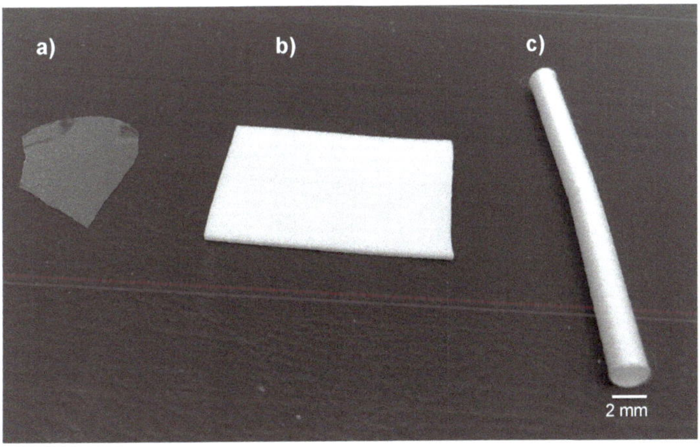

Fig. 11: Silk-based materials: (a) film; (b) sponge sheet; (c) artificial vessel.

4. Large-scale Rearing of Transgenic Silkworms in a New Industrial Sericulture

For the mass production of recombinant proteins or silks produced by transgenic silkworms, the regulations for the handling of genetically modified organisms (GMOs) should be followed. The regulation of GMOs, including transgenic silkworms, will differ between countries. In many countries, it is necessary to clear the regulations of the "Act on the Conservation and Sustainable Use of Biological Diversity through Regulations on the Use of Living Modified Organisms" (commonly known as the "Cartagena Protocol") for rearing genetically modified silkworms. In Japan, rearing with prevention of the dispersal of transgenic silkworms (type 2 use) and rearing without preventing their dispersal (type 1 use) are both in use.

In Japan, rearing by the protocol of type 2 use is already established for the practical production of transgenic silkworms by some companies and farmer associations. In addition, establishment of the protocol of type 1 use has been tried for many years. Finally, a silk-raising farmer in Japan began rearing transgenic silkworms that produce green fluorescent silks, according to the protocol of type 1 use in 2017 (Fig. 12). In India, an attempt to rear a virus-resistant transgenic silkworm has been made by silk-raising farmers (Subbaiah et al. 2013).

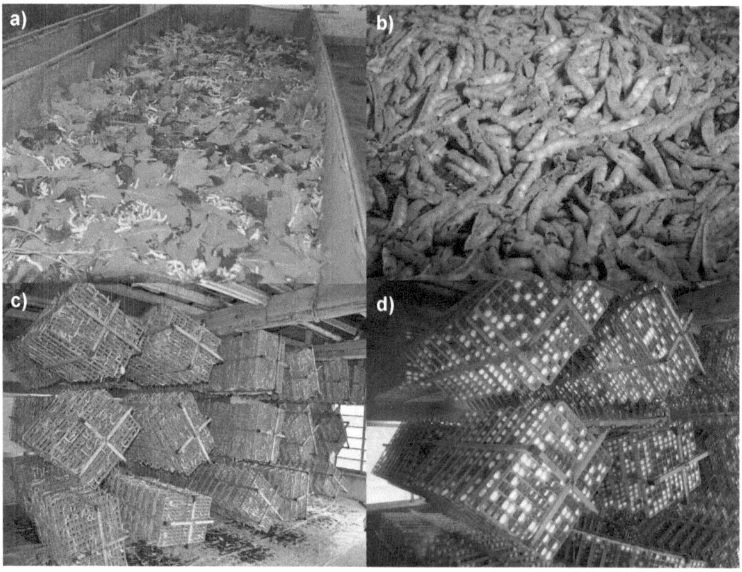

Fig. 12: Rearing of transgenic silkworm in a farmer's house: (a) (c) under natural light; (b) (d) under a blue light source with a yellow filter.

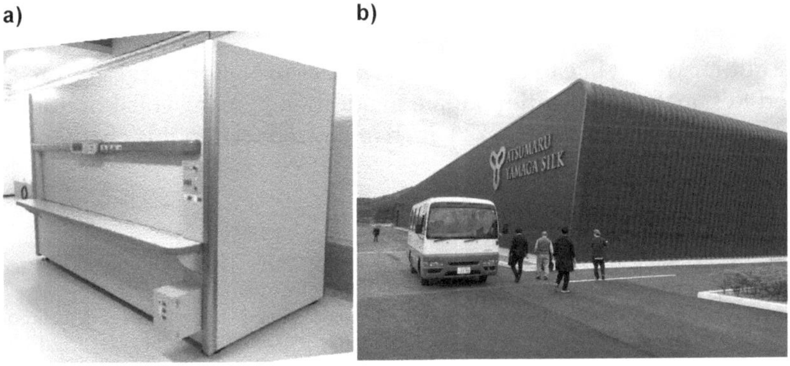

Fig. 13: A new rearing system and a large-scale silkworm factory: (a) A rearing system developed by Shinryo Corporation, Japan; (b) "Large-scale year-round aseptic sericulture plant" in Yamaga City, Kumamoto, Japan.

Shinryo Corporation, in Japan, developed a new mass rearing apparatus for transgenic silkworms (Fig. 13a). The apparatus can rear between 20,000 and 60,000 individuals with uniform temperature and humidity. Another Japanese company, Atsumaru Holdings Corporation, completed a large-scale year-round aseptic sericulture plant in 2017 at a cost of $21 million (Fig. 13b). The plant can rear a maximum of 560,000 silkworms at one time all year round. Thus, a new industrial sericulture with transgenic silkworms has begun.

5. Concluding Remark

This chapter introduced recent advances in the development of basic technologies and the applications of transgenic silkworms. The revolutionary transformation of the sericulture industry has been occurring via the development of medicinal products, new types of silk, and the new sericultural approaches. Silkworms will become more important bio-factories, because they can reduce emission of carbon dioxide in the production of useful proteins and fibers. Transgenic silkworms require a long lead time because the establishment of a transgenic line takes a few generations, but the silkworm is easily reproducible after the line has been established and shows stable expression of recombinant proteins. On the other hand, recombinant baculovirus-infected silkworm can produce useful proteins in a short time, although all individuals need to be inoculated with virus in each generation. The two production systems leadings to useful materials from silkworms are both very useful and complement each other. Thus, silkworms using new biotechnologies will construct a new silk road in the near future.

Acknowledgments

The author would like to thank the colleagues of National Agriculture and Food Research Organization Japan, in addition to many collaborators, for their contribution to the works which are introduced in this chapter.

References

Banno, Y., T. Shimada, Z. Kajiura and H. Sezutsu. 2010. The silkworm-an attractive BioResource supplied by Japan. Exp. Anim. 59(2): 139–146.

Daimon, T., T. Kiuchi and Y. Takasu. 2014. Recent progress in genome engineering techniques in the silkworm, *Bombyx mori*. Dev. Growth Differ. 56(1): 14–25. Doi: 10.1111/dgd.12096.

Horn, C., B. Jaunich and E. A. Wimmer. 2000. Highly sensitive, fluorescent transformation marker for Drosophila transgenesis. Dev. Genes Evol. 210(12): 623–629.

Iizuka, M., S. Ogawa, A. Takeuchi, S. Nakakita, Y. Kubo, Y. Miyawaki, J. Hirabayashi and M. Tomita. 2009. Production of a recombinant mouse monoclonal antibody in transgenic silkworm cocoons. FEBS J. 276(20): 5806–5820. Doi: 10.1111/j.1742-4658.2009.07262.x.

Iizuka, T., H. Sezutsu, K. Tatematsu, I. Kobayashi, N. Yonemura, K. Uchino, K. Nakajima, K. Kojima, C. Takabayashi, H. Machii, K. Yamada, H. Kurihara, T. Asakura, Y. Nakazawa, A. Miyawaki, S. Karasawa, H. Kobayashi, J. Yamaguchi, N. Kuwabara, T. Nakamura, K. Yoshii and T. Tamura. 2013. Colored fluorescent silk made by transgenic silkworms. Advanced Functional Materials. 23(42): 5232–5239. Doi: 10.1002/adfm.201300365.

Imamura, M., J. Nakai, S. Inoue, G. X. Quan, T. Kanda and T. Tamura. 2003. Targeted gene expression using the GAL4/UAS system in the silkworm *Bombyx mori*. Genetics. 165(3): 1329–1340.

International Silkworm Genome, C. 2008. The genome of a lepidopteran model insect, the silkworm *Bombyx mori*. Insect Biochem. Mol. Biol. 38(12): 1036–1045. Doi: 10.1016/j.ibmb.2008.11.004.

Itoh, K., I. Kobayashi, S. Nishioka, H. Sezutsu, H. Machii and T. Tamura. 2016. Recent progress in development of transgenic silkworms overexpressing recombinant human proteins with therapeutic potential in silk glands. Drug Discov. Ther. 10(1): 34–39. Doi: 10.5582/ddt.2016.01024.

Kojima, K., Y. Kuwana, H. Sezutsu, I. Kobayashi, K. Uchino, T. Tamura and Y. Tamada. 2007. A new method for the modification of fibroin heavy chain protein in the transgenic silkworm. Biosci. Biotechnol. Biochem. 71(12): 2943–2951.

Kuwana, Y., H. Sezutsu, K. Nakajima, Y. Tamada and K. Kojima. 2014. High-toughness silk produced by a transgenic silkworm expressing spider (Araneus ventricosus) dragline silk protein. PLoS One. 9(8): e105325. Doi: 10.1371/journal.pone.0105325.

Maeda, S. 1989. Gene transfer vectors of a baculovirus, *Bombyx mori*, and their use of expression of foreign genes in insect cells. pp. 167–181. *In*: Jun Mitsuhashi (ed.). Invertebrate Cell System Applications. CRC Press. 1.

Matsumoto, Y., M. Ishii, K. Ishii, W. Miyaguchi, R. Horie, Y. Inagaki, H. Hamamoto, K. Tatematsu, K. Uchino, T. Tamura, H. Sezutsu and K. Sekimizu. 2014. Transgenic silkworms expressing human insulin receptors for evaluation of therapeutically active insulin receptor agonists. Biochem. Biophys. Res. Commun. 455(3-4): 159–164. Doi: 10.1016/j.bbrc.2014.10.143.

Nakade, S., T. Tsubota, Y. Sakane, S. Kume, N. Sakamoto, M. Obara, T. Daimon, H. Sezutsu, T. Yamamoto, T. Sakuma and K. T. Suzuki. 2014. Microhomology-mediated end-joining-dependent integration of donor DNA in cells and animals using TALENs and CRISPR/Cas9. Nat. Commun. 5: 5560. Doi: 10.1038/ncomms6560.

Nikaido, Y., A. Kurosawa, H. Saikawa, S. Kuroiwa, C. Suzuki, N. Kuwabara, H. Hoshino, H. Obata, S. Saito, T. Saito, H. Osada, I. Kobayashi, H. Sezutsu and S. Takeda. 2015. *In vivo* and *in vitro* evaluation of novel mu-opioid receptor agonist compounds. Eur. J. Pharmacol. 767: 193–200. Doi: 10.1016/j.ejphar.2015.10.025.

Osanai-Futahashi, M., T. Ohde, J. Hirata, K. Uchino, R. Futahashi, T. Tamura, T. Niimi and H. Sezutsu. 2012. A visible dominant marker for insect transgenesis. Nat. Commun. 3: 1295. Doi: 10.1038/ncomms2312.

Quan, G. X., T. Kanda and T. Tamura. 2002. Induction of the white egg 3 mutant phenotype by injection of the double-stranded RNA of the silkworm white gene. Insect Mol. Biol. 11(3): 217–222.

Sajwan, S., Y. Takasu, T. Tamura, K. Uchino, H. Sezutsu and M. Zurovec. 2013. Efficient disruption of endogenous *Bombyx* gene by TAL effector nucleases. Insect Biochem. Mol. Biol. 43: 17–23. Doi: 10.1016/j.ibmb.2012.10.011.

Sakudoh, T., H. Sezutsu, T. Nakashima, I. Kobayashi, H. Fujimoto, K. Uchino, Y. Banno, H. Iwano, H. Maekawa, T. Tamura, H. Kataoka and K. Tsuchida. 2007. Carotenoid silk coloration is controlled by a carotenoid-binding protein, a product of the Yellow blood gene. Proc. Natl. Acad. Sci. U S A. 104(21): 8941–8946. Doi: 10.1073/pnas.0702860104.

Sakurai, T., H. Mitsuno, S. S. Haupt, K. Uchino, F. Yokohari, T. Nishioka, I. Kobayashi, H. Sezutsu, T. Tamura and R. Kanzaki. 2011. A single sex pheromone receptor determines chemical response specificity of sexual behavior in the silkmoth *Bombyx mori*. PLoS Genet. 7(6): e1002115. Doi: 10.1371/journal.pgen.1002115.

Saotome, T., H. Hayashi, R. Tanaka, A. Kinugasa, S. Uesugi, K. Tatematsu, H. Sezutsu, N. Kuwabara and T. Asakura. 2015. Introduction of VEGF or RGD sequences improves revascularization properties of *Bombyx mori* silk fibroin produced by transgenic silkworm. Journal of Materials Chemistry B. 3(35): 7109–7116. Doi: 10.1039/c5tb00939a.

Sato, M., H. Kitani and K. Kojima. 2017. Development and validation of scFv-conjugated affinity silk protein for specific detection of carcinoembryonic antigen. Sci. Rep. 7(1): 16077. Doi: 10.1038/s41598-017-16277-6.

Subbaiah, E. V., C. Royer, S. Kanginakudru, V. V. Satyavathi, A. S. Babu, V. Sivaprasad, G. Chavancy, M. Darocha, A. Jalabert, B. Mauchamp, I. Basha, P. Couble and J. Nagaraju. 2013. Engineering silkworms for resistance to baculovirus through multigene RNA interference. Genetics. 193(1): 63–75. Doi: 10.1534/genetics.112.144402.

Tada, M., K. Tatematsu, A. Ishii-Watabe, A. Harazono, D. Takakura, N. Hashii, H. Sezutsu and N. Kawasaki. 2015. Characterization of anti-CD20 monoclonal antibody produced by transgenic silkworms (*Bombyx mori*). MAbs. 7(6): 1138–1150. Doi: 10.1080/19420862.2015.1078054.

Takasu, Y., I. Kobayashi, K. Beumer, K. Uchino, H. Sezutsu, S. Sajwan, D. Carroll, T. Tamura and M. Zurovec. 2010. Targeted mutagenesis in the silkworm *Bombyx mori* using zinc finger nuclease mRNA injection. Insect Biochem. Mol. Biol. 40(10): 759–765. Doi: 10.1016/j.ibmb.2010.07.012.

Takasu, Y., S. Sajwan, T. Daimon, M. Osanai-Futahashi, K. Uchino, H. Sezutsu, T. Tamura and M. Zurovec. 2013. Efficient TALEN construction for *Bombyx mori* gene targeting. PLoS One. 8(9): e73458. Doi: 10.1371/journal.pone.0073458.

Takasu, Y., I. Kobayashi, T. Tamura, K. Uchino, H. Sezutsu and M. Zurovec. 2016. Precise genome editing in the silkworm *Bombyx mori* using TALENs and ds- and ssDNA donors—A practical approach. Insect Biochem. Mol. Biol. 78: 29–38. Doi: 10.1016/j.ibmb.2016.08.006.

Tamura, T., C. Thibert, C. Royer, T. Kanda, E. Abraham, M. Kamba, N. Komoto, J. L. Thomas, B. Mauchamp, G. Chavancy, P. Shirk, M. Fraser, J. C. Prudhomme and P. Couble. 2000. Germline transformation of the silkworm *Bombyx mori* L. using a piggyBac transposon-derived vector. Nat. Biotechnol. 18(1): 81–84. Doi: 10.1038/71978.

Tamura, T., N. Kuwabara, K. Uchino, I. Kobayashi and T. Kanda. 2007. An improved DNA injection method for silkworm eggs drastically increases the efficiency of producing transgenic silkworms. J. Insect Biotechnol. Sericol. 76: 155–159. Doi: 10.11416/jibs.76.3_155.

Tan, A., H. Tanaka, T. Tamura and T. Shiotsuki. 2005. Precocious metamorphosis in transgenic silkworms overexpressing juvenile hormone esterase. Proc. Natl. Acad. Sci. U S A. 102(33): 11751–11756. Doi: 10.1073/pnas.0500954102.

Tatematsu, K., I. Kobayashi, K. Uchino, H. Sezutsu, T. Iizuka, N. Yonemura and T. Tamura. 2010. Construction of a binary transgenic gene expression system for recombinant protein production in the middle silk gland of the silkworm *Bombyx mori*. Transgenic Res. 19(3): 473–487. Doi: 10.1007/s11248-009-9328-2.

Teulé, F., Y. G. Miao, B. H. Sohn, Y. S. Kim, J. J. Hull, M. J. Jr. Fraser, R. V. Lewis and D. L. Jarvis. 2012. Silkworms transformed with chimeric silkworm/spider silk genes spin composite silk fibers with improved mechanical properties. Proc. Natl. Acad. Sci. U S A. 109(3): 923–928. Doi: 10.1073/pnas.1109420109.

Tomita, M., H. Munetsuna, T. Sato, T. Adachi, R. Hino, M. Hayashi, K. Shimizu, N. Nakamura, Y. Tamura and K. Yoshizato. 2003. Transgenic silkworms produce recombinant human type III procollagen in cocoons. Nat. Biotechnol. 21(1): 52–56. Doi: 10.1038/nbt771.

Tomita, M., R. Hino, S. Ogawa, M. Iizuka, T. Adachi, K. Shimizu, H. Sotoshiro and K. Yoshizato. 2007. A germline transgenic silkworm that secretes recombinant proteins in the sericin layer of cocoon. Transgenic Res. 16(4): 449–465. Doi: 10.1007/s11248-007-9087-x.

Tomita, M. 2011. Transgenic silkworms that weave recombinant proteins into silk cocoons. Biotechnol. Lett. 33(4): 645–654. Doi: 10.1007/s10529-010-0498-z.

Tsubota, T. and H. Sezutsu. 2017. Genome editing of silkworms. Methods Mol. Biol. 1630: 205–218. Doi: 10.1007/978-1-4939-7128-2_17.

Tsubota, T., Y. Takasu, K. Uchino, I. Kobayashi and H. Sezutsu. 2017. TALEN-mediated genome editing of the ku80 gene in the silkworm *Bombyx mori*. J. Insect Biotechnol. Sericol. 86: 9–16. Doi: 10.11416/jibs.86.1_009.

Wang, Y., Z. Li, J. Xu, B. Zeng, L. Ling, L. You, Y. Chen, Y. Huang and A. Tan. 2013. The CRISPR/Cas system mediates efficient genome engineering in *Bombyx mori*. Cell Res. 23(12): 1414–1416. Doi: 10.1038/cr.2013.146.

CHAPTER 9

Glycosyltransferase Expression in Silkworm and its Applications in Glycobiology

Makoto Ogata,[1,*] *Taichi Usui*[2] and *Enoch Y. Park*[3]

1. Introduction

Carbohydrates have attracted significant attention in recent years as they play an important role in both intracellular and extracellular molecular recognition and intermolecular signal transduction (Varki 1993). Carbohydrates are also closely involved in the interactions between viruses of infectious diseases and host cells (Varki 1993). If carbohydrates involved in viral infections can be artificially synthesized, a biological recognition element material can be developed that will be able to quickly and specifically detect pathogenic viruses such as influenza viruses. However, artificial synthesis of carbohydrates is very difficult and costly because they have complex structures and are rich in diversity unlike chain polymers such as nucleic acids and proteins. The two main enzyme synthesis methods of such biologically active

[1] Department of Applied Chemistry and Biochemistry, National Institute of Technology, Fukushima College, 30 Nagao, Iwaki, Fukushima 970-8034, Japan.
[2] Integrated Bioscience Research Division, Graduate School of Science and Technology, Shizuoka University, 836 Ohya, Suruga-ku, Shizuoka 422-8529, Japan.
[3] Research Institute of Green science and Technology, Shizuoka University, 836 Ohya, Suruga-ku, Shizuoka 422-8529, Japan.
* Corresponding author: ogata@fukushima-nct.ac.jp

oligosaccharide chains are one that uses the transfer and condensation reaction of a glycoside hydrolase and the other that uses the synthesis reaction of a glycosyltransferase. The first method is a simple method, but the reaction specificity of enzymes and yield are low, therefore, its use is limited to the synthesis of saccharides with a degree of polymerization between 2 to 3. On the other hand, the second method is expected to be a versatile method of glycan synthesis with reaction specificity and high yield, given that the supply of donor substrate for the enzyme and the nucleic acid is sufficient in quantity. Recently, we have succeeded in easily preparing large quantities of glycans that have complex structures and are involved in viral infections by a using a chemo-enzymatic synthesis method which combines the commonly used organic chemistry method for glycan synthesis and an enzyme synthesis method that uses recombinant glycosyltransferases expressed in silkworms (Ogata et al. 2009c). The present paper reports the production of various recombinant glycosyltransferases using silkworms and the development of functional glycan material using these glycosyltransferases.

2. Glycosyltransferase Expression in Silkworm

Proteins derived from higher organisms often require post-translational modifications, by glycans for example, and proteins expressed in a bacterial system with incomplete post-translational modifications often have no biological activity. Since expression systems using insect cells and larva have a similar structure to that of animal cells, they are widely used for the expression of proteins derived from eukaryotic cells (Verma et al. 1998). We have successfully developed *Bombyx mori* nucleopolyhedrovirus (BmNPV) bacmid, which is a shuttle vector of *Escherichia coli* and silkworm, and have applied it to the production of recombinant proteins using silkworms (Motohashi et al. 2005). The advantage of this silkworm-BmNPV bacmid system is that it allows for gene transfer by directly inoculating bacmid DNA produced in *E. coli* into larvae of silkworm. This enables convenient and rapid production of recombinant proteins having physiological activity in large quantities.

We attempted to construct an enzyme library by expressing a synthesizing enzyme, referred to as glycosyltransferase, for a receptor molecule that is specifically recognized when pathogenic viruses adhere to or infect cells, using the silkworm-BmNPV bacmid expression system. As an example, we report here on the expression of α2,6-sialyltransferase (ST6GalI), a key enzyme, essential for the synthesis of receptor molecules recognized by human influenza virus. First, the transmembrane region was removed from the ST6GalI gene and amplified by polymerase chain reaction using rat liver cDNA as a template, following which, a fusion gene containing a silkworm-derived bombyxin secretory signal sequence and a purification tag sequence was constructed. A recombinant BmNPV bacmid was manufactured by inserting this fusion gene, which was then injected into the larvae of five-year-old silkworms. The larvae were reared for approximately one week under the temperature range of 25–27°C, and the hemolymph was sampled at different time points. The results of measurement of the activity in the haemolymph confirmed the desired glycosyltransferase activity in the haemolymph approximately 100 hours after the recombinant BmNPV bacmid was injected, as well as a maximum activity concentration of enzymes at around 160 hours (Fig. 1). Interestingly, differences in the enzyme expression level were observed depending on the type of tag sequence introduced into recombinant ST6GalI for purification: when His, Strep, and FLAG tags were used, the highest activity concentration (2.0 U/mL at 6.5 dpi) of enzymes in the haemolymph was obtained with the FLAG tag (Fig. 1). The silkworm haemolymph, containing recombinant ST6GalI, can be purified in 2 stages, ammonium sulfate fractionation and affinity column chromatography in order to obtain 2.2 mg of purified protein (enzyme activity recovery rate 64%) from 4.5 mL of the silkworm larval haemolymph (Ogata et al. 2009c). The results indicate that approximately 200 μg of purified recombinant protein can be expressed in each larva of silkworm. Recently, we have further succeeded in increasing the expression of recombinant ST6GalI

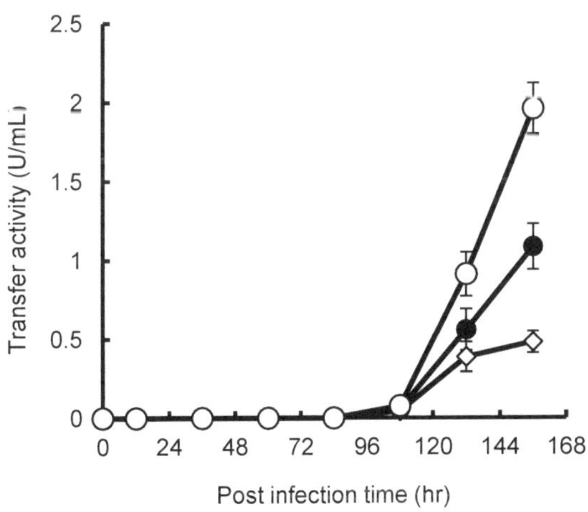

Fig. 1: Expression levels of rST6GalIs in silkworm larval hemolymph. Unfilled diamonds, His-ST6GalI; filled circles, Strep-ST6GalI; unfilled circles, FLAG-ST6GalI. Three type BmNPV bacmids were injected directly into 1st day of fifth-instar silkworm larvae. ST6GalI activities were determined by sialyltransferase activity assay described in Ogata et al. (2010b).

to approximately twice the level by modifying the promoter sequence of BmNPV bacmid (Kato et al. 2012). ST6GalI derived from mammals has been very expensive and difficult to obtain or prepare in large quantities. In this study, the BmNPV bacmid system utilizing silkworm larvae was demonstrated to be useful as an expression system of recombinant ST6GalI in the biologically active state.

The recombinant glycosyltransferase expression technology utilizing the BmNPV bacmid system can be applied not only to ST6GalI, but also to glycosyltransferases derived from eukaryotes. So far, the following four types of glycosyltransferases in addition to ST6GalI have been successfully expressed: β1,4-galactosyltransferase (β1,4-GalTI), β1,3-*N*-acetylglucosaminyltransferase (β1,3-GnTII), α1,3-fucosyltransferase (FUT6) and α2,3-sialyltransferase (ST3GalIII). This enables various physiologically active glycans to be prepared in large quantities.

3. Applications on Glycobiology

3.1 Synthesis of Artificial Sialoglycopolypeptide Utilizing Glycosyltransferases Expressed in Silkworms and Applications in the Detection of Influenza Virus

Interactions between carbohydrates and carbohydrate-binding proteins (lectin) play a key role in various biological processes (Varki 1993). This carbohydrate-lectin interaction often comes into play when viruses attach and infect host cells (Varki 1993). In the case of influenza virus, a lectin called hemagglutinin (HA), protruding as spikes on the surface of the virus, specifically recognizes glycans on the surface of host cells and binds to the cells, leading to infection (Suzuki et al. 1986). The structure of glycans involved in infections also varies depending on the host; for example, if we compare avian and human influenza viruses, the binding mode of sialic acid at the non-reducing end and even the internal glycan structure are considerably different (Ogata et al. 2009b).

We report here the synthesis of glycopolymers (glycopolypeptide) with the ability to bind to influenza virus, using glycosyltransferase expressed in silkworm larvae. Specifically, glycopolypeptides were synthesized by targeting the respective glycan structures recognized by the human influenza virus (Neu5Acα2,6(Galβ1,4GlcNAc)$_n$β1-R) (Fig. 2; Ogata et al. 2009b), avian influenza virus (Neu5Acα2,3Galβ1,4(6-sulfo)GlcNAcβ1-R) (Fig. 3; Ogata et al. 2014) and equine influenza virus (Neu5Gcα2,3Galβ1,4Glc/GlcNAcβ1-R) (Fig. 4; Ogata et al. 2017). As an example, a synthesis of glycopolypeptide containing the sialylated heptasaccharide, having the ability to bind to human influenza virus,

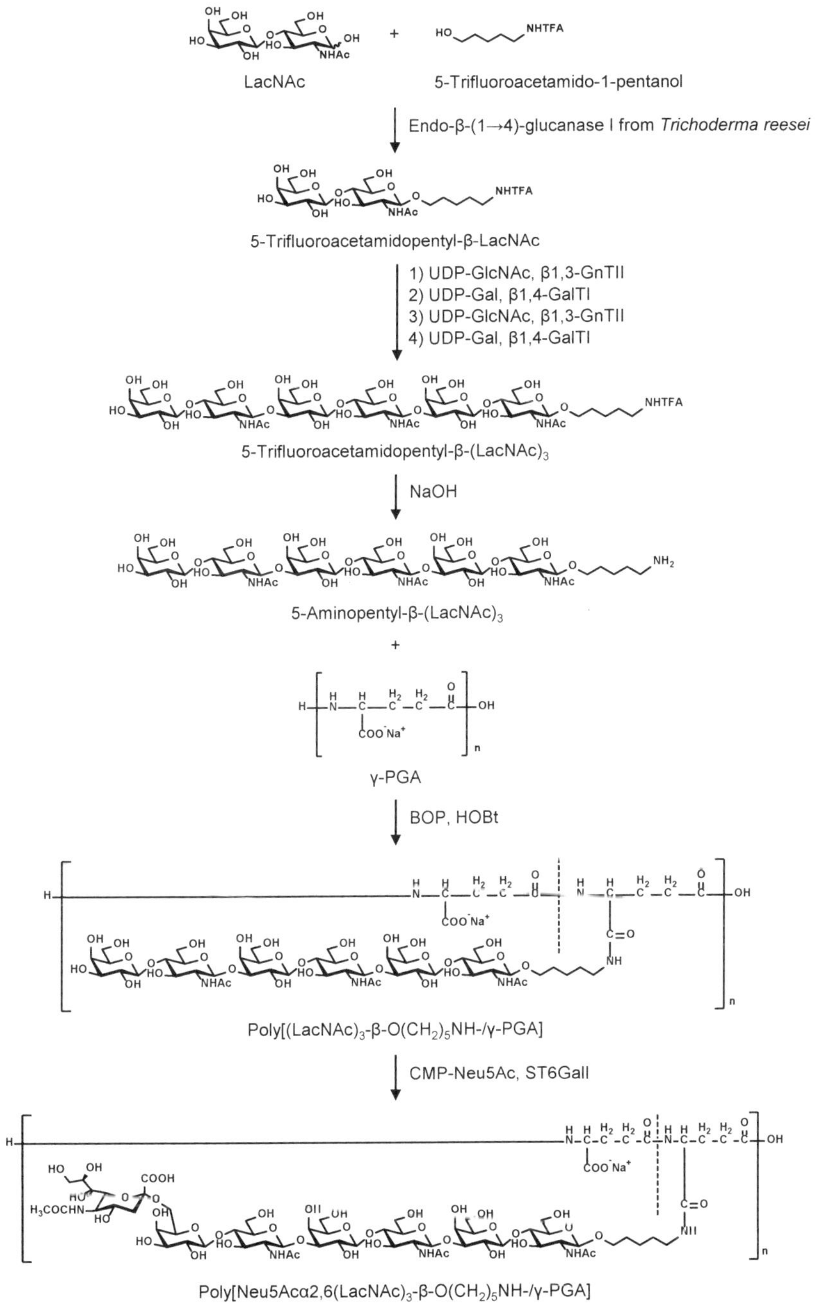

Fig. 2: Chemo-enzymatic synthesis of sialoglycopolypeptide for potentially blocking human influenza virus infection using glycosyltransferases expressed in BmNPV bacmid-injected silkworm larvae.

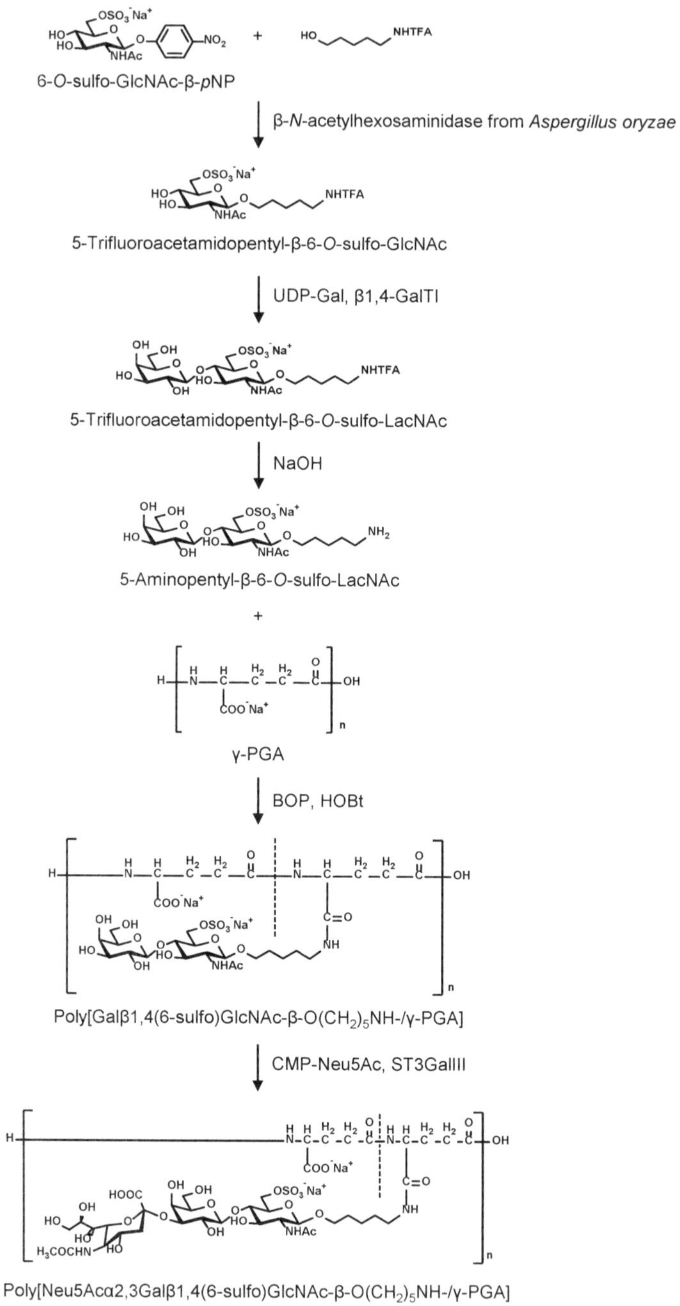

Fig. 3: Chemo-enzymatic synthesis of sulfated sialoglycopolypeptide as hemagglutination inhibitors against avian influenza virus using glycosyltransferases expressed in BmNPV bacmid-injected silkworm larvae.

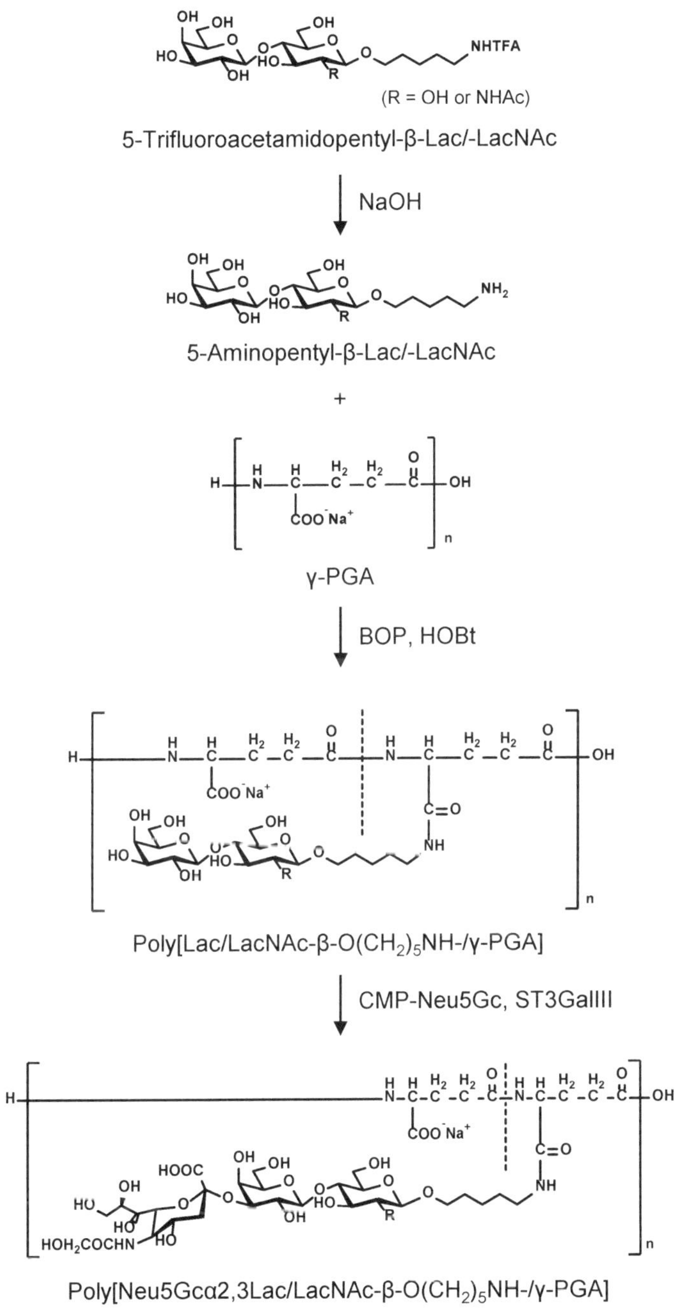

Fig. 4: Chemo-enzymatic synthesis of NeuGc-carrying sialoglycopolypeptide as effective inhibitors equine influenza virus hemagglutination using glycosyltransferases expressed in BmNPV bacmid-injected silkworm larvae.

is described below (Fig. 2). 5-Trifluoroacetamidopentyl-β-$(LacNAc)_3$ containing three repeated LacNAc units was synthesized using one-pot enzymatic synthesis without undergoing the protection-deprotection process, by repeated reaction of 5-trifluoroacetamidopentyl-β-LacNAc, an acceptor substrate, with two types of glycosyltransferases (β1,3-GnTII and β1,4-GalTI) and their corresponding donor substrates (UDP-GlcNAc and UDP-Gal). An asialoglycopolypeptide was further synthesized by multivalent introduction of this glycoside into γ-polyglutamic acid (γ-PGA) produced by *Bacillus subtilis* var. natto. Additionally, we succeeded in obtaining sialoglycopolypeptide containing Neu5Acα2,6$(LacNAc)_3$ conveniently and at high yields by using this asialoglycopolypeptide as the acceptor, CMP-β-Neu5Ac as the donor and ST6GalI as the enzyme source.

We have experimentally demonstrated that this sialoglycopolypeptide can become a powerful infection inhibitor against a target influenza virus (Ogata et al. 2009a, 2009b). For example, in a virus infection inhibition test (focus-forming assay) using Madine-Darby canine kidney (MDCK) cells, glycopolypeptides containing the sialylated trisaccharide (Neu5Acα2,3LacNAc) indicated a 50% infection inhibitory concentration of 10 nmol/L against avian influenza virus [A/Duck/HongKong/313/4/78 (H5N3)]. In addition, glycopolypeptides containing the sialylated heptasaccharide [Neu5Acα2,6$(LacNAc)_3$] showed an extremely strong inhibitory activity of 60 fmol/L against human influenza virus [A/WSN/33 (H1N1)]. These sialoglycopolypeptides show equivalent infection inhibitory effects against other target influenza viruses. They are powerful and structure selective infection inhibitors that are unmatched in the world. The infection inhibitory effect of glycopolypeptides containing the sialylated heptasaccharide has also been demonstrated in *in vivo* tests using mice, in which their protective effect against fatal human influenza virus [A/WSN/33 (H1N1)] infection has been confirmed (Fig. 5; Ogata et al. 2009b). The strong binding affinity and binding specificity of sialoglycopolymer materials to viruses can be utilized and applied to virus detection, in addition to infection inhibitors.

These glycopolypeptides are composed of hydrophilic glycan moiety and hydrophobic γ-PGA-derived main chain moiety, giving them the characteristics of an amphiphilic molecule. This allows for these glycopolypeptides to be easily fixed to a hydrophobic substrate, such as a polystyrene plate. In addition, the glycopolymers can be fixed to a universal plate by irradiating ultraviolet rays for approximately 1 minute after adding the glycocluster aqueous solution. The glycan fixed plates are then blocked with bovine serum albumin, and the specimen is added and washed, following which, they are treated with the primary

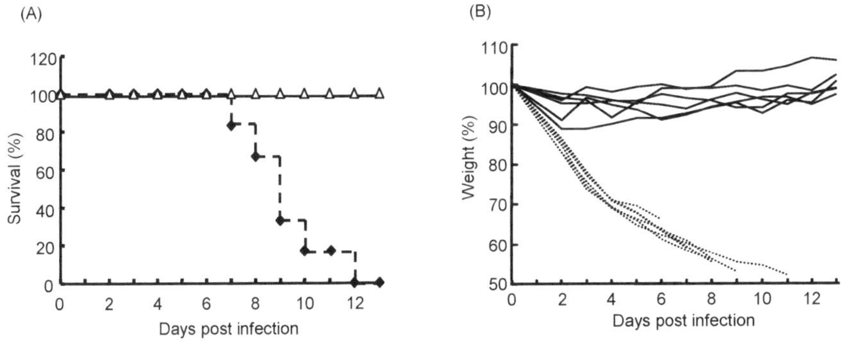

Fig. 5: Effect of glycopolypeptides influenza virus [A/WSN/33 (H1N1)] infection in mice was performed as described in Ogata et al. (2009b). Graph (A) indicates Kaplan-Meier survival curves (*P < 0.01, Log-rank test). Mice were intranasally inoculated with viral suspension and asialo- (filled diamonds) or the sialylated heptasaccharide (unfilled triangles) glycopolypeptides at a dose of 3 nmol/kg. Graph (B) shows body weight dynamics of mice. Asialo- (dotted line) or the sialylated heptasaccharide (solid line) glycopolypeptides.

and secondary antibodies and the absorbance is measured after adding the colorimetric substrate solution, thereby enabling the development of virus diagnostic technology based on the principle of enzyme-linked immunosorbent assay (ELISA) method (Ogata et al. 2007). This virus evaluation system has been used by many researchers as a simple and highly sensitive method for the detection of influenza virus and the analysis of carbohydrate recognition specificity (Dong et al. 2013, Hidari et al. 2008). Recently, this ELISA method was combined with the real-time PCR method in order to establish a method of directly determining the glycan recognition specificity of a virus within 4 hours, using a clinical sample taken from birds (Takahashi et al. 2013). In recent years, the highly pathogenic H5N1 avian influenza virus has become prevalent in livestock, especially in Asia; consequently, infections in humans are increasing year after year. There is a concern that recurrence of such infections will lead to mutations in the sugar-binding specificity of viruses and new viruses may emerge that can cause a pandemic. The present virus identification method can also be used as an analytical technique for easy monitoring of mutations in influenza viruses that require close monitoring. The method of synthesizing glycans with glycosyltransferase is not limited to glycans involved in viral infections but can also be used in the development of various glycan materials. Specifically, we have previously reported the use of glycosyltransferases expressed in silkworm as a powerful tool in the synthesis of enzyme activity measurement substrates (Fig. 6; Ogata et al. 2010b) and in the synthesis of glycoclusters as high-affinity cross-linker against lectin (Fig. 7; Ogata et al. 2012).

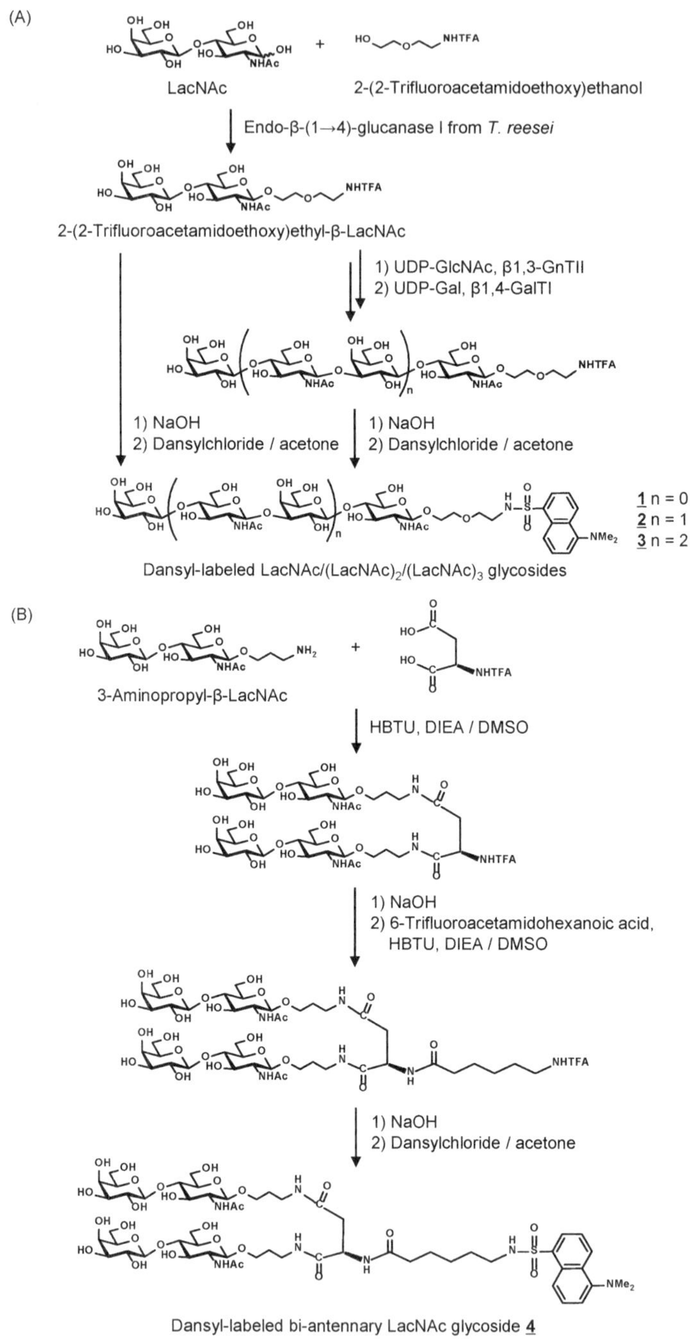

Fig. 6: Chemo-enzymatic synthesis of fluorescent labeled glycoside as acceptor substrate of sialyltransferases using recombinant glycosyltransferase expressed in silkworm larvae.

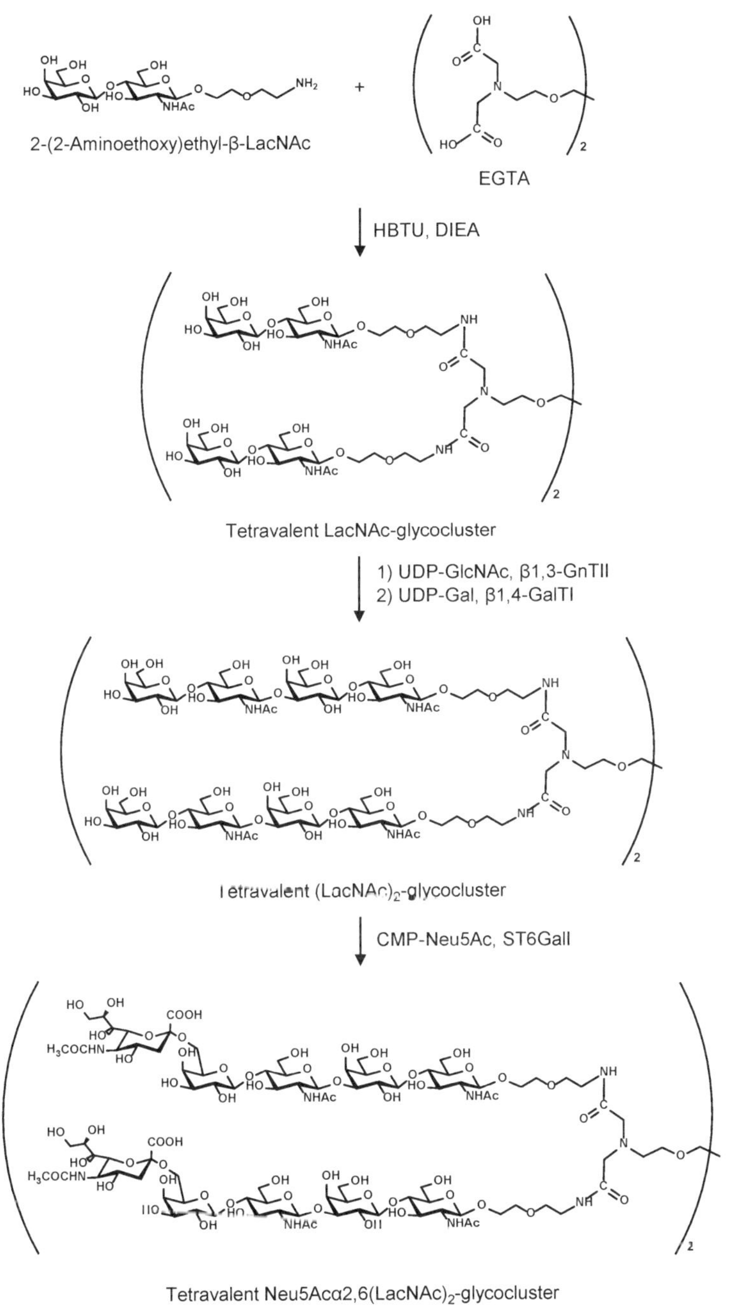

Fig. 7: Chemo-enzymatic synthesis of tetravalent sialo-glycocluster as high-affinity cross-linker against *Sambucus sieboldiana* agglutinin using recombinant glycosyltransferase expressed in silkworm larvae.

3.2 Synthesis of Fluorescent Labeled Glycosides as Acceptor Substrates for Sialyltransferases Utilizing Glycosyltransferases Expressed in Silkworms

Generally, the sialyltransferase assay using fluorescent labeled small galactoside as an acceptor substrate with CMP-β-Neu5Ac as a donor has been widely used to estimate the enzyme activity. However, the fluorescent substrate sometimes shows poor substrate recognition for the target enzyme and is troubled with low solubility in a buffer solution (> 10 mM). We designed and developed a chemo-enzymatic method for the synthesis of a series of dansyl-labeled glycosides carrying LacNAc repeats, which are useful as acceptor substrates for α2,6- and α2,3-sialyltransferase assays. The fluorescent labeled glycoside consists of three parts: glycan, in glycon site and spaced-linker and fluorescent group in aglycon site. Figure 6A represents the pathway from LacNAc via three steps: (1) enzymatic glycosylation of LacNAc to alkanol spacer; (2) enzymatic sugar elongation by consecutive additions of β1,4 linked Gal and β1,3 linked GlcNAc to the resulting LacNAc glycosides; (3) detrifluoroacylation followed by introduction of dansyl group into aglycon moiety. This led to the synthesis of dansyl-labeled di-, tetra- and hexa-saccharide glycosides (**1**, **2**, and **3**) carrying single, tandem, and triplet LacNAc repeats. In the first step, enzymatic *O*-glycosylation was carried out by a cellulase-mediated condensation reaction between LacNAc and alkanol (Ogata et al. 2010a). This has been shown to be a useful method for obtaining hydrophilic spacer-*O*-linked LacNAc glycoside. As an alternative design, a dansyl-labeled bi-antennary LacNAc glycoside (**4**) was synthesized as a mimetic molecule of *N*-linked glycan, such as asialo-α_1-acid glycoprotein. Figure 6B represents the pathway from 3-aminopropyl-β-LacNAc via three steps: (1) chemical coupling with dicarbonic acid; (2) spacer elongation of the resulting bi-antennary glycoside; (3) detrifluoroacylation, followed by the introduction of dansyl group into aglycon moiety.

The kinetic parameters for the transfer reaction of synthesized dansyl-labeled glycoside acceptors to a CMP-β-Neu5Ac donor mediated by α2,6- and α2,3-sialyltransferase were measured using the HPLC-fluorescence method. For α2,6-sialyltransferase with dansyl-labeled glycosides carrying LacNAc repeats, the K_m value of **1** (K_m = 0.61 mM) was 2.3- and 2.6-fold higher than that of **2** and **3**, respectively. Moreover, the V_{max}/K_m value of **1** was 5- and 7-fold higher than that of **2** and **3**, respectively, and decreased in a glycan length-dependent manner **1** > **2** > **3**. Interestingly, the enzyme bound preferentially to short glycan with single LacNAc. This was also the case for α2,3-sialyltransferase. The present dansyl-labeled glycosides **1**~**3** show high solubility (> 100 mM) in a buffer solution. In addition, bi-antennary LacNAc glycoside **4** indicated the best K_m values (Ogata et al.

2010b) of 0.24 mM for α2,6-sialyltransferase, but the V_{max}/K_m value was 81% of compound **1**. From a practical viewpoint, the new assay, developed using compound **1**, is more reliable and displayed greater sensitivity in comparison to conventional substrates.

The present synthetic acceptor substrates for sialyltransferase are expected to be useful as analytical tools for sialyltransferase expression and elucidation of physiological functions.

3.3 Design and Synthesis of Tetravalent Sialoglycoclusters as High-affinity Cross-Linker against Sambucus sieboldiana Agglutinin Utilizing Glycosyltransferases Expressed in Silkworms

In the field of glycobiology, despite the very low one-to-one binding affinity between glycans and proteins, the "glycocluster effect" (in which the binding affinity dramatically increases due to the formation of an aggregate structure at the cell surface layer) is well-known (Lee et al. 1983). This concept is an important design guideline in the development of glycan and antiviral drugs. The authors have previously established a method for synthesizing divalent ligands with high affinity for *Triticum vulgaris* agglutinin (WGA) and the ability to form cross-linked complexes (Misawa et al. 2008). The ability of glycans to cluster even at low valency was successfully increased by arranging glyco-ligands appropriately. The authors designed a tetravalent glycan ligand, in which the binding affinity between glycans and target proteins was dramatically enhanced. Specifically, we focused on the chelating principle, by which a tetravalent carboxy group ligand of a common metal chelating agent {ethylene glycol bis(β-aminoethylether)-*N,N,N′,N′*-tetraacetate (EGTA)} captures metal ions like crab pincers, and synthesized a novel tetravalent glycocluster by introducing a sialoglyco glycoside {Neu5Acα2,6(Galβ1,4GlcNAc)$_2$β O(CH$_2$)$_2$O(CH$_2$)$_2$NH-} that was synthesized by using various glycosyltransferase to the tetravalent carboxy group (Fig. 7; Ogata et al. 2012). Binding characterization analysis of this tetravalent glycocluster with *Sambucus sieboldiana* agglutinin (SSA) that indicates sugar recognition specificity similar to human influenza virus was evaluated with isothermal titration calorimetry (ITC). The tetravalent glycocluster indicated a very strong binding dissociation constant (K_d = 34 nM) with a value 28 times stronger than that of monovalent Neu5Acα2,6Galβ1,4GlcNAc (Fig. 8). Interestingly, the number of SSA binding to one molecule of tetravalent glycocluster varied depending on the length of glycans introduced. Specifically, when a trisaccharide structure (Neu5Acα2,6Galβ1,4GlcNAcβ-) with short glycan length was introduced, the binding ratio of tetravalent glycocluster to SSA was 1 to 2, whereas the binding ratio was 1 to 3 when a long pentasaccharide structure {Neu5Acα2,6(Galβ1,4GlcNAc)$_2$β-}

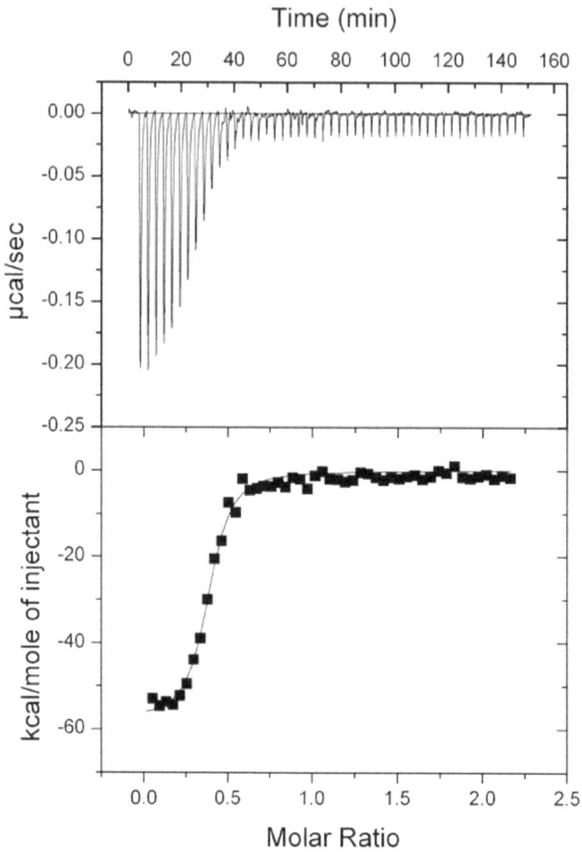

Fig. 8: ITC data for interaction of tetravalent Neu5Acα2,6(LacNAc)$_2$-glycocluster and SSA. (Top) Raw titration curve. (Bottom) Integrated titration curve. The solid line is the best-fit, which used a single-site model, $K_d = 34 \pm 5$ nM, $n = 0.38 \pm 0.01$.

was introduced. When morphological changes accompanying the cross-linking of tetravalent glycocluster were evaluated by quantitative precipitation assay and dynamic light scattering method, the formation of stoichiometric cross-linked complexes was observed in the concentration range supporting the ITC results. The particle diameter of these large cross-linked complexes formed between the tetravalent glycocluster and SSA was 600 to 800 nm (Fig. 9). Tetravalent glycoclusters were shown to not only increase the binding affinity due to the glycocluster effect but also to form a three-dimensional precipitate of cross-linked complexes with the target protein, where the length of glycans plays a significant role in the formation mode of cross-linked complexes (Fig. 10; Ogata et al. 2012). Application of the formation principle of cross-linked complexes by such

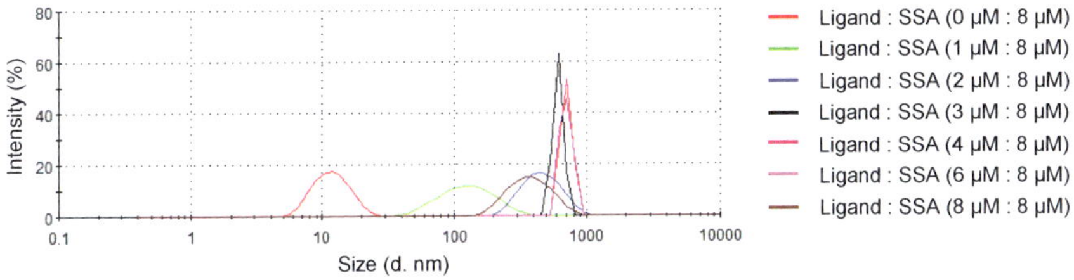

Fig. 9: DLS measurement for the complexes formed by tetravalent Neu5Acα2,6(LacNAc)$_2$-glycocluster and SSA. SSA solutions (16 μM, 50 μL) were mixed with sequential additions of 50 μL (0 ~ 16 μM) of tetravalent Neu5Acα2,6(LacNAc)$_2$-glycocluster in 10 mM PBS (pH 7.4). After an addition of ligand, the mixtures were each incubated for 1 hr at 4°C. DLS intensities were measured at 25°C using a Zetasizer Nano ZS.

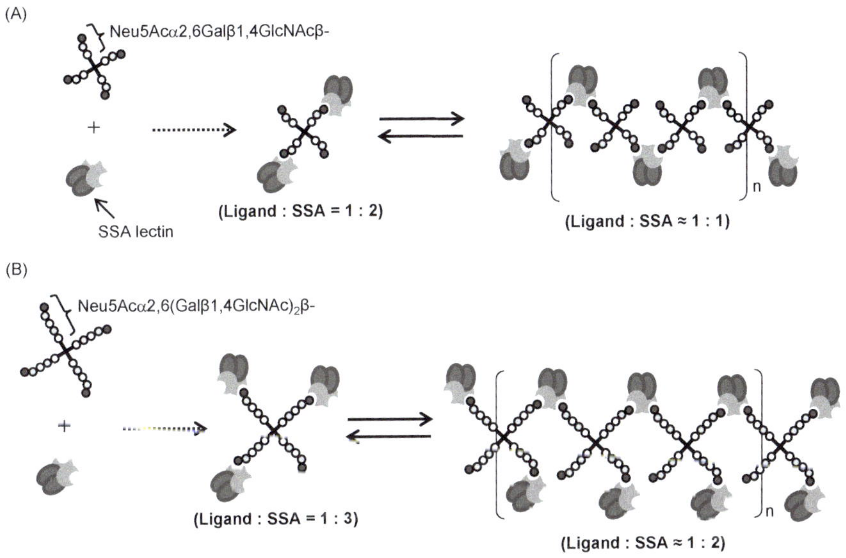

Fig. 10: Schematics of the tetravalent sialoglycoclusters/SSA lectin complexes formed by (A) tetravalent Neu5Acα2,6LacNAc-glycocluster and (B) tetravalent Neu5Acα2,6(LacNAc)$_2$-glycocluster.

glycan ligands to viral particles such as influenza virus instead of lectin can be expected to enable expansion of glycotechnology to a system that can be used to capture or detect viruses.

4. Concluding Remarks

We have demonstrated in our recent studies that the glycosyltransferase expression technology using silkworms can be a powerful tool in the synthesis of biologically active glycans having complex structures. In the future, we expect that these mass-produced carbohydrate molecules can be used to develop glycan materials with various functionalities.

Acknowledgements

We thank Prof. Takashi Suzuki (University of Shizuoka, Japan), associate Prof. Tadanobu Takahashi (University of Shizuoka, Japan), Prof. Kazuya I.P.J. Hidari (University of Aizu, Japan) and Dr. Takashi Yamanaka (Japan Racing Association, Japan) for influenza virus assays. These works were supported by a Grant-in-Aid for Young Scientists (B) (Nos. 22780101 and 24780111) and Scientific Research (C) (No. 26450147) from the Ministry of Education, Culture, Sports, Science, and Technology of Japan.

References

Dong, J., M. Harada, S. Yoshida, Y. Kato, A. Murakawa, M. Ogata, T. Kato, T. Usui and E. Y. Park. 2013. Expression and purification of bioactive hemagglutinin protein of highly pathogenic avian influenza A (H5N1) in silkworm larvae. J. Virol. Methods. 194: 271–276.

Hidari, K. I. P. J., T. Murata, K. Yoshida, Y. Takahashi, Y. H. Minamijima, Y. Miwa, S. Adachi, M. Ogata, T. Usui, Y. Suzuki and T. Suzuki. 2008. Chemo-enzymatic synthesis, characterization, and application of glycopolymers carrying lactosamine repeats as entry inhibitors against influenza virus infection. Glycobiology. 18: 779–788.

Kato, T., S. L. Manohar, S. Kanamasa, M. Ogata and E. Y. Park. 2012. Improvement of the transcriptional strength of baculovirus very late polyhedrin promoter by repeating its untranslated leader sequences and coexpression with the primary transactivator. J. Biosci. Bioeng. 113: 694–696.

Lee, Y. C., R. R. Townsend, M. R. Hardy, J. Lönngren, J. Arnarp, M. Haraldsson and H. Lönn. 1983. Binding of synthetic oligosaccharides to the hepatic Gal/GalNAc lectin. Dependence on fine structural features. J. Biol. Chem. 258: 199–202.

Misawa, Y., T. Akimoto, S. Amarume, T. Murata and T. Usui. 2008. Enzymatic synthesis of spacer-linked divalent glycosides carrying *N*-acetylglucosamine and *N*-acetyllactosamine: analysis of cross-linking activities with WGA. J. Biochem. 143: 21–30.

Motohashi, T., T. Shimojima, T. Fukagawa, K. Maenaka and E. Y. Park. 2005. Efficient large-scale protein production of larvae and pupae of silkworm by *Bombyx mori* nuclear polyhedrosis virus bacmid system. Biochem. Biophys. Res. Commun. 326: 564–569.

Ogata, M., T. Murata, K. Murakami, T. Suzuki, K. I. P. J. Hidari, Y. Suzuki and T. Usui. 2007. Chemo-enzymatic synthesis of artificial glycopolypeptides containing multivalent sialyloligosaccharides with a γ-polyglutamic acid backbone and their effect on inhibition of infection by influenza viruses. Bioorg. Med. Chem. 15: 1383–1393.

Ogata, M., K. I. P. J. Hidari, W. Kozaki, T. Murata, J. Hiratake, E. Y. Park, T. Suzuki and T. Usui. 2009a. Molecular design of spacer-N-linked sialoglycopolypeptide as polymeric inhibitors against influenza virus infection. Biomacromolecules. 10: 1894–1903.

Ogata, M., K. I. P. J. Hidari, T. Murata, S. Shimada, W. Kozaki, E. Y. Park, T. Suzuki and T. Usui. 2009b. Chemo-enzymatic synthesis of sialoglycopolypeptides as glycomimetics to block infection by avian and human influenza viruses. Bioconjugate Chem. 20: 538–549.

Ogata, M., M. Nakajima, T. Kato, T. Obara, H. Yagi, K. Kato, T. Usui and E. Y. Park. 2009c. Synthesis of sialoglycopolypeptide for potentially blocking influenza virus infection using a rat α2,6-sialyltransferase expressed in BmNPV bacmid-injected silkworm larvae. BMC Biotechnol. 9: 1–13.

Ogata, M., Y. Kameshima, T. Hattori, K. Michishita, T. Suzuki, H. Kawagishi, K. Totani, J. Hiratake and T. Usui. 2010a. Lactosylamidine-based affinity purification for cellulolytic enzymes EG I and CBH I from *Hypocrea jecorina* and their properties. Carbohydr. Res. 345: 2623–2629.

Ogata, M., T. Obara, Y. Chuma, T. Murata, E. Y. Park and T. Usui. 2010b. Molecular design of fluorescent labeled glycosides as acceptor substrates for sialyltransferases. Biosci. Biotechnol. Biochem. 74: 2287–2292.

Ogata, M., M. Yano, S. Umemura, T. Murata, E. Y. Park, Y. Kobayashi, T. Asai, N. Oku, N. Nakamura, I. Matsuo and T. Usui. 2012. Design and synthesis of high-avidity tetravalent glycoclusters as probes for *Sambucus sieboldiana* agglutinin and characterization of their binding properties. Bioconjugate Chem. 23: 97–105.

Ogata, M., H. Uzawa, K. I. P. J. Hidari, T. Suzuki, E. Y. Park and T. Usui. 2014. Facile synthesis of sulfated sialoglycopolypeptides with a γ-polyglutamic acid backbone as hemagglutinin inhibitors against influenza virus. J. Appl. Glycosci. 61: 1–7.

Ogata, M., A. Koizumi, T. Otsubo, K. Ikeda, M. Sakamoto, R. Aita, T. Kato, E. Y. Park, T. Yamanaka and K. I. P. J. Hidari. 2017. Chemo-enzymatic synthesis and characterization of *N*-glycolylneuraminic acid-carrying sialoglycopolypeptides as effective inhibitors against equine influenza virus hemagglutination. Biosci. Biotechnol. Biochem. 81: 1520–1528.

Suzuki, Y., Y. Nagao, H. Kato, M. Matsumoto, K. Nerome, K. Nakajima and E. Nobusawa. 1986. Human influenza A virus hemagglutinin distinguishes sialyloligosaccharides in membrane-associated gangliosides as its receptor which mediates the adsorption and fusion processes of virus infection. J. Biol. Chem. 261: 17057–17061.

Takahashi, T., T. Kawakami, T. Mizuno, A. Minami, Y. Uchida, T. Saito, S. Matsui, M. Ogata, T. Usui, N. Sriwilaijaroen, H. Hiramatsu, Y. Suzuki and T. Suzuki. 2013. Sensitive and direct detection of receptor binding specificity of highly pathogenic avian influenza A virus in clinical samples. PLoS ONE. 8: e78125.

Varki, A. 1993. Biological roles of oligosaccharides: All of the theories are correct. Glycobiology. 3: 97–130.

Verma, R., E. Boleti and A. J. George. 1998. Antibody engineering comparison of bacterial, yeast, insect and mammalian expression system. J. Immunol. Methods. 216: 165–181.

CHAPTER 10

Silkworm Expression of Cell Surface Molecules

Shunsuke Kita

1. Introduction

Baculovirus—insect cell expression vector system (BEVS) is one of the useful recombinant protein expression systems that uses baculovirus, which has a strong polyhedrin promoter and is, therefore, suitable for the production of a large amount of recombinant proteins. There are two types of baculovirus that are generally utilized in BEVS. The first, *Autographa californica* nuclear polyhedrosis virus (AcMNPV), is widely used in this system as a commercially available baculovirus and is famous for its application in recombinant expression systems, in combination with insect cell lines such as sf9, sf21 and High5 cells. The other is *Bombyx mori* nuclear polyhedrosis virus (BmNPV), which infects silkworm and its cell lines. The silkworm is a very attractive host for protein production, for the following reasons: (1) Simple feeding and low cost: Large-scale and time-consuming cell cultivation is not necessary, (2) Only a few pieces of indispensable equipment are needed: Essentially one incubator for breeding silkworms, (3) High-level protein expression: The very strong polyhedrin promoter and the high expression state are ready for producing silk proteins (fibroins) (4) the intramolecular disulfide bond formation and the post-translational modifications are similar to those

Center for Research and Education on Drug Discovery and Laboratory of Biomolecular Science, Faculty of Pharmaceutical Sciences, Hokkaido University, Sapporo 060-0812, Japan.
E-mail: kita@pharm.hokudai.ac.jp

in mammals. The AcMNPV BEVS employs a very useful bacmid DNA, comprising of the baculoviral DNA genome with the *Escherichia coli* origin and transposition sequences (Luckow et al. 1993). Since the replication of bacmid DNA and the transposition of the desired genes of the transfer vector onto the bacmid DNA can occur in *E. coli*, the time-consuming preparation of recombinant viruses is not required. The BmNPV bacmid DNA can be mixed with a lipid reagent and directly injected into the silkworm for the infection (Motohashi et al. 2005). Therefore, this bacmid system for silkworms is quite useful. Using the BmNPV bacmid-silkworm expression system, a human G-protein coupled receptor (GPCR), the nociception receptor, was expressed in the membranes of fat bodies and on the viral surface (Kajikawa et al. 2009).

In this chapter, a preparation of human killer cell immunoglobulin-like receptor 2DL1 (KIR2DL1) in silkworm is described in detail. It is generally difficult to obtain a sufficient amount of immune cell surface receptors, since these receptors are highly glycosylated and are necessary for proper folding and stability in many cases; therefore, an expression system that is easy-to-use and produces a high expression level of posttranslationally-modified receptors has been eagerly awaited. Using the baculovirus—silkworm expression system, 0.2 mg of KIR2DL1 proteins was easily obtained from 1 silkworm larva and the purified proteins showed uniform *N*-linked glycosylation (Sasaki et al. 2009). Through the preparation and functional analysis of KIR2DL1 proteins, the advantages of the silkworm expression system for the preparation of human proteins are shown.

2. Silkworm Expression of KIR2DL1

Killer cell Immunoglobulin-like receptors (KIRs) are immune cell surface receptors expressed by natural killer (NK) cells and certain populations of T cells. They consist of 15 functional inhibitory (KIR2DL and KIR3DL) and activating (KIR2DS and KIR3DS) receptors (Colonna 1996, Long and Rajagopalan 2000). KIRs are type I transmembrane glycoproteins, with two or three extracellular Ig-like domains in the extracellular region and a cytoplasmic signaling domain. The cytoplasmic domains of KIRs either have an immunoreceptor tyrosine-based inhibitory motif (ITIM) or a short tail with a charged lysine in their transmembrane domain, which couples to the immunoreceptor tyrosine-based activation motif (ITAM) containing adapter protein DAP12. The KIRs basically recognize the classical major histocompatibility class I (MHC-I) molecules using their extracellular regions. KIR2DL1 specifically binds to HLA-Cw14 (and its relatives) on target cells and suppresses the immune responses of NK cells and T cells upon binding to HLA-Cw4 (Maenaka et al. 1999, Wagtmann et al. 1995). MHC-I expression on tumor cells or virus-infected cells is lower than that of uninfected normal cells, although the presentation of the unusual peptide MHC-I activates T cells and results in the elimination of virus-infected cells and tumor cells. These cells, exhibiting reduced or no MHC-I expression, can potentially become susceptible to KIR-expressing immune cells because they lack KIR-mediated inhibitory signals. Genetic analyses clearly demonstrated that the KIR genes are associated with infectious diseases, autoimmune diseases and cancer (Carrington and Martin 2006, Tsuchiya et al. 2007). The accumulated evidence has revealed the physiological significance of KIR2DL1 in immune regulation.

3. Material and Methods

3.1 Construction of the BmNPV Bacmid Expressing KIR2DL1

The gene encoding the extracellular region of KIR2DL1 was obtained by PCR, using the following primers (forward, 5′-TTAGGATCCATGTCGCTCTTGGTCGTCAGC-3′; reverse, 5-TTAGA ATTCTTAATGATGATGATGATGATGGTGCAGGTGTCGGGGGTTA CC-3′) from pKMATHNK1 (Maenaka et al. 1999). The sequence encoding the hexahistidine tag was introduced at the 5′ end of the reverse primer for further purification. The KIR2DL1 gene was inserted into the donor plasmid pFastBac1 vector. The resultant plasmid was designated as pFastBac1/KIR2DL1. Transposition was carried out by transforming the plasmid pFastBac1/KIR2DL1 into BmDH10Bac cells (Motohashi et al. 2005). White kanamycin- and gentamycin-resistant colonies were selected, and then the BmNPV bacmid, designated as BmNPV bacmid/KIR2DL1, was isolated.

3.2 Expression of the KIR2DL1 Gene in Silkworm Larvae and Protein Purification

First day 5th instar larvae were used for protein production. Silkworm larvae were directly injected with the BmNPV bacmid/KIR2DL1. The BmNPV bacmid/KIR2DL1 (1.0 μg) was suspended in 3 μl of DMRIE-C reagent and the resultant mixture was injected into the dorsal side of the larvae. After the larvae were cultured for 144 h, the haemolymph was collected and immediately combined with sodium thiosulfate. The collected haemolymph was diluted in buffer A (20 mM Tris–HCl (pH 8.0), 100 mM NaCl). The recombinant proteins were purified by Ni Sepharose 6 Fast Flow resin (GE Healthcare).

3.3 SDS–PAGE and Western Blotting Analysis

In order to perform sodium dodecyl sulfate–polyacrylamide gel electrophoresis (SDS–PAGE), the eluted fractions were loaded on a 15–

25% slab gel, and then stained with Coomassie brilliant blue. For Western blotting, the proteins, separated by SDS–PAGE as described above, were transferred onto a polyvinylidene difluoride (PVDF) membrane. The membrane was incubated in PBS-T containing a 1/1000 dilution of the penta-His antibody (Qiagen) and then incubated with a horseradish peroxidase labeled mouse IgG antibody (Amersham Biosciences, Piscataway, NJ). After washing with PBS-T, the detection was carried out using an ECL Plus Western Blotting Detection System (GE Healthcare).

3.4 Surface Plasmon Resonance (SPR) and Circular Dichroism Spectroscopy (CD)

SPR measurements were performed using a BIAcore 2000 instrument (GE Healthcare). The HLA-Cw4 was immobilized and the binding assay was performed using serial dilutions of the KIR2DL1, ranging from 0.06 to 30 μM. The CD spectra of a 7.2 μM solution of KIR2DL1 were measured with a Jasco-J 720 spectropolarimeter.

3.5 Sugar Digestion and Profiling Analysis

The KIR2DL1 protein was digested with either Endo H (New England Biolabs, MA, USA) or PNGaseF (New England Biolabs) and then subjected to SDS–PAGE in order to determine the existence of *N*-glycan and its Endo H sensitive sugars. A detailed sugar composition analysis was performed by matrix-assisted laser desorption/ionization-time of flight mass spectrometry (MALDI-TOF-MS) after protease digestion. The KIR2DL1 protein (0.3 mg) was proteolyzed with a mixture of chymotrypsin and trypsin and was further digested with glycoamidase A, releasing the *N*-glycans. After the removal of the peptide materials, the reducing ends of the resultant *N*-glycans (PA-oligosaccharides) were derivatized with 2-aminopyridine (Wako, Osaka, Japan). The PA-glycan mixture was separated by an octadecyl silica (ODS) column (Shimadzu, Kyoto) and the individual fractions were subjected to MALDI-TOF-MS. The identification of *N*-glycan structures was based on GU and mass values in comparison to PA-glycans in the GALAXY database (http://www.glycoanalysis.info/galaxy2/ENG/systemin1.jsp). The *N*-glycans were confirmed by co-chromatography on the ODS column.

4. Results

4.1 Expression of KIR2DL1 in Silkworms by Direct Injection of its Bacmid DNA

The BmNPV bacmid/KIR2DL1, bearing the extracellular region (residues 1–224) of KIR2DL1 with a hexahistidine tag at the C-terminus, was successfully prepared (Fig. 1A). The obtained bacmid DNA (0.3 μg) was directly injected into the 5th instar silkworm body. Six days after the injection, the body haemolymph was collected and centrifuged at 10,000 *g* for 10 min, then 0.5 μl of the haemolymph was used for Western blotting. The Western blotting analysis using the anti-pentahistidine antibody clearly showed that the recombinant KIR2DL1 protein was expressed in the haemolymph of silkworm by the direct bacmid DNA injection method (Fig. 1B). To optimize the expression conditions of this system, different amounts of injected bacmid DNA were examined. The largest

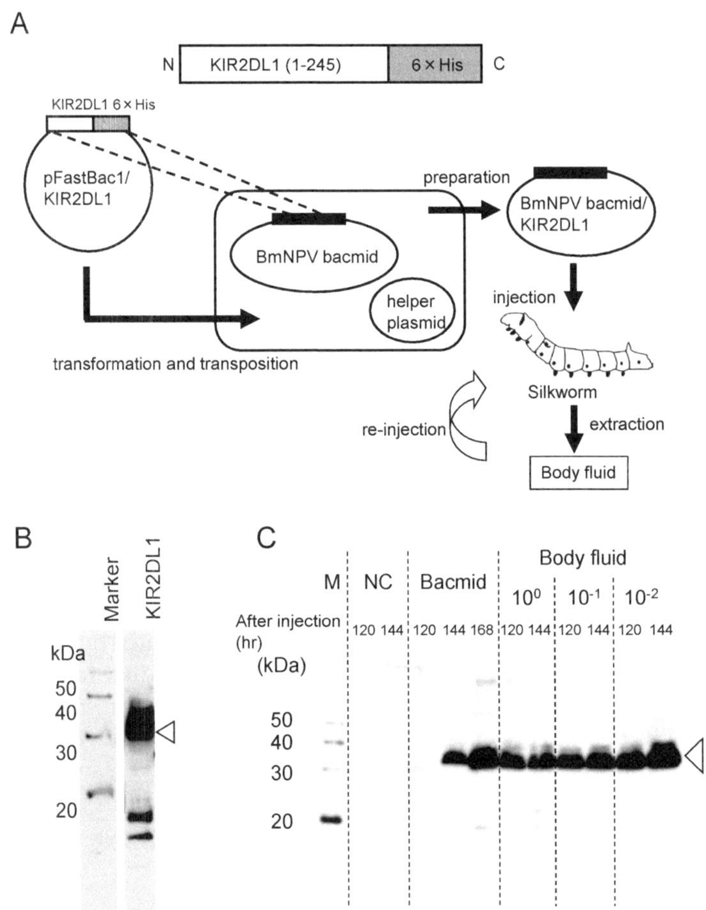

Fig. 1: (A) Construction of the recombinant BmNPV bacmid DNA and expression strategy using silkworm larvae. (B) Western blot analysis for detection of the expressed KIR2DL1 protein in the haemolymph of silkworms, using an anti-pentahistidine antibody. (C) Western blot analysis for the detection of the recombinant KIR2DL1 protein (open arrowhead) expressed by the injection of BmNPV bacmid DNA or body fluid solution. NC, negative control.

expression of the recombinant protein was obtained by injecting 1.0 µg of DNA. Furthermore, the body fluid derived from the silkworm larvae expressing the recombinant KIR2DL1 protein included a large amount of recombinant baculovirus. Thus, in terms of experimental cost and time to produce the protein, it is worth examining the effectiveness of directly injecting the body fluid into other larvae. The needles, which were briefly dipped into the body fluid solutions at several diluted concentrations, were used to penetrate the silkworm larvae, and the expression levels were evaluated by Western blotting using an anti-pentahistidine antibody. As clearly shown in Fig. 1C, even 100-times-diluted body fluid still exhibited sufficient infectivity (also confirmed by Western blotting using an anti-gp64 (BmNPV envelope protein) antibody) and conferred the stable expression of a large amount of the KIR2DL1 protein. Therefore, the injection of the body fluid including the recombinant viruses remarkably increases both cost-effectiveness and time-efficiency, when compared with that of bacmid DNA, which requires DNA preparation and lipid reagent for successful infection.

4.2 *Purification of KIR2DL1 Proteins from Hemolymph*

The haemolymph, including the recombinant KIR2DL1 protein, was collected using a needle to puncture the abdominal leg of silkworm larvae, then sodium thiosulfate was immediately added in order to prevent melanization. The haemolymph solution was subjected to Ni^{2+} affinity column chromatography for the purification. The SDS–PAGE displayed a single band of about 30 kDa of the KIR2DL1 protein in the fractions eluted with 100 or 150 mM imidazole, and the protein was further confirmed by Western blotting using an anti-pentahistidine antibody (Fig. 2). For biochemical characterizations, the protein was further purified by anion-exchange chromatography. The final yield from one larva was 0.2 mg of the purified protein 1 mg quantities of recombinant protein can easily be obtained from several larvae, therefore, this system can achieve high-level expression at a low cost and in a relatively short time period.

4.3 *Characterization of the Recombinant KIR2DL1 Expressed in Silkworms*

For the molecular characterization of the KIR2DL1 protein expressed in the haemolymph of silkworm, several biochemical analyses were performed. To determine whether the recombinant protein was *N*-glycosylated, it was treated with Endo H or PNGase F, which eliminate the *N*-linked sugars differently (PNGase F cleaves any kind of *N*-linked sugar except for α-1,3-fucose-containing glycans, while Endo H digests the high mannose and some hybrid oligosaccharides of *N*-linked sugars). The PNGase F

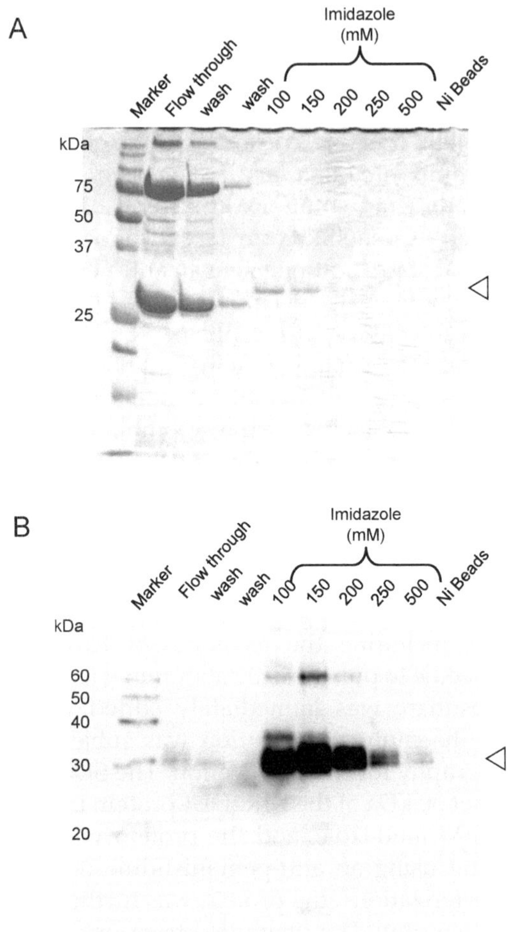

Fig. 2: Purification of the KIR2DL1 protein from the haemolymph. The proteins eluted from the Ni^{2+} affinity chromatography column were separated by 15–25% gradient SDS–PAGE (A) and subjected to a Western blotting analysis using an anti-pentahistidine antibody (B). The open arrowhead indicates the KIR2DL1 protein.

treatment shifted the band of the KIR2DL1 protein to a lower position in the SDS–PAGE gel, but the Endo H treatment did not (Fig. 3A). This result indicated that the obtained protein was *N*-glycosylated, but its *N*-linked sugars were not Endo H-sensitive. A further sugar composition analysis was performed, using a combination of HPLC and MS. This test revealed that the appended sugar of the protein consisted of only two kinds of small paucimannose-type sugars, Manα1-6Manβ1-4GlcNAcβ1-4(Fucα1-6)GlcNAc and Manα1-6Manβ1-4GlcNAcβ1-4GlcNAc (Fig. 3B). The former was significantly predominant (83–90%) over the latter (10–17%). Interestingly,

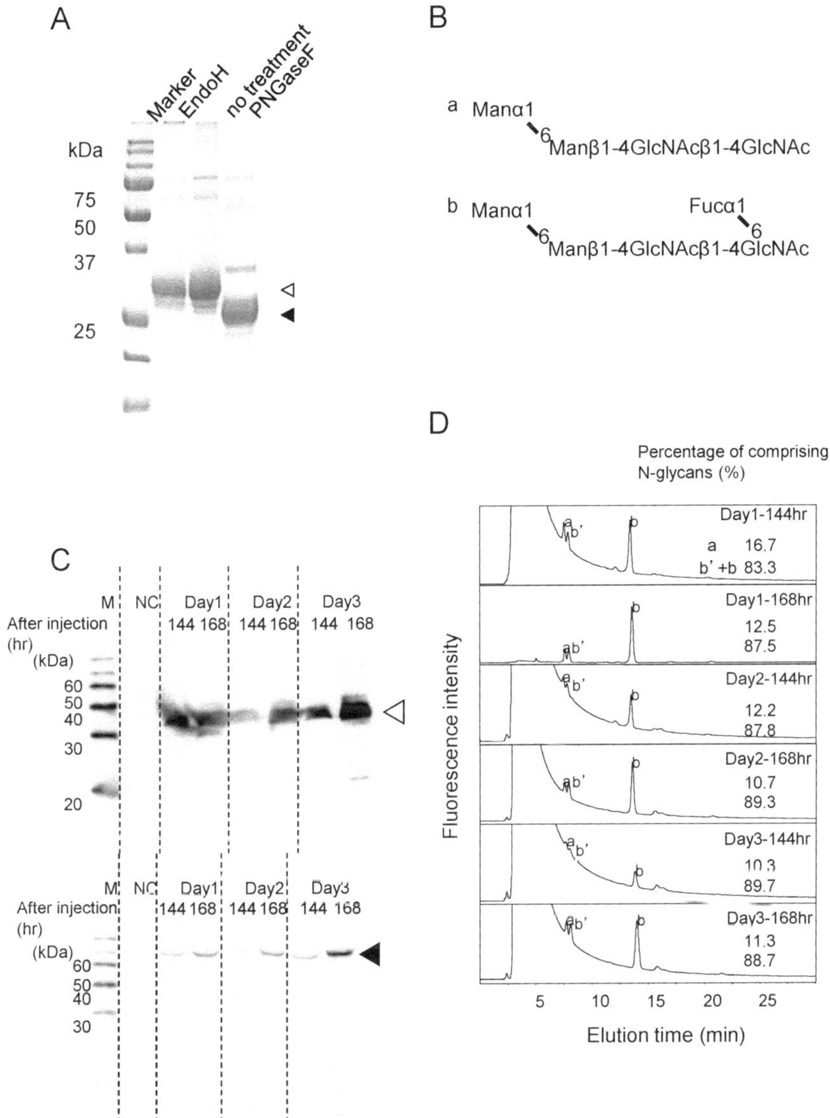

Fig. 3: (A) KIR2DL1 expressed in silkworm has an Endo H insensitive *N*-glycan modification. The open arrowhead indicates the glycosylated KIR2DL1 protein, and the filled arrowhead indicates the deglycosylated protein. (B) The structures of the PA-oligosaccharides obtained from the KIR2DL1 protein expressed in silkworm haemolymph. Western blot analysis (C) and *N*-linked sugar profile (D) of the recombinant KIR2DL1 expressed at different injection and extraction times. Day 1 indicates the first day of 5th instars. (C) The open arrowhead indicates the KIR2DL1 protein, and the filled arrowhead indicates gp64. (D) The structures of *N*-linked sugars were determined by HPLC and MS, as described under Materials and methods. In (D), b′ indicates epimerized b, which is a byproduct obtained during the 2-aminopyridine derivatization reaction.

the sugar composition of the paucimannose-type oligosaccharides was maintained under the different expression conditions, in terms of the injection times and expression levels (Fig. 3C–D). Although the sugar content was slightly different from that of the previously reported IgG antibody (Park et al. 2009), the relatively homogeneous and small sugar modifications seem potentially useful for structural studies. Next, we analyzed the binding ability of the recombinant KIR2DL1 protein to its ligand, HLA-Cw4, by surface plasmon resonance (SPR). As shown in Fig. 4, the purified KIR2DL1 specifically recognized HLA-Cw4, with a K_d of 8.6 ± 0.69 μM, although no binding to HLA-A11 and BSA was observed. Furthermore, the CD spectrum of the KIR2DL1 protein displayed the typical β-sheet secondary structure (data not shown), which corresponds to the β-sandwich feature in the crystal structure of KIR2DL1 (Fan et al. 1999). Taken together, these results clearly indicated that the silkworm could produce the KIR2DL1 protein with the proper structural and functional features.

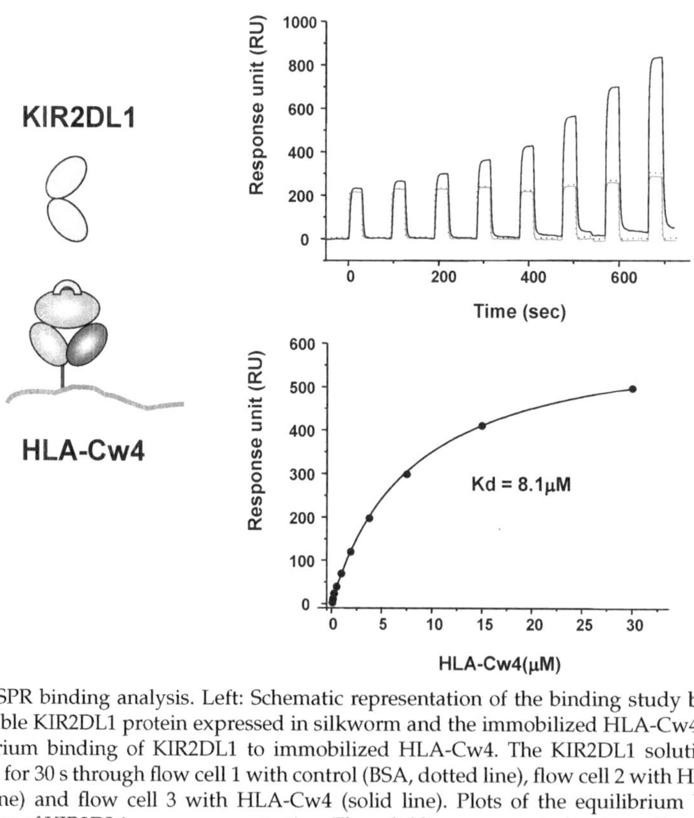

Fig. 4: SPR binding analysis. Left: Schematic representation of the binding study between the soluble KIR2DL1 protein expressed in silkworm and the immobilized HLA-Cw4. Right: Equilibrium binding of KIR2DL1 to immobilized HLA-Cw4. The KIR2DL1 solution was injected for 30 s through flow cell 1 with control (BSA, dotted line), flow cell 2 with HLA-A11 (gray line) and flow cell 3 with HLA-Cw4 (solid line). Plots of the equilibrium binding responses of KIR2DL1 versus concentration. The solid line represents direct non-linear fits of the 1:1 Langmuir binding isoform to the data. RU, response units.

5. Discussion

Immune cell surface receptors are highly glycosylated, thus, it is often difficult to express sufficient amounts of the recombinant proteins. Therefore, an expression system that is easy-to-use and produces a high expression level of posttranslationally-modified receptors has been eagerly awaited. In this chapter, the successful expression of human KIR2DL1 by a simple and rapid method, combining silkworm expression with the direct injection of the BmNPV bacmid DNA, is described in detail. A few silkworm larvae (costing only ~ 20 US dollars) are sufficient to produce ~ mg order of human KIR2DL1 proteins. As described above, the BmNPV baculovirus bacmid system (Motohashi et al. 2005) also has some remarkable advantages: (1) It is directly applicable to silkworm expression: The commercially available *A. californica* nucleopolyhedrovirus (AcMNPV) BEVS normally requires the large-scale cultivation of insect cell lines, (2) It is an easy and quick method: It requires only bacmid DNA, which can replicate in *E. coli*, thus, the site-specific transposition for introducing the target gene into bacmid DNA can be done in *E. coli*, while in contrast, a high-titer recombinant virus, together with proficient and time-consuming virus-handling techniques, are necessary for AcMNPV BEVS, (3) High biohazard safety: The bacmid DNA itself lacks infective activity (only gene transfer ability) and the baculoviruses are not infectious to mammalian cells. Here, using the above-mentioned BmNPV bacmid-silkworm expression system (Motohashi et al. 2005), the recombinant KIR2DL1 protein was produced in body fluid only 5–7 days after the injection of DNA into silkworms. The obtained body fluid was applied to a Ni^{2+} affinity column for purification, and 0.2 mg of the highly purified protein could be obtained from one larva. Furthermore, the recombinant baculoviruses, which are easily purified from the body fluid by simple ultracentrifugation steps, could be directly injected into other larvae in order to stably produce the KIR2DL1 protein (Fig. 1C). This technique is quite useful for the significant reduction of both cost and time. Finally, the standard method for silkworm expression and purification of recombinant cell surface receptors was established.

The sugar modifications are often important for functional and structural analyses, as well as for practical applications. The previous report of the silkworm expression of mouse interferon-β demonstrated that the expressed protein harbored variable sets of paucimannose-type *N*-linked sugars, including Manα1-6Manβ1-4GlcNAcβ1-4GlcNAc (6.4%) and Manα1-6(Manα1-3)Manβ1-4GlcNAcβ1-4(Fucα1–6)GlcNAc (1.1%) (Misaki et al. 2003). These paucimannose-type sugars are possibly derived from GlcNAcMan5GlcNAc2 by the mannosidase-mediated excision of α-linked mannosyl residues, followed by the GlcNAcase-mediated deletion of a β-1,2-linked terminal GlcNAc residue (Watanabe et al. 2002). In contrast, in the BmNPV bacmid-silkworm expression system, the *N*-linked sugar composition of the expressed KIR2DL1 protein is comprised of only two kinds of small paucimannose-type oligosaccharides, Manα1-6Manβ1-4GlcNAcβ1-4(Fucα1-6)GlcNAc (83–90%) and Manα1-6Manβ1-4GlcNAcβ1-4GlcNAc (10–17%). Similarly, the previous report of IgG expressed by the same system revealed that the IgG protein harbored two sugar types, Manα1-6Manβ1-4GlcNAcβ1-4(Fucα1-6)GlcNAc (77.5%) and Manα1-6(Manα1-3)Manβ1-4GlcNAcβ1-4(Fucα1-6)GlcNAc (12.7%). The major one, Manα1-6Manβ1-4GlcNAcβ1-4(Fucα1-6)GlcNAc, was maintained in the expression of different proteins, thus, small *N*-linked paucimannose-type sugars were essentially preferable, even though some composition differences were observed. Furthermore, the BmNPV bacmid-silkworm expression did not show any detectable differences in the sugar modifications under various conditions (Fig. 3D). These small and relatively homogeneous *N*-linked sugar modifications in the BmNPV bacmid-silkworm expression system have the potential to be applicable to structural and biochemical studies of immune cell surface receptors.

The example for preparation of KIR2DL1 protein tells us many technological aspects of the baculovirus—silkworm expression system. In addition to theses techniques, further useful options and available materials are described below. For the expressions of protease sensitive proteins, a protease deficient bacmid BmNPV-*CP*$^-$-*Chi*$^-$, which lacks cysteine protease and chitinase genes, is available. The successful expressions of several proteins using the protease deficient bacmid BmNPV-*CP*$^-$-*Chi*$^-$ were reported (Nagaya et al. 2004, Park et al. 2008), therefore, this bacmid is widely used for protein production. The growing temperature is also one of parameters for decreasing protease activities and controlling protein expression speed. The silkworm generally grows at around 25 degrees Celsius, and the growing temperature can be adjusted within the range of 20 to 30 degrees Celsius. At lower temperatures, silkworm grow slowly and the expression level is lower than that of silkworm at 25 degrees Celsius (data not shown). On the other hand, at higher temperatures, silkworm grow rapidly and the expression level is higher than that of silkworm at 25 degrees Celsius. If the target protein is sensitive to protease and is easily degraded, changing the growing temperature is worth trying.

6. Concluding Remark

This chapter describes successful examples for the preparation of human cell surface antigen using the baculovirus—silkworm expression system. The BmNPV system is rapid and easy to use in producing large amounts of human proteins, thus, more and more proteins will be expressed using this system in the near future. The application of this system is not limited to functional studies of human immune cell surface receptors. It can be extended to structural studies, high throughput drug screening and production of biological relevant proteins for vaccination.

References

Carrington, M. and M. P. Martin. 2006. The impact of variation at the KIR gene cluster on human disease. Curr. Top. Microbiol. Immunol. 298: 225–257.

Colonna, Marco. 1996. Natural killer cell receptors specific for MHC class I molecules. Curr. Opin. Immunol. 8(1): 101–107.

Fan, Q. R., L. Mosyak, D. N. Garboczi, C. C. Winter, N. Wagtmann, E. O. Long and D. C. Wiley. 1999. Structure of a human natural killer cell inhibitory receptor. Transplant. Proc. 31(4): 1871–1872.

Kajikawa, Mizuho, Kaori Sasaki, Yoshitaro Wakimoto, Masaru Toyooka, Tomoko Motohashi, Tsukasa Shimojima, Shigeki Takeda, Enoch Y. Park and Katsumi Maenaka. 2009. Efficient silkworm expression of human GPCR (Nociceptin Receptor) by a *Bombyx mori* Bacmid DNA system. Biochem. Biophys. Res. Commun. Elsevier Inc. 385(3): 375–379.

Long, Eric O. and Sumati Rajagopalan. 2000. HLA Class I recognition by killer cell Ig-like receptors. Semin. Immunol. 12(2): 101–108.

Luckow, V. A., S. C. Lee, G. F. Barry and P. O. Olins. 1993. Efficient generation of infectious recombinant baculoviruses by site-specific transposon-mediated insertion of foreign genes into a Baculovirus genome propagated in *Escherichia coli*. J. Virol. 67(8): 4566–4579.

Maenaka, K., T. Juji, T. Nakayama, J. R. Wyer, G. F. Gao, T. Maenaka, N. R. Zaccai et al. 1999. Killer cell immunoglobulin receptors and T cell receptors bind peptide-major histocompatibility complex class I with distinct thermodynamic and kinetic properties. J. Biol. Chem. 274(40): 28329–28334.

Misaki, Ryo, Hidekazu Nagaya, Kazuhito Fujiyama, Itaru Yanagihara, Takeshi Honda and Tatsuji Seki. 2003. N-Linked glycan structures of mouse interferon-β produced by *Bombyx mori* Larvae. Biochem. Biophys. Res. Commun. 311(4): 979–986.

Motohashi, Tomoko, Tsukasa Shimojima, Tatsuo Fukagawa, Katsumi Maenaka and Enoch Y. Park. 2005. Efficient large-scale protein production of larvae and pupae of silkworm by *Bombyx mori* nuclear polyhedrosis virus bacmid system. Biochem. Biophys. Res. Commun. 326(3): 564–569.

Nagaya, Hidekazu, Toshimichi Kanaya, Hiroki Kaki, Yoneko Tobita, Masashi Takahashi, Hitomi Takahashi, Yuichi Yokomizo and Shigeki Inumaru. 2004. Establishment of a large-scale purification procedure for purified recombinant bovine interferon-tau produced by a silkworm-baculovirus gene expression system. J. Vet. Med. Sci. 66(11): 1395–1401.

Park, Enoch Y., Takahiro Abe and Tatsuya Kato. 2008. Improved expression of fusion protein using a cysteine- protease- and chitinase-deficient *Bombyx mori* (Silkworm) multiple nucleopolyhedrovirus bacmid in silkworm larvae. Biotechnol. Appl. Biochem. 49(Pt 2): 135–140.

Park, Enoch Y., Motoki Ishikiriyama, Takuya Nishina, Tatsuya Kato, Hirokazu Yagi, Koichi Kato and Hiroshi Ueda. 2009. Human IgG1 expression in silkworm larval hemolymph using BmNPV bacmids and its N-linked glycan structure. J. Biotech. 139(1): 108–114.

Sasaki, Kaori, Mizuho Kajikawa, Kimiko Kuroki, Tomoko Motohashi, Tsukasa Shimojima, Enoch Y. Park, Sachiko Kondo, Hirokazu Yagi, Koichi Kato and Katsumi Maenaka. 2009. Silkworm expression and sugar profiling of human immune cell surface receptor, KIR2DL1. Biochem. Biophys. Res. Commun. 387(3): 575–580.

Tsuchiya, Naoyuki, Chieko Kyogoku, Risa Miyashita and Kimiko Kuroki. 2007. Diversity of human immune system multigene families and its implication in the genetic background of rheumatic diseases. Curr. Med. Chem. 14(4): 431–439.

Wagtmann, Nicolai, Sumati Rajagopalan, Christine C. Winter, Marta Peruui and Eric O. Long. 1995. Killer cell inhibitory receptors specific for HLA-C and HLA-B identified by direct binding and by functional transfer. Immunity. 3(6): 801–809.

Watanabe, Satoko, Takehiro Kokuho, Hitomi Takahashi, Masashi Takahashi, Takayuki Kubota and Shigeki Inumaru. 2002. Sialylation of N-Glycans on the recombinant proteins expressed by a baculovirus-insect cell system under β-N-Acetylglucosaminidase inhibition. J. Biol. Chem. 277(7): 5090–5093.

CHAPTER 11

Expression and Purification of Mitochondrial Membrane Protein OPA1 for Reconstitution of Membrane Fusion

Tadato Ban[1,*] and *Naotada Ishihara*[1,2]

1. Introduction

Mitochondria have a double membrane structure, composed of outer membrane (OM) and inner membrane (IM), and play a central role in ATP production, the phospholipid metabolism and apoptosis (Labbe et al. 2014, Mishra and Chan 2016). Mitochondria fuse together via OM fusion followed by IM fusion and form a mitochondrial network (Fig. 1A) (Ishihara et al. 2013). The membrane fusion is required for proper mitochondrial functions, and dysfunction of mitochondrial dynamics are often observed within neurodegenerative and metabolic diseases (Nunnari and Suomalainen 2012, Pernas and Scorrano 2016). In mammalian cells, IM-protein optic atrophy 1 (OPA1) is required for IM fusion (MacVicar and Langer 2016). Various spliced variants are expressed in human, and the variant 1 (960 residues) is processed to 873-residue

[1] Department of Protein Biochemistry, Institute of Life Science, Kurume University, 67 Asahimachi, Kurume, Fukuoka 830-011, Japan.
[2] Department of Biological Science, Graduate School of Science, Osaka University, 1-1 Machikaneyama, Toyonaka, Osaka 560-0043, Japan.
*Corresponding author: ban_tadato@kurume-u.ac.jp

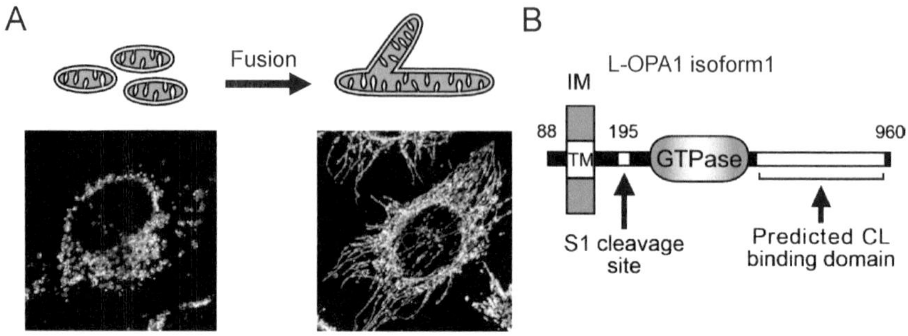

Fig. 1: Mitochondrial inner membrane fusion is regulated by GTPase OPA1. (A) Mitochondria fuse together and regulate the size, length and physiological function. (B) L-OPA1 has a transmembrane domain (TM) at the N-terminus, a GTPase domain, and a predicted phospholipid cardiolipin (CL)-binding domain at the C-terminus.

mature L-OPA1. L-OPA1 is anchored to IM by a transmembrane domain (TM) at the N-terminus and has a GTPase domain and a predicted phospholipid cardiolipin (CL) binding domain at the C-terminus (Fig. 1B). L-OPA1 is further proteolytically cleaved at the S1 cleavage site (Fig. 1B) and to form S-OPA1, which is lacking TM domain and modulates IM fusion (Anand et al. 2014, Ishihara et al. 2006, Mishra et al. 2014). CL is the most abundant negative phospholipid in IM, consisting of two phosphatidyl groups and four acyl chains. This unique dimeric structure confers an affinity to proteins, which can be essential for stabilizing higher order assemblies such as respiratory complex (Claypool 2009), regulating the spatio-temporal redistribution of proteins during apoptosis (Lutter et al. 2000), mitophagy (Chu et al. 2013) and mitochondrial dynamics (Choi et al. 2006). CL association triggers assembly of S-OPA1 into higher oligomers and enhances GTPase activity (Ban et al. 2010), indicating that CL cooperates with OPA1 in IM fusion. To elucidate the molecular mechanism of membrane fusion and to identify the essential components, *in vitro* reconstitution using recombinant proteins and model membranes have been performed (Wickner and Schekman 2008). Reconstitution approaches have particularly addressed the molecular mechanisms of various membrane fusion by SNAREs in secretory/endocytic membranes (Weber et al. 1998) and atlastin in endoplasmic reticulum membranes (Orso et al. 2009). However, due to the difficulty of preparing active recombinant L-OPA1, reconstituting approach of IM fusion has not been reported. Recently we established a method for expression and purification of human L-OPA1 using the silkworm expression system and developed *in vitro* assays (Ban et al. 2017). In this chapter, we describe the details of method and discuss the unique fusion reaction by OPA1.

2. Expression and Purification Recombinant Human L-OPA1

2.1 L-OPA1 Expression Using the BmNPV Bacmid-silkworm Expression System

For reconstitution of IM fusion *in vitro*, we prepared active recombinant L-OPA1 using the BmNPV bacmid-silkworm expression system. The details of silkworm expression systems are described in previous reports (Kajikawa et al. 2009). The cDNA corresponding to human L-OPA1 isoform1 was PCR-amplified from HeLa cells and then sub-cloned into the donor plasmid pFastBac Htb (10359016, Thermo Fisher Scientific). The donor plasmid encoding L-OPA1 transformed into *E. coli* BmDH10Bac. The transposition was confirmed by colony PCR with the universal primers M13 forward (-40) (5′-GTTTTCCCAGTCACGAC-3′) and M13 reverse (5′-CAGGAAACAGCTATGAC-3′). The recombinant BmNPV bacmid DNA was amplified and purified using a Qiafilter plasmid Midi Kit (No. 12243, QIAGEN). We obtained more than 100 μg of bacmid DNA from 50 ml culture. Purified bacmid-DNA was introduced into silkworm larva by *in vivo* lipofection using DMRIE-C (10459014, Thermo Fisher Scientific). We reared 60 silkworm larvae simultaneously, therefore 90 μg (1.5 μg × 60) of bacmid was mixed with 180 μl (3 μl × 60) of DMRIE-C and incubated for 45 min at room temperature, then diluted in 3000 μl (50 μl × 60) of sterilized water. Aliquots (50 μl) of bacmid-DMRIE-C mixture were directly injected into the dorsal side of a silkworm larva with 0.30 × 25 mm needle (No. 121109, Tochigi Seiko). Feeding was provided after 30 min incubation at room temperature after the injection. Silkworm larvae were fed on an artificial diet (silkmate 2S, Nihon Nosan Kogyo), followed by rearing for 7 days in a climate chamber (LPH-180-E, Nippon Medical & Chemical Instruments) under 25°C, 80% humidity. We reared 15 silkworm larvae in 22 × 16 × 6.6 cm box with breathing holes and feed them every morning as follows. For 15 silkworm larvae; day 1 (20 g), day 2 (30 g), day 3 (45 g), day 4 (50 g), day 5 (50 g), day 6 (35 g). The infected fat bodies were isolated manually and collected in an ice-cold PB buffer (50 mM NaPi, pH 8.0, 500 mM NaCl, 10% glycerol, 1 mM DTT, 1 mM PMSF, 0.5% sodium thiosulfate) (Fig. 2A). The collected fat bodies were homogenized in a 50 ml of glass-Teflon potter homogenizer with 10 strokes at 1300 rpm, then frozen in liquid nitrogen and stored at –80°C.

2.2 L-OPA1 Purification from Silkworm Fat Body

L-OPA1 was fractionated by centrifugation and solubilization with detergent (Fig. 2B). We stored the isolated fat bodies from 15 silkworm in a 50 ml conical tube and handled 4 tubes at once, except during sonication.

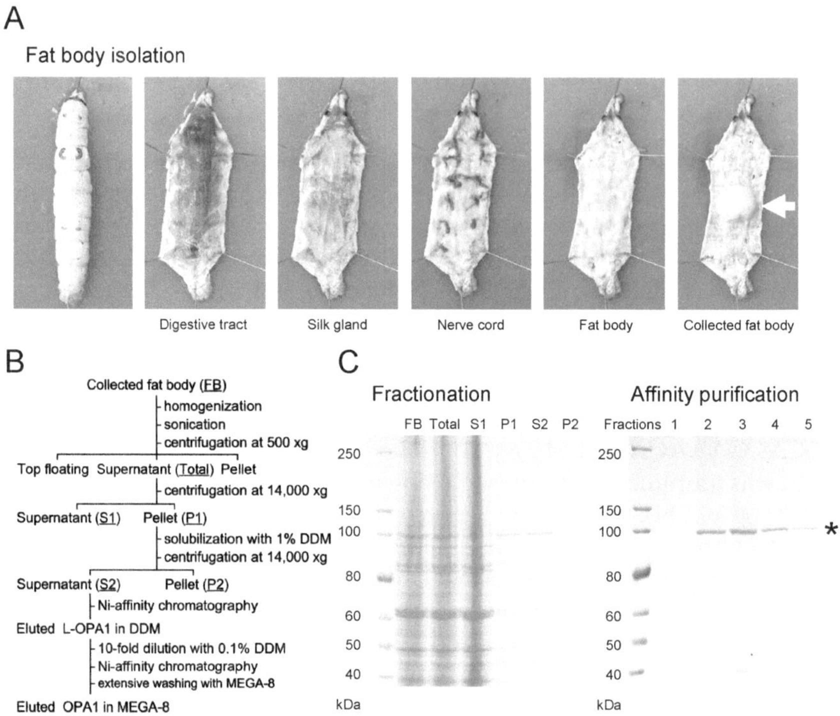

Fig. 2: Preparation of recombinant human L-OPA1 using silkworm larvae. (A) The fat bodies expressing L-OPA1 were collected manually. Arrow indicates collected fat body. (B) Procedures for the fractionation and purification of recombinant L-OPA1 from fat bodies. L-OPA1 was solubilized by DDM and purified by Ni-affinity chromatography. (C) SDS-PAGE analysis and staining with Coomassie blue of fractionated and purified L-OPA1. Asterisk indicates purified L-OPA1.

The frozen fat bodies were thawed on ice and transferred to a 50 ml glass beaker, and then sonicated on ice using a sonicator (W-385 with flat tip, Astrason) at output level 9, cycle time 2, and duty cycle 50 for 10 min (FB in Fig. 2B). The disrupted fat bodies were transferred to a new 50 ml conical tube and centrifuged at 500 xg for 5 min at 4°C. After removal of top floating, supernatant was gently decanted to a new 50 ml conical tube (Total in Fig. 2B). The supernatant was further centrifuged at 14,000 xg for 1 hr at 4°C, and the supernatant (S1 in Fig. 2B) was decanted from tube in order to harvest the pellet fraction (P1 in Fig. 2B). For solubilization of L-OPA1 from the pellet fraction, 40 ml of ice-cold PB buffer containing 10 mM imidazole and 1% dodecyl maltoside (DDM) (D316, Dojindo) was added to the tube and the pellet was suspended gently and incubated for 2 hr at 4°C with gentle agitation. The solubilized pellet was centrifuged at 14,000 xg for 30 min at 4°C and the supernatant (S2 in Fig. 2B) was transferred to a new 50 ml tube. After addition of 750 µl (bed volume) of Ni-Sepharose 6 fast flow beads (17531801, GE Healthcare), the sample was incubated overnight at 4°C with gentle agitation. After centrifugation at 50 xg for 1 min at 4°C, supernatant was removed and Ni- Sepharose 6 beads were washed with 1 ml of ice-cold PB buffer containing 10 mM imidazole and 1% DDM three times as same. When we handled 4 tubes at once, Ni-sepharose 6 beads were suspended with ice-cold PB buffer containing 10 mM imidazole and 1% DDM and carefully applied into Econo-column (No.737-1007; Bio-rad) from the four 50 ml conical tubes. Ni-Sepharose 6 beads were washed with 16 ml of ice-cold RB buffer (50 mM Tris, pH 7.4, 300 mM NaCl, 10% glycerol, 1 mM DTT) containing 10 mM imidazole and 0.1% DDM, followed by further washing with 16 ml of ice-cold RB buffer containing 20 mM imidazole and 0.1% DDM. L-OPA1 was eluted 8 times with 1.5 ml of ice-cold RB buffer containing 250 mM imidazole and 0.1% DDM. The eluted L-OPA1 fractions were analyzed by SDS-PAGE and Coomassie blue staining (Fig. 2C). L-OPA1 concentration was determined by Bradford assay. In usual case, the total protein yield was more than 2 mg from 60 silkworm larvae. The eluted L-OPA1 fractions were frozen in liquid nitrogen and stored at –80°C.

3. Membrane Fusion and Liposome Binding Assay

3.1 Preparation of L-OPA1 Proteoliposome

To better understand the function of L-OPA1, we reconstituted *in vitro* membrane fusion with proteoliposomes containing L-OPA1. Proteoliposomes were prepared on the basis of a detergent-dialysis method (Fig. 3A) (Ban et al. 2017, Mima and Wickner 2009). Because DDM is not appropriated for detergent-dialysis methods due to its low critical micelle concentration (CMC), DDM was replaced with MEGA-8 (M014; Dojindo), which has higher CMC (Fig. 2B). After 10-fold dilution with RB buffer containing 0.1% DDM, purified L-OPA1 was applied into Econo-column with 1 ml (bed volume) Ni-Sepharose 6 by gravity flow at 4°C. L-OPA1 bound to the Ni-sepharose 6 column was washed with 20 column volumes of RB buffer containing 2.5% MEGA-8, and L-OPA1 was eluted with RB buffer containing 2.5% MEGA-8 and 250 mM imidazole (Fig. 2C). The eluted L-OPA1 fractions were frozen in liquid nitrogen and stored at –80°C. For the preparation of liposomes, the following lipids were purchased from Avanti Polar Lipids. 1-palmitoyl-2-oleoyl-*sn*-glycero-3-phosphocholine (POPC) (No. 850457), 1-palmitoyl-2-oleoyl-*sn*-glycero-3-phosphoethanolamine (POPE) (No. 850757), 1-palmitoyl-2-oleoyl-*sn*-glycero-3-phosphate (POPA) (No. 840857), 1-palmitoyl-2-oleoyl-*sn*-glycero-3-phospho-L-serine (POPS) (No. 840034), 1-palmitoyl-2-oleoyl-*sn*-glycero-3-phospho-(1′-rac-glycerol) (POPG) (No. 840457), L-α-phosphatidylinositol (soy-PI) (No. 840044)

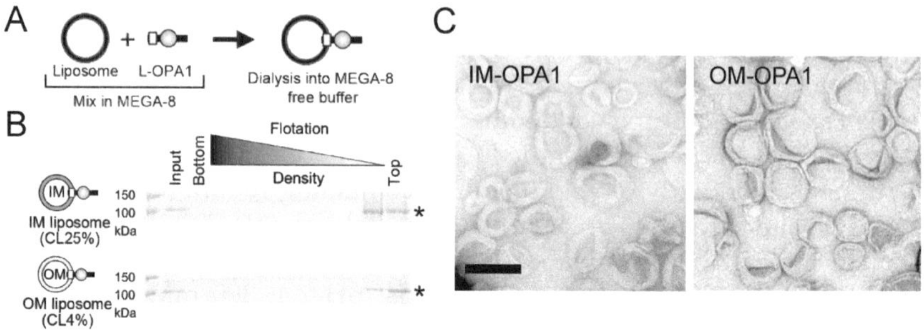

Fig. 3: Preparation of proteoliposomes containing L-OPA1. (A) Recombinant L-OPA1 and IM or OM liposomes were mixed in the presence of detergent, and proteoliposomes were reconstituted by dialysis. The lipid composition of IM liposome (25% CL, 42% PC, 25% PE, 8% PI), OM liposome (4% CL, 60% PC, 27% PE, 9% PI). For fluorescence-labelled or biotin-modified liposome, 3% of PE were replaced, respectively. (B, C) Proteoliposomes were analyzed by density gradient flotation and negative staining electron microscopy. Asterisk indicates L-OPA1. Scale bar represents 100 nm.

and 1′,3′-bis[1,2-dioleoyl-*sn*-glycero-3-phospho]-*sn*-glycerol ($CL(18:1)_4$) (No. 710335). The lipid mixtures for IM liposome or OM liposome were designed from the lipid compositions of mitochondrial inner or outer membranes (Ardail et al. 1990). The lipid compositions of liposomes were as follows (percent-molar): "IM" liposomes ($CL(18:1)_4$:POPC:POPE:Soy-PI; 25:42:25:8), "OM" liposomes ($CL(18:1)_4$:POPC:POPE:Soy-PI; 4:60:27:9). The indicated lipids in a chloroform solution were mixed in glass tube and dried with a N_2 gas stream. To remove chloroform, the lipid mixtures were further dried in a vacuum for 30 min. The dried lipid films were dissolved in an RB buffer containing MEGA-8 at a concentration of 5%. After purified L-OPA1 was mixed with detergent-lipid mixtures in a PCR tube, the detergent-lipid-protein mixtures containing 3.2% MEGA-8, 2 mM lipids and 2 μM L-OPA1 were incubated for 2 hr with gentle rotation at 4°C. For removal of MEGA-8, these mixtures were transferred to a dialysis button (HR3-328; Hampton Research) and then dialyzed against RB buffer at 4°C, resulting in the formation of proteoliposomes containing L-OPA1.

To analyze the efficiency of L-OPA1 incorporation into liposomes, we performed the density gradient flotation assay. The dialysate was mixed with an equal volume of 80% Histodenz (D2158, Sigma-Aldrich) in RB buffer and transferred to an 11 × 34 mm ultra-clear tube (No. 343778, Beckman Instruments). The sample solution was layered with 30% Histodenz in RB buffer to 2 ml. After the addition of 200 μl RB buffer, the sample solution was centrifuged at 214,000 xg for 2 hr in a swinging bucket rotor (TLS-55, Beckman Instruments) at 4°C. Ten fractions were collected at different points from the top to the bottom of the tube and analyzed by SDS-PAGE and Coomassie blue staining. L-OPA1 was observed in top fractions, indicating that L-OPA1 was integrated into the both IM and OM liposomes (Fig. 3B). We further examined the proteoliposomes containing L-OPA1 by negative staining electron microscopy. Aliquots of L-OPA1 proteoliposomes were applied to carbon-collodion coated 400 mesh copper grids (6512, Nisshin EM) and stained with 2% uranyl acetate. Images were acquired using a transmission electron microscope (H-7650; Hitachi) at 100 kV and showed spherical unilamellar vesicles with a diameter of ~ 100 nm and without remarkable tubulation (Fig. 3C), suggesting intact integrity and morphology for both L-OPA1 proteoliposomes containing IM or OM lipid.

3.2 Membrane Fusion Assay

Membrane fusion assay was performed by a well-known fluorescence energy transfer-based method (Fig. 4A) (Scott et al. 2003, Weber et al. 1998). For the preparation of fluorescent labeled donor liposomes, 3% of POPE in IM or OM lipids were replaced with 1.5% (N-(7-nitrobenz-2-oxa-1,3-diazol-4-yl)-1,2-dihexadecanoyl-*sn*-glycero-3-phosphoethanolamine) (NBD-PE) (N360; Molecular Probes) and 1.5% lissamine rhodamine B 1,2-dihexadecanoyl-*sn*-glycero-3-phosphoethanolamine (Rh-PE) (L1392; Molecular Probes), and proteoliposomes were prepared as described above. A sample mixture of 18 µl containing donor L-OPA1 proteoliposomes (L-OPA1 and liposomes at a concentration of 200 nM and 200 µM) and non-fluorescent acceptor L-OPA1 proteoliposomes (L-OPA1 and liposomes at a concentration of 500 nM and 500 µM) was incubated in a black 384-well plate (No. 3676, Corning) for 10 min at 30°C using a fluorescence plate reader (Infinite F200 Pro; Tecan). Reactions were initiated by the addition of 2 µl of GTP/$MgCl_2$ at a concentration of 5 mM at 30°C. NBD fluorescence (λ excitation = 460 nm, λ emission = 540 nm) was measured at 30-s intervals. The data were normalized to maximum NBD fluorescence using the equation; Lipid mixing (%) = $(F_t-F_0)/(F_{max}-F_0) \times 100$, in which *Ft* is the NBD fluorescence during the measurement, F_0 is the initial NBD fluorescence just after the addition of GTP/$MgCl_2$ and F_{max} is the NBD fluorescence after the addition of 4% of 2 µl Triton X-100 to the reaction mixture.

We examined whether recombinant L-OPA1 could mediate membrane fusion. When L-OPA1 proteoliposomes containing IM lipids were used as both donor and acceptor, clear membrane fusion was observed in the presence of GTP (Fig. 4A). In contrast, fusion was not observed for protein-free liposomes even in the presence of GTP (Fig. 4A). L-OPA1-induced membrane fusion was dependent on GTP hydrolysis (Ban et al. 2017), indicating that L-OPA1 alone can cause the membrane fusion via GTP hydrolysis. When L-OPA1 was excluded from acceptor liposomes, efficient membrane fusion was still observed (Fig. 4A), suggesting

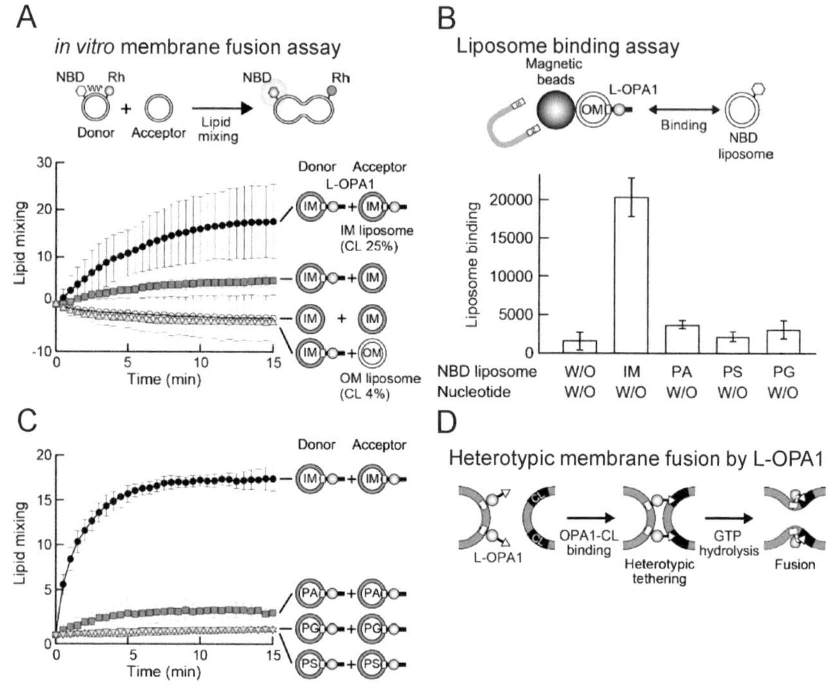

Fig. 4: *In vitro* fusion and liposome binding assay with L-OPA1 proteoliposome. (A) L-OPA-mediated membrane fusion *in vitro*. After addition of GTP/$MgCl_2$, membrane fusion was measured by FRET-based lipid mixing. (B) Specific interaction between L-OPA1 and CL was analyzed by liposome binding assay. (C) Another negative lipids were analyzed. POPA, POPS, POPG was used instead of CL. (D) Model for L-OPA1-mediated membrane fusion. L-OPA1 directly binds to CL in the absence of GTP for tethering of the opposite membrane, and fusion is completed by GTP hydrolysis. All data are the average of three independent measurements and error bars represent SD.

that L-OPA1 on the one side is sufficient for fusion. However, such a heterotypic fusion was not observed for OM liposome containing only 4% CL (Fig. 4A), suggesting that CL is also an essential component for L-OPA1-mediated membrane fusion.

3.3 Liposome Binding Assay

To uncover CL function in L-OPA1-mediated fusion, we performed the liposome binding assay (Fig. 4B) (Ban et al. 2017, Sugiura and Mima 2016). For immobilization of L-OPA1 proteoliposomes containing OM lipids on the surface of avidin-coated magnetic beads (Dynabeads M-280, Thermo Fisher Scientific), 3% POPE in OM lipids were replaced with 1,2-dioleoyl-sn-glycero-3-phosphoethanolamine-N-(biotinyl) (biotinyl-PE) (No. 870282, Avanti Polar Lipids). After addition of 5 μl of magnetic beads, 50 μl of L-OPA1 proteoliposomes containing OM lipids (L-OPA1 and liposomes at a concentration of 150 nM and 150 μM) in RB buffer were incubated for 30 min at room temperature with gentle rotation. NBD-labelled liposomes were prepared by replacement of 1.5% POPE in each liposome with NBD-PE. NBD-labelled liposomes (150 μM) were mixed with the sample mixture and incubated in the absence of GTP/ $MgCl_2$ for 30 min at room temperature with gentle rotation. To quantify the interaction between L-OPA1 and NBD-labelled liposomes, magnetic beads were pulled down with a magnet and washed with RB buffer three times, followed by resuspension in RB buffer containing 0.4% Triton X-100, and then NBD fluorescence (λ excitation = 460 nm, λ emission = 540 nm) was measured by plate leader.

We observed IM liposomes binding efficiently to L-OPA1 proteoliposomes in the absence of GTP (Fig. 4B). In contrast, when CL of IM liposome was replaced with another negative phospholipid such as POPA, POPS or POPG, liposome binding was not observed (Fig. 4B), indicating that L-OPA1 has specific affinity to CL. Membrane fusion was not observed in L-OPA1 proteoliposome containing other negative phospholipids (Fig. 4C). Note that L-OPA1-mediated liposome binding did not require GTP-binding and hydrolysis, but L-OPA1-mediated membrane fusion required GTP hydrolysis. These results indicate that CL-to-L-OPA1 pairing causes membrane tethering, independently of GTP, although GTP hydrolysis is needed for the subsequent fusion (model in Fig. 4D). In addition, the C-terminal deletion mutant (Δ 581–940) of L-OPA1 showed severe reduction of both liposome binding and membrane fusion, even though this mutant retains GTPase activity (Ban et al. 2017), suggesting that the domain next to the GTPase domain of L-OPA1 is essential for CL binding (Fig. 1B).

4. Conclusion

To evaluate L-OPA1-mediated mitochondrial inner membrane fusion we have developed methods to express and purify active human OPA1 using the BmNPV bacmid-silkworm expression system. By combining recombinant L-OPA1 and liposome-based *in vitro* assay, we uncovered the fusion machinery in IM. In contrast to all previously reported intracellular membrane fusion machinery, such as SNAREs (Weber et al. 1998), mitofusin on OM fusion (Koshiba et al. 2004) and atlastin (Orso et al. 2009), in which trans-protein complex between the opposite membranes are required for tethering steps and fusion, L-OPA1 on one side of the membrane and CL on the other side are sufficient for the membrane fusion (model in Fig. 4D). Interestingly, the cell-free mitochondrial fusion assay showed that the yeast OPA1 ortholog Mgm1 on both sides of the membrane forms a trans-complex during mitochondrial fusion (Meeusen et al. 2004), indicating that mammalian mitochondria have a unique fusion machinery. Thus, the reconstituting approach by the present methods will provide critical insight into the mechanisms of mitochondrial membrane fusion. The silkworm-expression system might have advantages for the expression of large membrane proteins in comparison with the bacterial expression system and will have applicability for analysis of mitochondrial dynamics regulated by membrane proteins.

Acknowledgements

We thank A. Ichimura (Kurume University) for EM observations, Dr. T. Oka (Rikkyo University) and Dr. K. Maenaka (Hokkaido University) for advice on the silkworm expression system, Dr. J. Mima (Osaka University) for advice on the preparation of assays with proteoliposomes

References

Anand, R., T. Wai, M. J. Baker, N. Kladt, A. C. Schauss, E. Rugarli and T. Langer. 2014. The i-AAA protease YME1L and OMA1 cleave OPA1 to balance mitochondrial fusion and fission. J. Cell. Biol. 204(6): 919–929.

Ardail, D., J. P. Privat, M. Egret-Charlier, C. Levrat, F. Lerme and P. Louisot. 1990. Mitochondrial contact sites. Lipid composition and dynamics. J. Biol. Chem. 265(31): 18797–18802.

Ban, T., J. A. Heymann, Z. Song, J. E. Hinshaw and D. C. Chan. 2010. OPA1 disease alleles causing dominant optic atrophy have defects in cardiolipin-stimulated GTP hydrolysis and membrane tubulation. Hum. Mol. Genet. 19(11): 2113–2122.

Ban, T., T. Ishihara, H. Kohno, S. Saita, A. Ichimura, K. Maenaka, T. Oka, K. Mihara and N. Ishihara. 2017. Molecular basis of selective mitochondrial fusion by heterotypic action between OPA1 and cardiolipin. Nat. Cell Biol. 19(7): 856–863.

Choi, S. Y., P. Huang, G. M. Jenkins, D. C. Chan, J. Schiller and M. A. Frohman. 2006. A common lipid links Mfn-mediated mitochondrial fusion and SNARE-regulated exocytosis. Nat. Cell Biol. 8(11): 1255–1262.

Chu, C. T., J. Ji, R. K. Dagda, J. F. Jiang, Y. Y. Tyurina, A. A. KapraJlov, V. A. Tyurin, N. Yanamala, I. H. Shrivastava, D. Mohammadyani, K. Z. Q. Wang, J. Zhu, J. Klein-Seetharaman, K. Balasubramanian, A. A. Amoscato, G. Borisenko, Z. Huang, A. M. Gusdon, A. Cheikhi, E. K. Steer, R. Wang, C. Baty, S. Watkins, I. Bahar, H. Bayir and V. E. Kagan. 2013. Cardiolipin externalization to the outer mitochondrial membrane acts as an elimination signal for mitophagy in neuronal cells. Nat. Cell Biol. 15(10): 1197–1205.

Claypool, S. M. 2009. Cardiolipin, a critical determinant of mitochondrial carrier protein assembly and function. Biochim. Biophys. Acta. 1788(10): 2059–2068.

Ishihara, N., Y. Fujita, T. Oka and K. Mihara. 2006. Regulation of mitochondrial morphology through proteolytic cleavage of OPA1. EMBO J. 25(13): 2966–2977.

Ishihara, N., H. Otera, T. Oka and K. Mihara. 2013. Regulation and physiologic functions of GTPases in mitochondrial fusion and fission in mammals. Antioxid. Redox. Signal. 19(4): 389–399.

Kajikawa, M., K. Sasaki, Y. Wakimoto, M. Toyooka, T. Motohashi, T. Shimojima, S. Takeda, E. Y. Park and K. Maenaka. 2009. Efficient silkworm expression of human GPCR (nociceptin receptor) by a *Bombyx mori* bacmid DNA system. Biochem. Biophys. Res. Commun. 385(3): 375–379.

Koshiba, T., S. A. Detmer, J. T. Kaiser, H. Chen, J. M. McCaffery and D. C. Chan. 2004. Structural basis of mitochondrial tethering by mitofusin complexes. Science. 305(5685): 858–862.

Labbe, K., A. Murley and J. Nunnari. 2014. Determinants and functions of mitochondrial behavior. Annu. Rev. Cell Dev. Biol. 30: 357–391.

Lutter, M., M. Fang, X. Luo, M. Nishijima, X. Xie and X. Wang. 2000. Cardiolipin provides specificity for targeting of tBid to mitochondria. Nat. Cell Biol. 2(10): 754–761.

MacVicar, T. and T. Langer. 2016. OPA1 processing in cell death and disease-the long and short of it. J. Cell Sci. 129(12): 2297–2306.

Meeusen, S., J. M. McCaffery and J. Nunnari. 2004. Mitochondrial fusion intermediates revealed *in vitro*. Science. 305(5691): 1747–1752.

Mima, J. and W. Wickner. 2009. Complex lipid requirements for SNARE- and SNARE chaperone-dependent membrane fusion. J. Biol. Chem. 284(40): 27114–27122.

Mishra, P., V. Carelli, G. Manfredi and D. C. Chan. 2014. Proteolytic cleavage of Opa1 stimulates mitochondrial inner membrane fusion and couples fusion to oxidative phosphorylation. Cell Metab. 19(4): 630–641.

Mishra, P. and D. C. Chan. 2016. Metabolic regulation of mitochondrial dynamics. J. Cell Biol. 212(4): 379–387.

Nunnari, J. and A. Suomalainen. 2012. Mitochondria: In sickness and in health. Cell. 148(6): 1145–1159.

Orso, G., D. Pendin, S. Liu, J. Tosetto, T. J. Moss, J. E. Faust, M. Micaroni, A. Egorova, A. Martinuzzi, J. A. McNew and A. Daga. 2009. Homotypic fusion of ER membranes requires the dynamin-like GTPase atlastin. Nature. 460(7258): 978–983.

Pernas, L. and L. Scorrano. 2016. Mito-morphosis: Mitochondrial fusion, fission and cristae remodeling as key mediators of cellular function. Annu. Rev. Physiol. 78: 505–531.

Scott, B. L., J. S. Van Komen, S. Liu, T. Weber, T. J. Melia and J. A. McNew. 2003. Liposome fusion assay to monitor intracellular membrane fusion machines. Methods Enzymol. 372: 274–300.

Sugiura, S. and J. Mima. 2016. Physiological lipid composition is vital for homotypic ER membrane fusion mediated by the dynamin-related GTPase Sey1p. Sci. Rep. 6: 20407.

Weber, T., B. V. Zemelman, J. A. McNew, B. Westermann, M. Gmachl, F. Parlati, T. H. Söllner and J. E. Rothman. 1998. SNAREpins: Minimal machinery for membrane fusion. Cell. 92(6): 759–772.

Wickner, W. and R. Schekman. 2008. Membrane fusion. Nat. Struct. Mol. Biol. 15(7): 658–664.

CHAPTER 12

Virus-like Particles Expression in Silkworms

Vipin Kumar Deo[1,*] and *Enoch Y. Park*[2]

1. Introduction

Humans have always been fascinated with new materials (organic and inorganic) and have consistently produced novel materials. In fact, human development in each century has been characterized by the development of materials such as copper (Copper age), iron and steel (Iron age), etc. These materials have been processed in order to produce a wide variety of items with different applications. In the present century and in the near future, organic material of a biological nature can provide the next new materials which can help humankind make advances in science and society. Different kinds of materials of biological origin (biomaterials) are used increasingly in day-to-day life, for example in cosmetic materials, fabrics and eco-friendly materials for human use (non-medicinal and medicinal). Since long ago, sea sponges have been used primarily as scrubs in baths owing to their unique absorption properties, plus their natural aroma and aesthetic appeal. Recently, many applications in various healthcare and medicinal fields have also been discovered. Sponges produce

[1] Laboratory of Biotechnology, College of Global-Interdisciplinary Studies, Shizuoka University, Shizuoka 422-8529, Japan.
[2] Laboratory of Biotechnology, Research Institute of Green Science and Technology, Shizuoka University, 836 Ohya Suruga-ku, Shizuoka 422-8529, Japan.
E-mail: park.enoch@shizuoka.ac.jp
* Corresponding author: deo.vipin.kumar@shizuoka.ac.jp

unique antibiotics based upon their flora and fauna and some, such as the Okinawa marine sponge *Theonella* sp. Collected from Ie island, have been found to be suitable in treating various forms of cancers and other bacterial and viral infections (Donia and Hamann 2003, Vinothkumar and Parameswaran 2013).

Progress in biology is, in a way, related to the size of different materials being used, as shown in Fig. 1A. Similarly, viruses have been known to humans for a long time, as they are a cause of much sorrow due to their high mortality rates across continents. With advancements in science, viruses can now be isolated and classified very precisely and distinctively. As a result, clear understanding of the virus and its mechanism of entry into the cells became possible, leading researchers to focus on the architecture of the virus (Ludwig and Wagner 2007). Now we have considerable X-ray crystallography data of different viruses in bound and un-bounded form to ligands giving a detailed general outlook (morphology, presence/absence of lipid layer, size, etc.). The above information, coupled with data provided by cryo and transmission electron microscopy, respectively, has presented a more holistic summary of viruses (Akarsu et al. 2003, Deo et al. 2015, Harris et al. 1999, Latham and Galarza 2001, Nandhagopal et al. 2004, Scheifele et al. 2007). Viruses have been classified based on the genetic material content, RNA or DNA. The genetic material can be single strand or double strand, based on which, the classification of viruses takes

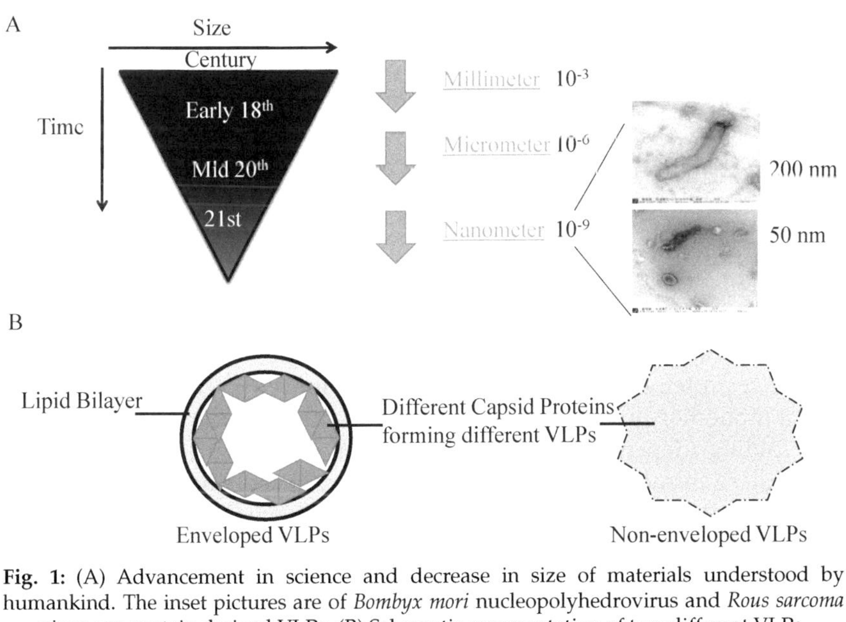

Fig. 1: (A) Advancement in science and decrease in size of materials understood by humankind. The inset pictures are of *Bombyx mori* nucleopolyhedrovirus and *Rous sarcoma* virus gag protein derived VLPs. (B) Schematic representation of two different VLPs.

place (using public database like PUBMED) (Dong et al. 2017). The initial studies or interest in virology was for developing vaccine candidates but, of late VLPs have also shown potential as a drug delivery system (Kan-Davelaar et al. 2014, Keswani et al. 2013, Pokorski and Steinmetz 2011). The focus of this chapter is virus-like particles (VLPs), which cannot be classified as a virus but have the dual potential of being a vaccination candidate and being efficient drug delivery systems. The characteristics, expression and purification of VLPs in silkworms is the primary focus of this chapter.

VLPs have been expressed in different expression systems like plant (Santi et al. 2006), bacterial, yeast, animal cells, insect cells and silkworms, using appropriate expression vectors (Kim and Kim 2017). The choice of expression system depends upon a number of variables, such as the type of VLPs (enveloped or non-enveloped), capsid protein expression and its post-translational modifications in host cells and finally the protein trafficking (localization of VLPs) of the proteins for VLPs production. Different expression systems produce VLPs with different production efficiencies, and no known universal expression system is available for VLPs. Here we shall briefly discuss the suitability of different expression systems and compare them with VLPs expression in silkworms.

2. Enveloped and Non-enveloped Coat Proteins Role in VLPs Formation

Viruses are extremely deadly, owing to their nanoscale size and the fact that they have evolved to specifically target receptors on the mammalian cells (Chen et al. 2006, Singh et al. 2006, Molino and Wang 2014). If VLPs are to be made, then they can, in theory, retain all the properties of the virus but not cause any infection due to lack of lethal genetic material (Akarsu et al. 2003, Deo et al. 2015, Harris et al. 1999, Latham and Galarza 2001, Nandhagopal et al. 2004, Scheifele et al. 2007). Such a concept has been tested and many researchers have found the idea very attractive for further research as a vaccination and drug delivery systems platform (Kan-Davelaar et al. 2014, Krishnamachari et al. 2011).

Coat proteins are the backbone on which a virus structure is based; it is a monomer protein with the ability to self-assemble. The expression and folding (dependent upon the minimum energy (ΔG)) of the protein takes place intracellularly, guided by the signal sequence encoded in its sequence (Pokorski and Steinmetz 2011). Thus, completely folded coat protein is assembled inside the cell and, based upon whether a virus has a lipid layer or not, translocated to plasma membrane or kept in the endoplasmic reticulum, respectively (Nadaraia-Hoke et al. 2013, Parent

2011, Wang et al. 2010, Wills et al. 1994). Self-assembly takes place in the cell and the genetic material is packaged during the structure assembly assisted by nucleo-capsid proteins. Their role is restricted to packaging of the genetic material, which we shall not discuss as this chapter covers only the role of capsid proteins involved in VLP formation. The translocation of protein is driven by the presence of signal peptide, which is present in the original virus coat protein. Hence, the assembly of coat protein is self-driven, independent from viral genetic material and this property makes the coat protein a very attractive material for making the next generation of biomaterial. An ideal biomaterial should self-assemble, and coat proteins are the closest approximation of this idea. When the coat protein alone expressed using plants, yeast, bacterial, mammalian and baculovirus expression systems, it produced VLPs (Fuenmayor et al. 2017, Liu et al. 2013). Since they represent the parent virus closely, they activate the immune system without any chance of causing the infection to spread. The immune system activation by VLPs is still under research but from the known facts available, it is understood to activate both the innate and adaptive immune systems (Fiers et al. 2004, Ludwig and Wagner 2007). Such approach has been verified and is very beneficial in developing the next generation of vaccines (Cid-Arregui et al. 2003, Kang et al. 2012, Noad and Roy 2003, Quan et al. 2008, Zhang et al. 2014).

Due to presence and absence of lipid bilayer, the VLPs as shown in Fig. 1B can be broadly classified into two groups; enveloped VLPs and non-enveloped VLPs. The presence and absence of lipid bilayers depends upon the signal sequence associated with the capsid protein and membrane associated amino acid region (hydrophobic), plus capsid protein localization plays an important role too. Most of the enveloped VLP capsid proteins translocate to the plasma membrane or are localized in the ER-Golgi region during the protein trafficking. As a result, during the budding of VLPs they are enveloped with lipid bilayers present in those regions, whereas the non-enveloped VLPs are localized in cytosols and then released outside either by exocytosis or the disruption of the cell integrity (like cell death). The presence of a lipid bilayer has merits, such as the ability of the lipid bilayer to display other proteins on its surface, packaging of the VLPs using large unilamellar vesicles. The demerits of enveloped VLPs are that other proteins are displayed on the membrane controlled by the host cell. In addition, the purification process of enveloped VLPs is hindered by the presence of a lipid bilayer, as rigorous pH, salt or ions exchange-based columns cause damage to the integrity of the VLP's macromolecular structure. Similarly, the non-enveloped VLPs have merits, like being easy to purify using the rigorous methods mentioned above. The prominent demerit associated with non-enveloped VLPs is that they do not have scope for surface modification, as a result their scope of applications is restricted to vaccination only. If other applications are to be considered, then the exposed coat protein surface needs to be engineered either by chemical modifications (after expression and purification) or by producing a recombinant coat protein fused with a target. Both approaches can cause serious damage to the coat protein structure stability, hence, non-enveloped VLPs cannot be efficiently produced without danger to the VLPs structure itself.

Presence and absence of a lipid bilayer on VLPs is an important factor during the VLPs expression in any expression system. Various researchers have used different expression systems as mentioned above and from the literature review; it is observed that the non-enveloped VLPs are best suited for production in plants, yeast and bacterial expression systems. The primary reason being that the VLPs localize in the Vacuoles, Cytosols and Endoplasmic reticulum or Golgi bodies, respectively, after undergoing the necessary post-translational modifications. The specific reason for localization to these compartments in different expression systems is related to the amino acid sequence of the capsid proteins, which form non-enveloped VLPs. Pupae of Silkworms are ideal for production of non-enveloped VLPs as they have high fat body content. As a result of the high fat body content, the cell machinery expresses proteins and they are localized intracellularly.

3. Development of Proteins Expression System in Silkworms

In this section, we shall focus on Bacmid based expression in silkworms or insect cells. The earliest development in the use of insect cell expression systems began with the discovery of baculovirus. However, the process of producing recombinant baculovirus was very tedious and time-consuming, leading to research into modifications in the baculovirus genome and insertion of a blue/white screening system. Later adoption of the transposon-based transfer of cDNA into the baculovirus genome increased the efficiency and screening of recombinant baculovirus. Finally, Bacmid (a shuttle vector with baculovirus genome and *ori*) can replicate in bacteria was developed which brought rapid development in expression of foreign protein in silkworms and insect cells. All this knowledge was applied in producing bacmid (Motohashi et al. 2005) for *Bombyx mori* Nucleopolyhedrovirus (BmNPV), leading to a robust system for the expression and purification of protein from silkworms, as shown in Fig. 2. This simplifies the cloning in of any cDNA using modern recombinant cloning techniques and then screening and amplifying the bacmid in bacteria. The recombinant bacmid can then easily be transfected with high efficiency in silkworms. The whole process takes less than 1 month and can be scaled up to an industrial level.

The life cycle of silkworm is as shown schematically in Fig. 3, and it requires a 65% humidity chamber with ambient temperature of 25°C for proper growth. Within a space of 2–3 weeks, the silkworms can be injected with bacmids and proteins purified efficiently.

Expression of VLPs in silkworms requires minimalistic equipment for protein expression, as shown in Fig. 4. Since there are no known diseases or infections in silkworms which can affect humans, they are completely safe to handle. A recombinant bacmid carrying the cDNA for the capsid protein (enveloped or non-enveloped) is injected into a silkworm and then fed and reared for 5 days post injection. Finally, the haemolymph is collected from silkworm; this will be used to obtain VLPs.

As explained schematically in Fig. 4 group antigen protein is used as a model for VLPs formation. The gag protein is a conserved capsid protein

Early 90s	**Discovery and characterization of Baculovirus** ↓
Middle of 90s	**Modification in screening of recombinant Baculovirus** ↓ **Discovery of Transposons**
Late 90s	**Development of Bacmid for Baculovirus based infection** ↓
Early of 2000	**Development of Transposons based insect cell expression system** ↓ **Developed first BmNPV based bacmid, Increased the host range of AcNPV bacmid, Created protease deficient bacmid based insect expression system** ↓
Present	**Present day Stable insect cell and bacmid based expression system** ↓ **Using bacmid based expression system to produce VLPs**

Fig. 2: Schematic representation of events leading to development of bacmid expression system for silkworms.

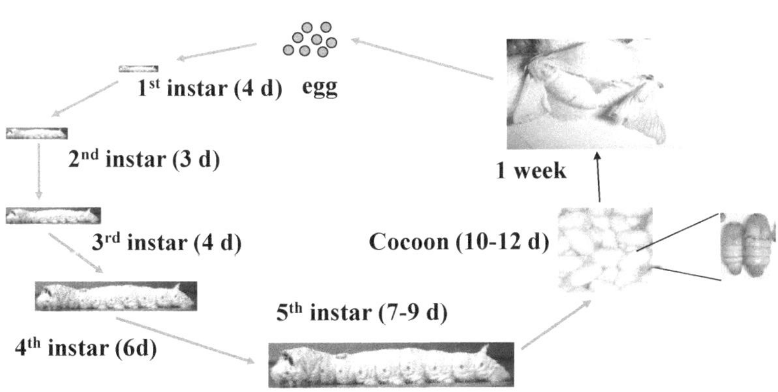

Fig. 3: Schematic representation of different development stages of silkworms.

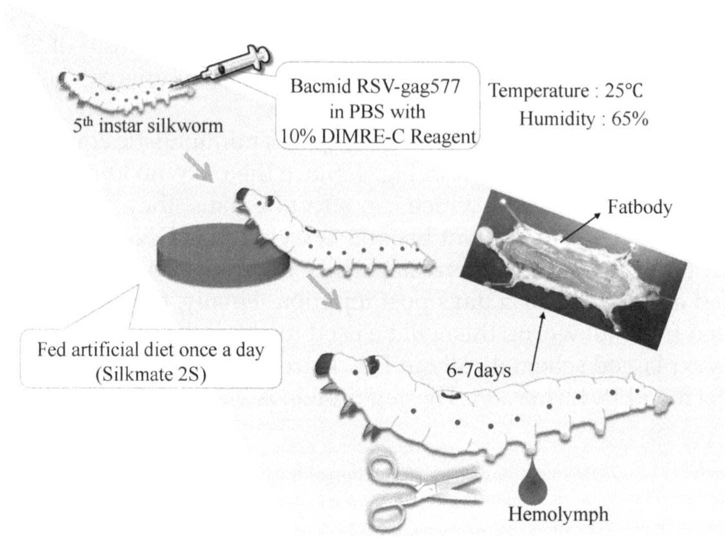

Fig. 4: Schematic representation of expression of proteins in silkworms.

belonging to the family of Retroviridae *Rous Sarcoma* virus (Nandhagopal et al. 2004, Parent 2011, Scheifele et al. 2007, Xiang et al. 1997, Yu et al. 2001). In order to express gag protein in the silkworms, the gag cDNA has to be inserted into a bacmid which, in turn, can be used to transfect insect cell or silkworm. Bacmid contain transposons flanking the p10/ Polyhedrin/immediate early or other (the promoters can be changed as per requirements) promotor with a cassette to insert foreign cDNA in frame for expression and are commercially available (Deo et al. 2006, 2011). The bacmid transfected into the silkworms transposes the cassette into the host genome where the transcription of cDNA takes place. Once the translation and post-translation modifications of the protein are completed, the protein is ready for VLP formation. VLP formation takes place inside the cell and, depending upon the signal peptide of coat protein, migrates towards either plasma membrane, endoplasmic reticulum or cytoplasm of the cell, respectively. For example, each gag protein, after being processed independently, moves towards the plasma membrane owing to the signal peptide and attaches to the plasma membrane (Zimmerberg and Kozlov 2005). The gag protein (as well as other coat proteins) has different regions, Matrix (MA) associated region (interacts with lipid bilayer), capsid (CA) associated region (interacts with other capsid associated region) and the protease cleavage rich region (interacts with the genetic material) (Ghanam et al. 2012, Wills et al. 1994, Yu et al. 2001). The protease region is redundant and can be removed since VLPs are designed to be empty cages for carrying drugs or dyes for DDS, or serve as a vaccination platform. In case of gag protein, CA-CA interaction assists in binding all the gag proteins to prepare the VLP shell structure during budding. Each monomeric protein aggregates at the plasma membrane, leading to a decrease in surface tension of the membrane. This causes the membrane to bleb, as shown in the Fig. 5 (Bennett and Tieleman 2013, Chang et al. 2008, Drin and Antonny 2010, Maeda et al. 2001, Zimmerberg and Kozlov 2005). The budding process and size (diameter) of the VLP formation is dependent upon the surface tension. The size also depends upon the protein displayed on the lipid bilayer. The presence of bilayer is beneficial to the VLPs as it can be used to display different proteins on its surface, thus also making it an appropriate tool for display of different proteins. In the model case given, three different proteins are anchored (Chang et al. 2008, Deo et al. 2016, Metzner et al. 2008, Paulick and Bertozzi 2008). The selection of appropriate anchors is very important, while constructing the recombinant bacmid the signal sequence for the anchor can be incorporated. Not all known anchors can be used, only the ones which can be processed by the cell machinery available in silkworms. Further research is needed in order to elucidate which anchors are available in silkworms.

The size of VLPs expressed in silkworms depends upon the coat protein, the anchors used and the number of proteins displayed. We have observed that there is a size shift in VLPs to larger or smaller size (diameter in nanometer), depending upon the anchor used. For example, glycophosphoinositol or Hemagglutinin trans membrane anchors causes increase or decrease, respectively, in the size of VLPs compared to VLPs not displaying any protein. It is still not known why the size of VLPs is

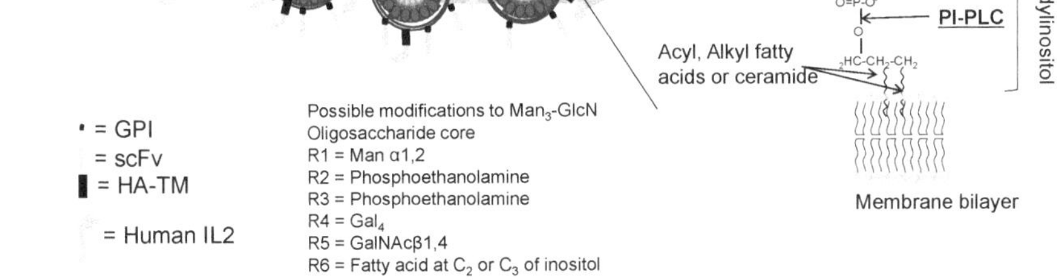

Fig. 5: Schematic representation of co-expression of gag, short chain fragment region (scFv) and human interleukin-2 proteins in silkworms. All the components of VLPs assemble at the plasma membrane including two different anchors glycophosphoinositol (GPI) and hemagglutinin trans-membrane region (HA-TM).

related to the above criteria, but our hypothesis is that the surface tension of the lipid bilayer is affected during the budding process of VLPs. This step is vital for VLP expression in silkworms and the choice of anchors and the number of different proteins to be displayed on a single VLP is limited as a result.

4. VLPs as a Platform for Vaccination

VLPs formed using coat proteins (owing to their self-folding ability), with enveloped and non-enveloped VLPs produced in silkworms as shown in Table 1, can be considered as prime candidates for vaccination platforms. Owing to the presence of the lipid bilayer in enveloped VLPs, the immune system is unable to recognize the capsid protein, however, the antigenic protein displayed on the surface serves as the primary antigen. As a result, enveloped VLPs displaying the antigen are the ideal candidate. These can be produced by co-expressing 2 or more proteins in silkworms, as mentioned above. Due to the absence of lipid bilayer in non-enveloped VLPs, the VLP itself serves as the antigen. All the exposed surface of the non-enveloped VLPs are a prospective target of the immune system. VLPs are recognized by both innate and adaptive immune systems in animals and humans alike. Another advantage is that in order to initiate the immune system, no added adjuvants are required. Many researchers have shown that in animal experiments, the efficacy of the VLPs produced in silkworms makes them good candidates for vaccination platforms.

Coat proteins (either one or more) form a stable structure. The understanding of this phenomenon leads to the question: Can two or more coat proteins (from different virus families) join together to

Table 1: Different capsid proteins expressed in silkworms to produce VLPs (enveloped/non-enveloped).

Name of Capsid Protein to Produce VLPs	Virus Family (Enveloped/Non-enveloped)	Reference
VP2	*Parvoviridae* (non-enveloped)	Feng et al. 2011
GAG	*Retroviridae* (enveloped)	Deo et al. 2011
S1, S3	*Reoviridae* (non-enveloped)	Chakrabarti et al. 2010
GAG (Rous Sarcoma), M1 (Influenza)	*Retroviridae* (enveloped)	Deo et al. 2015
L1	*Papillomaviridae* (non-enveloped)	Paliniyadi et al. 2012
VP60	*Caliciviridae* (non-enveloped)	Zheng et al. 2016
M, E	*Flaviviridae* (non-enveloped)	Matsuda et al. 2017

(Note) E: Envelope protein, GAG: group antigen protein, L: Late protein, M1: Matrix protein, S: Spike protein, VP: viral protein.

form new VLPs? This approach has wide applications as a universal vaccination platform against one or multiple viruses (Fiers et al. 2004, Solórzano et al. 2010, Zheng et al. 2013). Several coat proteins, such as M1 and gag, are known to interact with each other and form VLPs when co-expressed in silkworms (Deo et al. 2015). Both the proteins belong to two different viruses, Influenza and Rous sarcoma virus, respectively, and they interact with each other to form VLPs. The capsid-associated region of gag interacts with M1 to form VLPs which resemble both gag and M1-based VLPs, respectively, with characteristic changes, for example, in size (smaller). A new biomaterial such as this can serve as a novel biomaterial for vaccination.

5. VLPs Application in Drug Delivery Systems

Ailment or disease is treated with different active pharmaceutical ingredients (APIs) or drugs (e.g., chemical molecules, antibodies, peptides) in modern medicine (Hoelder et al. 2012, Mihich and Ehrke 2000, Sim and Radvanyi 2014). These drugs undergo rigorous research and development, followed by stringent clinical trials (very time-consuming and expensive). As a result, there are various health regulatory bodies controlling each step of drug development. The rules and standards of these regulatory bodies differ from country to country. Despite this lengthy process, there is a large repertoire of drugs around the world. All these drugs have working parameters, wherein the concentrations at which they work are limited, plus, they are toxic to normal cells. This major issue can be controlled with proper use of drug delivery systems (DDS). As a result, a lot of research and development is taking place in this field. VLPs (both enveloped and non-enveloped) are empty cages which can be packaged using different techniques, thus providing a viable next generation drug delivery platform.

The efficiency of APIs/drugs is directly related to the efficiency of the DDS platform, as shown in Fig. 6A. Generally, the DDS transports the drug to the afflicted area with more accuracy and is as a result less toxic. The specificity of the DDS is influenced by many factors, such as the targeting of specific receptors and affinity for specific cells or tissues. A non-specific DDS transports the APIs/drugs into a host circulatory system and stays active for certain period of time (depending upon its formulation). The APIs/drugs should then reach the necessary target within that time. The concentration has to reach a critical amount in terms of the body weight (proportional to the amount of fluid in the host) before the drug starts showing effect upon the target. Since they are non-specific in nature, they tend to harm the healthy cells as well as the affected cells, causing harmful toxicity. Ideally, when a DDS is very efficient (say 99%), then the amount

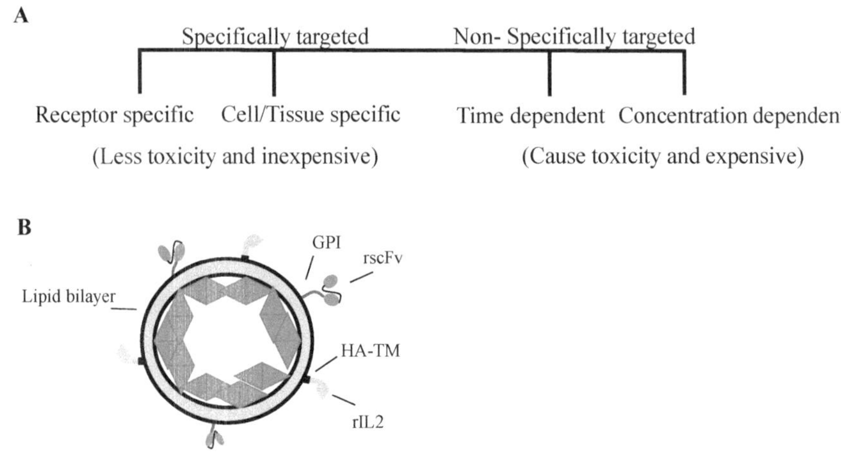

Fig. 6: (A) Broad classification of different types of DDS based on the their specific targeting. VLPs fall under specifically targeted DDS. (B) Schematic representation of enveloped VLPs displaying two different proteins with DDS application. Such VLPs can deliver drugs on target site plus induce innate immune system by attracting macrophages using interleukins chemokine properties.

of APIs/drugs needed for therapy/treatment is very little, reducing toxic side effects and bring down the overall cost of the treatment of the disease many-fold. The above-described DDS is further complicated by the mode of delivery (for specific & non-specific) which can be either oral, injection, nasal spray or an ointment (for adsorption through epidermal layer), while the target is in a tissue or cell. Finally, the hurdle of half-life (time the APIs/drugs spend in the host system) is very important too. A short and stable half-life is ideal, during which the target should be achieved. VLPs (both enveloped and non-enveloped) satisfy the necessity for ideal DDS. Owing to their small size, VLPs can easily diffuse through epithelial cells of the blood circulatory system and other tissues; this phenomenon is called effect of size. This property, coupled with the ability to home in on specific targets via the use of peptides displayed on VLPs, antibodies or receptors, as shown in the Fig. 6B, make VLPs ideal DDS. Using different chemokines combined with packaged APIs/drugs which can be delivered on-site can reduce the side effects and also improve the therapy (Arenberg et al. 2000, Robertson 2002, Roda et al. 2006, Loetscher et al. 1996, Ward et al. 1998). Researchers have also shown the potential of combining the chemokines, a part of the innate immune system, with drugs in order to target cancers efficiently. Such innovative new therapy models are possible due to the existence of VLPs. The packaging of enveloped VLPs can be done using methods like the fusion of VLPs with large uni-lamellar vesicles that carry the drugs, electroporation or any technique which allows the lipid bilayer to be penetrated and have material inserted inside it (Fukushima et al. 2008, Kamiya et al. 2010, Keswani et al. 2013, Sarker et al. 2014, Tsumoto and Yoshimura 2009). Packaging the non-enveloped VLPs with drugs is more of a challenge because they have no lipid bilayer. In order to package non-enveloped VLPs with drugs, exploiting the hydrophobic/hydrophillic nature of proteins in VLPs structure or the ionic charge of the proteins is the best option. There is still not a well-established protocol for packaging of drugs on VLPs, and currently no known VLPs packaged with drugs are being used commercially. The most important need is efficient and large-scale production of VLPs. Silkworm-based expression system has increased the enveloped VLPs production, but many hurdles still remain, like purity of VLPs, costs and versatility to produce different VLPs from other capsid proteins.

6. Conclusion

In this chapter, we briefly highlighted the role of new a nano-biomaterial called VLPs, as shown in Fig. 7. There properties are unique due to their nano-scale structure, and the ability to modify their surface makes them ideal for DDS and vaccination platforms. Utilizing silkworm as an expression factory for VLPs is suitable, since all the necessary post-translational modifications necessary for proper folding of capsid protein for formation of VLPs can be performed. The ease of handling and scaling-up of VLP expression in silkworm make it ideal for commercial application. A large-scale research, focusing on using VLPs in vaccinations, has generated valuable information and several non-enveloped VLPs, as shown in Table 1. For example, GARDASIL (human papillomavirus 9-valent VLP vaccine) has already been commercialized, and development

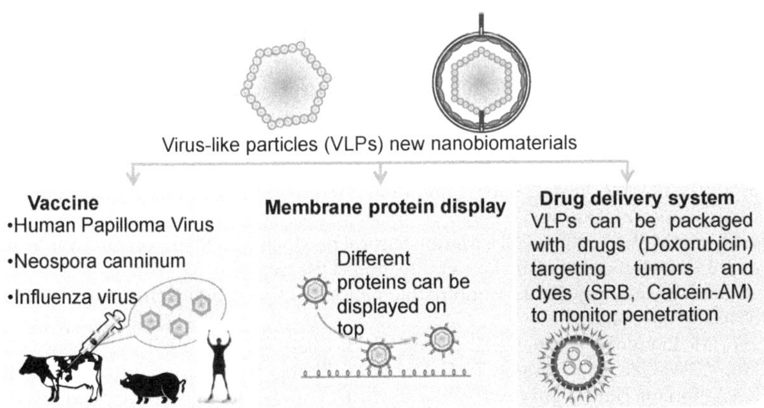

Fig. 7: Schematic representation of VLPs (enveloped and non-enveloped applications). Various proteins have been displayed on VLPs and they can be packaged with different drugs and for vaccinations too.

of other products for animal care is in progress at various levels. Further work is necessary in order to increase production of VLPs and improve the purification yield so as to develop VLPs as a DDS platform.

References

Akarsu, H., W. P. Burmeister, C. Petosa, I. Petit, C. W. Müller, R. W. Ruigrok and F. Baudin. 2003. Crystal structure of the M1 protein-binding domain of the influenza A virus nuclear export protein (NEP/NS2). The EMBO J. 22(18): 4646–4655.

Arenberg, D. A., M. P. Keane, B. DiGiovine, S. L. Kunkel, S. R. Strom, M. D. Burdick, M. D. Iannettoni and R. M. Strieter. 2000. Macrophage infiltration in human non-small-cell lung cancer: The role of CC chemokines. Cancer Immunol. Immunother. CII 49(2): 63–70.

Bennett, W. F. and P. Tieleman. 2013. Computer simulations of lipid membrane domains. Biochimica et Biophysica Acta (BBA)—Biomembranes. 1828: 1765–1776.

Chakrabarti, M., S. Ghorai, S. K. K. Mani and A. K. Ghosh. 2010. Molecular charecterization of genome segments 1 and 3 encoding two capsid proteins of Antheraea mylitta cytoplasmic polyhedrosis virus. J. Virol. 7: 181–192.

Chang, D. -K., S. -F. Cheng, E. Kantchev, C. -H. Lin and Y. -T. Liu. 2008. Membrane interaction and structure of the transmembrane domain of influenza hemagglutinin and its fusion peptide complex. BMC Biol. 6(2): 1–12.

Chen, C., M. -C. Daniel, Z. T. Quinkert, M. De, B. Stein, V. D. Bowman, P. R. Chipman, V. M. Rotello, C. C. Kao and B. Dragnea. 2006. Nanoparticle-templated assembly of viral protein cages. Nano Lett. 6(4): 611–615.

Cid-Arregui, A., V. Juárez and H. Z. Hausen. 2003. A synthetic E7 gene of human papillomavirus type 16 that yields enhanced expression of the protein in mammalian cells and Is useful for DNA immunization studies. J. Virol. 77(8): 4928–4937.

Deo, V. K., M. Hiyoshi and E. Y. Park. 2006. Construction of hybrid *Autographa californica* nuclear polyhedrosis bacmid by modification of p143 helicase. J. Virol. Methods. 134(1-2): 212–216.

Deo, V. K., Y. Tsuji, T. Yasuda, T. Kato, N. Sakamoto, H. Suzuki and E. Y. Park. 2011. Expression of an RSV-gag virus-like particle in insect cell lines and silkworm larvae. J. Virol. Methods. 177(2): 147–152.

Deo, V. K., T. Kato and E. Y. Park. 2015. Chimeric virus-like particles made using GAG and M1 capsid proteins providing dual drug delivery and vaccination platform. Mol. Pharm. 12(3): 839–845.

Deo, V. K., T. Kato and E. Y. Park. 2016. Virus-like particles displaying recombinant short-chain fragment region and interleukin 2 for targeting colon cancer tumors and attracting macrophages. J. Pharm. Sci. 105(5): 1614–1622.

Dong, R., H. Zheng, K. Tian, S. -C. C. Yau, W. Mao, W. Yu, C. Yin, C. Yu, R. L. He, J. Yang and S. S. T. Yau. 2017. Virus database and online inquiry system based on natural vectors. Evol. Bioinform. Online. 13: 1–7.

Donia, M. and M. T. Hamann. 2003. Marine natural products and their potential applications as anti-infective agents. The Lancet Infectious Diseases. 3(6): 338–348.

Drin, G. and B. Antonny. 2010. Amphipathic helices and membrane curvature. FEBS Lett. 584(9): 1840–1847.

Feng, H., M. Liang, H. -L. Wang, T. Zhang, P. -S. Zhao, X. -J. Shen, R. -Z. Zhang, G. -Q. Hu, Y. -W. Gao, C. -Y. Wang, T. -C. Wang, W. Zhang, S. -T. Yang and X. -Z. Xia. 2011. Recombinant canine parvovirus-like particles express foreign epitopes in silkworm pupae. Vet. Microbiol. 154(1-2): 49–57.

Fiers, W., M. D. Filette, A. Birkett, S. Neirynck and W. M. Jou. 2004. A 'universal' human influenza A vaccine. Virus Res. 103: 173–176.

Fuenmayor, J., F. Gòdia and L. Cervera. 2017. Production of virus-like particles for vaccines. New Biotechnol. 39(Pt B): 174–180.

Fukushima, H., M. Mizutani, K. Imamura, K. Morino, J. Kobayashi, K. Okumura, K. Tsumoto and T. Yoshimura. 2008. Development of a novel preparation method of recombinant proteoliposomes using baculovirus gene expression systems. J. Biochem. 144(6): 763–770.

Ghanam, R. H., A. B. Samal, T. F. Fernandez and J. S. Saad. 2012. Role of the HIV-1 matrix protein in gag intracellular trafficking and targeting to the plasma membrane for virus assembly. Front. Microbiol. 3: 55.

Grgacic, E. V. L. and D. A. Anderson. 2006. Virus-like particles: Passport to immune recognition. Methods. 40(1): 60–65.

Harris, A., B. Sha and M. Luo. 1999. Structural similarities between influenza virus matrix protein M1 and human immunodeficiency virus matrix and capsid proteins: An evolutionary link between negative-stranded RNA viruses and retroviruses. The J. Gen. Virol. 80(Pt 4): 863–869.

Hoelder, S., P. A. Clarke and P. Workman. 2012. Discovery of small molecule cancer drugs: successes, challenges and opportunities. Mol. Oncol. 6(2): 155–176.

Kamiya, K., K. Tsumoto, S. Arakawa, S. Shimizu, I. Morita, T. Yoshimura and K. Akiyoshi. 2010. Preparation of connexin43-integrated giant liposomes by a baculovirus expression-liposome fusion method. Biotechnol. Bioeng. 107(5): 836–843.

Kang, S. -M., J. -M. Song, F. -S. Quan and R. Compans. 2009. Influenza vaccines based on virus-like particles. Virus Res. 143: 140–146.

Kang, S. -M., M. -C. Kim and R. Compans. 2012. Virus-like particles as universal influenza vaccines. Expert Rev. Vaccines. 11(8): 995–1007.

Kan-Davelaar, H. E. V., J. C. M. Hest, J. J. L. M. Cornelissen and M. S. T. Koay. 2014. Using viruses as nanomedicines. Br. J. Pharmacol. 171(17): 4001–4009.

Keswani, R. K., I. M. Pozdol and D. W. Pack. 2013. Design of hybrid lipid/retroviral-like particle gene delivery vectors. Mol. Pharm. 10(5): 1725–1735.

Kim, H. J. and H. -J. J. Kim. 2017. Yeast as an expression system for producing virus-like particles: What factors do we need to consider? Lett. Appl. Microbiol. 64(2): 111–123.

Krishnamachari, Y., S. M. Geary, C. D. Lemke and A. K. Salem. 2011. Nanoparticle delivery systems in cancer vaccines. Pharm. Res. 28(2): 215–236.

Latham, T. and J. M. Galarza. 2001. Formation of wild-type and chimeric influenza virus-like particles following simultaneous expression of only four structural proteins. J. Virol. 75(13): 6154–6165.

Liu, F., X. Wu, L. Li, Z. Liu and Z. Wang. 2013. Use of baculovirus expression system for generation of virus-like particles: Successes and challenges. Protein Expr. Purif. 90: 104–116.

Loetscher, B. P., M. Seitz, M. Baggiolini and B. Moser. 1996. Interleukin-2 regulates CC chemokine receptor expression and chemotactic responsiveness in T lymphocytes. J. Exp. Med. 184(2): 569–577.

Ludwig, C. and R. Wagner. 2007. Virus-like particles-universal molecular toolboxes. Curr. Opin. Biotechnol. 18(6): 537–545.

Maeda, J., J. F. Repass, A. Maeda and S. Makino. 2001. Membrane topology of coronavirus E protein. Virology. 281(2): 163–169.

Matsuda, S., R. Nerome, K. Maegawa, A. Kotaki, S. Sugita, K. Kawasaki, K. Kuroda, R. Yamaguchi, T. Takasaki and Kuniaki Neromea. 2017. Development of a Japanese encephalitis virus-like particle vaccine in silkworms using codon-optimised prM and envelope genes. Heliyon. 3(4): e00286.

Metzner, C., B. Salmons, W. Günzburg and J. Dangerfield. 2008. Rafts, anchors and viruses—a role for glycosylphosphatidylinositol anchored proteins in the modification of enveloped viruses and viral vectors. Virology. 382(2): 125–131.

Mihich, E. and M. J. Ehrke. 2000. Anticancer drugs plus cytokines: Immunodulation based therapies of mouse tumors. Int. J. Immunopharmacol. 22(12): 1077–1081.

Molino, N. M. and S. W. Wang. 2014. Caged protein nanoparticles for drug delivery. Curr. Opin. Biotechnol. 28: 75–82.

Motohashi, T., T. Shimojima, T. Fukagawa, K. Maenaka and E. Y. Park. 2005. Efficient large-scale protein production of larvae and pupae of silkworm by *Bombyx mori* nuclear polyhedrosis virus bacmid system. Biochem. Biophys. Res. Commun. 326(3): 564–569.

Nadaraia-Hoke, S., D. V. Bann, T. L. Lochmann, N. Gudleski-O'Regan and L. J. Parent. 2013. Alterations in the MA and NC domains modulate phosphoinositide-dependent plasma membrane localization of the Rous sarcoma virus gag protein. J. Virol. 87(6): 3609–3615.

Nandhagopal, N., A. A. Simpson, M. C. Johnson, A. B. Francisco, G. W. Schatz, M. G. Rossmann and V. M. Vogt. 2004. Dimeric Rous sarcoma virus capsid protein structure relevant to immature gag assembly. J. Mol. Biol. 335: 275–282.

Noad, R. and P. Roy. 2003. Virus-like particles as immunogens. Trends Microbiol. 11(9): 438–444.

Palaniyandi, M., T. Kato and E. Y. Park. 2012. Expression of human papillomavirus 6b L1protein in silkworm larvae and enhanced greenfluorescent protein displaying on its virus-like particles. SpringerPlus. 1: 29–35.

Parent, L. J. 2011. New insights into the nuclear localization of retroviral gag proteins. Nucleus (Austin, Tex.). 2(2): 92–97.

Paulick, M. G. and C. R. Bertozzi. 2008. The glycosylphosphatidylinositol anchor: A complex membrane-anchoring structure for proteins. Biochem. 47: 6991–7000.

Pokorski, J. K. and N. Steinmetz. 2011. The art of engineering viral nanoparticles. Mol. Pharm. 8(1): 29–43.

Quan, F. S., D. Steinhauer, C. Huang, T. M. Ross, R. W. Compans and S. M. Kang. 2008. A bivalent influenza VLP vaccine confers complete inhibition of virus replication in lungs. Vaccine. 26: 3352–3361.

Robertson, M. J. 2002. Role of chemokines in the biology of natural killer cells. J. Leukoc. Biol. 71(2): 173–183.

Roda, J. M., R. Parihar, C. Magro, G. J. Nuovo, S. Tridandapani and W. E. Carson. 2006. Natural killer cells produce T cell-recruiting chemokines in response to antibody-coated tumor cells. Cancer Res. 66(1): 517–526.

Santi, L., Z. Huang and H. Mason. 2006. Virus-like particles production in green plants. Methods. 40(1): 66–76.

Sarker, S. R., R. Hokama and S. Takeoka. 2014. Intracellular delivery of universal proteins using a lysine headgroup containing cationic liposomes: deciphering the uptake mechanism. Mol. Pharm. 11: 164–174.

Scheifele, L. Z., S. P. Kenney, T. M. Cairns, R. C. Craven and L. J. Parent. 2007. Overlapping roles of the Rous sarcoma virus gag p10 domain in nuclear export and virion core morphology. J. Virol. 81(19): 10718–28.

Sim, G. C. and L. Radvanyi. 2014. The IL-2 cytokine family in cancer immunotherapy. Cytokine & Growth Factor Rev. 25(4): 377–390.

Singh, P., G. Destito, A. Schneemann and M. Manchester. 2006. Canine parvovirus-like particles, a novel nanomaterial for tumor targeting. J. Nanobiotechnol. 4(2): 1–11.

Solórzano, A., J. Ye and D. R. Pérez. 2010. Alternative live-attenuated influenza vaccines based on modifications in the polymerase genes protect against epidemic and pandemic flu. J. Virol. 84(9): 4587–4596.

Tsumoto, K. and T. Yoshimura. 2009. Chapter 5 recombinant proteoliposomes prepared using baculovirus expression systems. Methods Enzymol. Vol. 465: 95–109.

Vinothkumar, S. and P. S. Parameswaran. 2013. Recent advances in marine drug research. Biotechnol. Adv. 31(8): 1826–1845.

Wang, D., A. Harmon, J. Jin, D. H. Francis, J. Christopher-Hennings, E. Nelson, R. C. Montelaro and F. Li. 2010. The lack of an inherent membrane targeting signal is responsible for the failure of the matrix (M1) protein of influenza A virus To bud into virus-like particles. J. Virol. 84(9): 4673–4681.

Ward, S. G., K. Bacon and J. Westwick. 1998. Chemokines and T lymphocytes more than an attraction. Immunity. 9(1): 1–11.

Wills, J. W., C. E. Cameron, C. B. Wilson, Y. Xiang, R. P. Bennett and J. Leis. 1994. An assembly domain of the Rous sarcoma virus gag protein required late in budding. J. Virol. 68(10): 6605–6618.

Yan, X., T. W. Ridky, N. K. Krishna and J. Leis. 1997. Altered Rous sarcoma virus gag polyprotein processing and its effects on particle formation. J. Virol. 71(3): 2083–2091.

Yu, F., J. Swati, M. Yu, K. Richard, S. Martha and V. Vogt. 2001. Characterization of Rous sarcoma virus gag particles assembled *in vitro*. J. Virol. 75(6): 2753–2764.

Zhang, N., J. Shibo and L. Du. 2014. Current advancements and potential strategies in the development of MERS-CoV vaccines. Expert Rev. Vaccines. 13(6): 761–774.

Zheng, M., J. Luo and Z. Chen. 2013. Development of universal influenza vaccines based on influenza virus M and NP genes. Infection. 42(2): 251–262.

Zheng, X., S. Wang, W. Zhang, X. Liu, Y. Yi, S. Yang, X. Xia, Y. Li and Z. Zhang. 2016. Development of a VLP-based vaccine in silkworm pupae against rabbit hemorrhagic disease virus. Int. Immunopharm. 40: 164–169.

Zimmerberg, J. and M. Kozlov. 2005. How proteins produce cellular membrane curvature. Nature Reviews Mol. Cell Biol. 7(1): 9–19.

CHAPTER 13

Gene Delivery Based on *Bombyx mori* Nucleopolyhedrovirus (BmNPV)

*Tatsuya Kato** and *Enoch Y. Park**

1. Introduction

In general, baculoviruses have been utilized in the production of recombinant proteins in insect cells and insects. In addition, it has been reported that baculoviruses can enter mammalian cells without any amplification and allow foreign genes to be expressed in mammalian cells using mammalian promoters (Hofmann et al. 1995). Now, using this property of baculoviruses, new gene delivery technology has been developed. In most cases, *Autographa californica* multiple nucleopolyhedrovirus (AcMNPV) was used to transduce foreign genes into mammalian cells and organs (Hitchman et al. 2011). Recently, it was reported that *Bombyx mori* nucleopolyhedrovirus (BmNPV) can be used for the transduction of foreign genes into mammalian cells (Imai et al. 2016, Kato et al. 2016, Kenoutis et al. 2006), which opens a new utilization of BmNPV. In this chapter, we focus on gene delivery into mammalian cells and tissues using baculoviruses; we also introduce the methodology and discuss future possibilities (Fig. 1).

Laboratory of Biotechnology, Research Institute of Green Science and Technology, Shizuoka University, 836 Ohya Suruga-ku, Shizuoka 422-8529, Japan.
* Corresponding author: kato.tatsuya@shizuoka.ac.jp; park.enoch@shizuoka.ac.jp

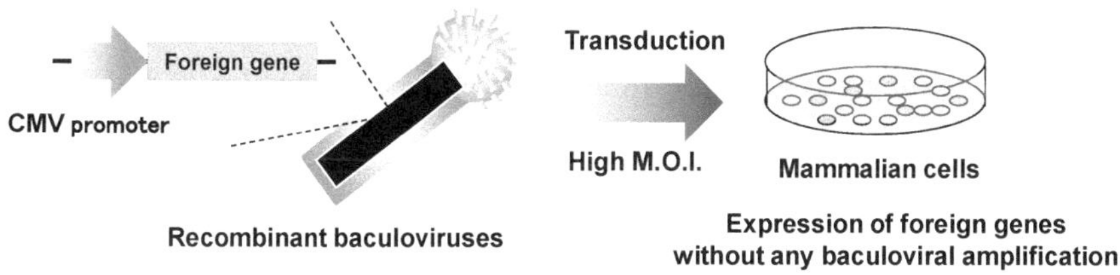

Fig. 1: Outline of transgene expression in mammalian cells via baculoviral transduction.

2. AcMNPV

2.1 Transduction of AcMNPV into Mammalian Cells

AcMNPV infects lepidopteran species, such as *A. californica* and *Trichoplusia ni*, but does not infect other hosts, such as mammals. However, when a mammalian expression cassette (composed of a mammalian promoter, a foreign gene and a mammalian terminator) is in the genome of AcMNPV, the foreign gene is expressed in mammalian cells during the transduction of the AcMNPV. In addition, the expression of the foreign gene is achieved without the baculoviral amplification in mammalian cells. This point is a merit of the recombinant protein production using baculoviruses because the contamination of baculoviruses in recombinant proteins purified from insect cell culture is a huge problem for its clinical applications. Thompson et al. reported the investigation of influenza virus-like particle (VLP) production in Sf-9 cells and human embryonic kidney (HEK) 293 cells using recombinant baculoviruses (Thompson et al. 2015). In this study, the gene coding each component of influenza VLP (hemagglutinin, neuraminidase and M1 protein) was inserted into the genome of one AcMNPV under the control of cytomegarovirus (CMV) promoter. Influenza VLPs were expressed in HEK293 cells using this recombinant AcMNPV at a multiplicity of infection (M.O.I.) 60 and purified using a density gradient ultracentrifuge. The expression level of influenza VLPs in HEK293 cells is lower than that in Sf-9 cells, but its VLPs purified from HEK293 cell did not contain any baculoviral contamination. However, in nature, baculoviruses immunized into mice induce the adaptive immune response (Hervas-Stubbs et al. 2007). In addition, influenza VLPs purified from insect cells protected mice from the lethal challenge of influenza virus compared to that purified from mammalian cells transfected with an expression plasmid (Margine et al. 2012). Margine et al. (2012) showed that the baculoviruses contaminated in purified influenza VLPs led to the activation of the innate immune response. This immunogenicity is still a problem, although baculoviruses are safe to mammals.

2.2 Development of DNA Vaccine-based on AcMNPV

Taking advantage of its immunogenicity, AcMNPV has been developed as a carrier of DNA vaccines to infectious viruses and parasites. Baculoviruses have an adjuvant activity and enhance humoral and cellular responses in addition to the innate immune response (Abe et al. 2005, Hervas-Stubbs et al. 2007, Suzuki et al. 2010). Ge et al. (2016) constructed recombinant AcMNPV having F protein gene or HN protein gene from Newcastle disease virus (NDV) under the CMV promoter control in order to prevent chickens from being affected by this virus infection. This recombinant AcMNPV allowed each protein to be expressed in chicken embryo fibroblasts. In addition, the immunization of this recombinant AcMNPV induced the production of NDV-specific antibodies, neutralizing antibodies and some cytokines (IFN-γ, IL-2, IL-4) without any adjuvant. In this paper, the display of vesicular stomatitis virus G protein (VSV-G) on the surface of recombinant AcMNPV and the insertion of woodchuck hepatitis virus post-transcriptional regulatory element and the inverted terminal repeats (ITRs) from adeno-associated virus were carried out in order to improve the expression of each protein in chicken cells. Especially, the display of envelope proteins from some viruses, including VSV-G, was adopted in order to improve the transduction efficiency of recombinant baculoviruses into non-host cells. In most studies on the vaccine development, using recombinant baculoviruses as a DNA vaccine, this improvement was carried out. VSV-G-displaying pseudotype AcMNPV, containing hemagglutinin (HA) gene from influenza virus under the control of CMV promoter, was constructed to use this pseudotype AcMNPV as a vaccine to influenza virus (Wu et al. 2009). Immunization of this recombinant AcMNPV induced humoral and cellular immune response (HA-specific antibody production and IFN-γ expression), leading to the prevention of influenza virus infection in mice and chickens. Using human endogenous retrovirus envelope protein (HERV env), HERV env-display on the surface of AcMNPV led to the significant enhancement of human papillomavirus 16 (HPV16) L1 protein expression in Huh7 cells and NIH3T3 cells by the recombinant AcMNPV (Lee et al. 2010). In addition, the induction of HPV L1-specific antibodies production in mice immunized with the HERV env-displaying AcMNPV was comparable to that in mice immunized with Gardasil® and its IFN-γ production was 450-fold higher than that in mice immunized with Gardasil®. In this study, human elongation factor promoter 1α was used for the expression of HPV16 L1. This HERV env-displaying AcMNPV was also applied to the development of influenza virus vaccines based on VLPs composed of HA, M1 protein and neuraminidase (Gwon et al. 2016). Granulocyte-macrophage colony-stimulating factor was displayed on the surface of this HERV env-displaying AcMNPV in order to enhance its adjuvant activity (Choi et al. 2015).

By using baculoviruses and surface display technology, recombinant AcMNPV with a mammalian expression cassette can be improved for vaccine development. Baculovirus surface display technology has been already established (Grabherr and Ernst 2010, Mäkelä and Oker-Blom 2006). Some transmembrane proteins can be displayed on the surface of baculoviruses without any modification. The transmembrane domain of glycoprotein 64 (GP64) from baculoviruses can be used for the display of recombinant proteins as an anchor to the envelope of baculoviruses. The gene of *Plasmodium berghei* circumsporozoite protein (PbCSP), fused with the transmembrane domain of GP64 from AcMNPV, was inserted under the control of polyhedrin and CMV promoters in AcMNPV genome and PbCSP-displaying AcMNPV with the mammalian PbCSP expression cassette was constructed (Yoshida et al. 2009). Immunization of this recombinant AcMNPV in mice induced both Th1 and Th2 responses and protected mice from *P. berghei* infection. The authors mentioned that this AcMNPV can be used as a hybrid vaccine having the capacity of both sub-unit and DNA vaccines. CMV early enhancer-chicken beta actin (CAG) promoter was also used for the *Plasmodium* antigen expression using AcMNPV (Iyori et al. 2013). The baculovirus-based vaccine has been applied to the prevention of white spot syndrome virus (WSSV) infection in shrimps (Cho et al. 2017, Syed and Kwang 2011).

2.3 Gene Delivery using AcMNPV

AcMNPV has been also used for gene delivery and gene therapy by exploiting its capacity to transduce foreign genes into mammalian cells (Kwang et al. 2016). Apart from human viral vectors such as adenoviruses, adeno-associated viruses and retroviruses, which have been conventionally used for the gene therapy, baculoviruses cannot be amplified in mammalian cells and pre-existing immunity to baculoviruses has not been observed in mammals (Ahi et al. 2011, Strauss et al. 2007). In addition, retroviral vectors are particularly in danger of causing aberrant transformations by insertional mutagenesis, baculoviruses, however, have no risk of this mutagenesis (Schambach et al. 2013). Therefore, baculoviruses have been developed as a carrier of foreign genes and RNAs for gene therapy.

AcMNPVs have a strong tropism to hepatocytes *in vitro*, but the gene delivery using recombinant AcMNPVs to livers *in vivo* failed because baculoviruses were inactivated by serum complement components (Hofmann et al. 1995, Sandig et al. 1996). However, the gene delivery to liver tissues *ex vivo* was achieved using recombinant AcMNPVs. Based on these results, several papers show that eye, brain, testis and central nerve system are favorable targets for gene therapy using baculoviruses because these are immune-privileged tissues (Haeseleer et al. 2001, Kinnunen et al. 2009, Sarkis et al. 2000, Tani et al. 2003, Wang et al. 2005). To escape the host's complement system, baculoviruses were shielded by some proteins using the baculovirus surface display technology. Decay-accelerating factor (DAF) was displayed on the surface of recombinant AcMNPVs for the construction of the complement-resistant gene transfer vector (Hüser et al. 2001, Iyori et al. 2017). DAF is the best candidate to escape the inactivation of recombinant AcMNPVs by the complement system among a family of complement receptors and regulatory proteins. DAF protected recombinant AcMNPVs from inactivation by the complement *in vitro* and the production of inflammatory cytokines by macrophages treated with DAF-displaying recombinant AcMNPVs was restricted compared to that by control recombinant AcMNPVs (Kaikkonen et al. 2010). Intraportal immunization of these DAF-displaying recombinant AcMNPVs in mice improved the gene delivery to liver tissues compared to the control.

2.4 Treatment of Cancers using AcMNPV

AcMNPVs have been also applied to the treatment of cancers. To express suicide genes and anti-tumor genes in cancer cells specifically, AcMNPVs have been used as a carrier of foreign genes. Herpex simplex virus, thymidine kinase (TK), which is well-known as a suicide protein, was expressed in glioblastoma cells using a recombinant AcMNPV under the control of a glioblastoma-specific promoter to suppress the growth of these cells (Balani et al. 2009). In this study, human high mobility group box2 promoter was used to express TK in glioblastoma cells because this protein was over-expressed in glioblastomas. The constructed recombinant AcMNPV also suppressed the growth of human glioblastoma cells in xenografted rats by its intra-tumor injection, leading to the longevity of xenografted rats. Interferon-β expressed in mouse Lewis lung carcinoma cells by a recombinant AcMNPV inhibited the proliferation of the cells *in vitro* and suppressed metastases in lungs of mice *in vivo* (Lykhova et al. 2015). miRNA expression using recombinant baculoviruses was also carried out for the cancer therapy. MicroRNA 122 (miR-122), which is known as a tumor suppressor for hepatocellular carcinoma, was expressed in HCC cells using a recombinant AcMNPV together with miR-151 sponge, which can sequester miR-151 over expressing HCC cells (Chen et al. 2015). In this study, a *Sleeping Beauty* transposase was co-expressed in order to prolong the expression of the miRNA and miRNA sponge by the integration of the expression cassettes into the host's chromosomes.

To extend the expression of transgenes by recombinant baculoviruses, Epstein-Barr nuclear antigen 1 and the origin of viral DNA replication, which are essential for the episomal maintenance of the Epstein-Barr virus genome, were co-expressed (Shan et al. 2006). Transduction of foreign genes into stem cells can be also achieved by AcMNPV for the cancer therapy. A recombinant AcMNPV allowed the expression of TK In human bone marrow mesenchymal stem cells (MSCs) and these TK-transduced MSCs, injected into tumor-xenografted mice, suppressed the tumor growth, leading to the longevity of the mice compared to control tumor-xenografted mice (Bak et al. 2010). Similarly, TK was expressed in induced pluripotent stem cell-derived neural stem cells using a recombinant AcMNPV and the injection of these TK-expressing cells into the contralateral hemisphere in a mouse intracranial human glioma xenograft mice inhibit the tumor growth *in vivo* (Lee et al. 2011).

2.5 Tissue Engineering using AcMNPV

Several studies show the possibilities for a baculovirus to be used in tissue engineering. It was reported that a recombinant baculovirus mediates the transgene expression in rat chondrocytes cultivated in a rotating-shaft bioreactor with its normal differentiation capacity (Ho et al. 2004, Chen et al. 2006). Based on these results, a recombinant AcMNPV containing the mammalian expression cassette of bone morphogenetic protein 2 was transduced into de-differentiated rabbit chondrocytes *ex vivo* and mature cartilaginous constructs were produced. These constructs implanted into cartilage-defective rabbits repaired their osteochondral defects (Chen et al. 2009). In addition, adipose-derived stem cells expressing BMP-6 and transforming growth factor-β (TGF-β) by recombinant AcMNPVs were cultivated in a scaffold for their 3D cultivation and the implantation of this construct into cartilage-defective rabbits led to the *in vivo* regeneration of hyaline cartilage (Lu et al. 2013). In this study, to prolong its expression, flippase recombinase was co-expressed in ASC.

3. BmNPV

3.1 Transduction of BmNPV into Mammalian Cells

BmNPV infects silkworms, *B. mori*, but not *A. californica* and *Trichoplusia ni*. As well as AcMNPV, BmNPV cannot replicate in mammalian cells and infect mammals. Kenoutis et al. (2016) reported that BmNPV carried GFP gene to HEK293 cells for its expression at M.O.I. 500 and GFP expression was observed at low M.O.I. in the presence of trichostatin A (TSA) which is an inhibitor of histone deacetylases and Dulbecco's phosphate-buffered saline. Imai et al. (2016) showed that the EGFP expression level was low at M.O.I. 500 when a recombinant BmNPV containing EGFP gene under the CAG promoter control was transduced into HEK293T cells. However, using PBS and a histone deacetylase inhibitor, sodium butyrate, during the exposure of BmNPV, the transduction efficiency was enhanced. In our paper, the expression of EGFP was not observed in HEK293T cells when a recombinant BmNPV containing EGFP gene under the control of CMV promoter was transduced into the cells at M.O.I. 150 or 300 without the use of PBS and a histone deacetylase inhibitor (Fig. 2; Kato et al. 2016). A histone deacetylase inhibitor inhibits the expression of a foreign gene in Sf-9 cells using a recombinant AcMNPV possessing a gene expression cassette composed of a foreign gene and p10 promoter (Peng et al. 2007). On the other hand, the histone deacetylase inhibitor enhances the transgene expression in mammalian cells by recombinant AcMNPVs under the control of mammalian promoters (Condreay et al. 1999). As well as AcMNPV, histone deacetylase inhibitors including TSA and sodium butylate are one of valuable tools for enhancing the expression of foreign genes in mammalian cells using recombinant BmNPVs. In the case of AcMNPVs, microtubule depolymerizing agents such as nocodazole and vinblastine also improve the AcMNPV-mediated transgene expression in mammalian cells, while actin-polymerization inhibitors such as cytochalasin D and latrunculin inhibited the transport of baculoviral nucleocapsids to the host's nuclei (Salminen et al. 2005). These results suggest that microtubules may be a barrier to the transport of nucleocapsids of baculoviruses and actin is important in introducing the nucleocapsids into the nuclei in cells.

In addition, a culture medium seems to also be a crucial factor in transducing foreign genes into mammalian cells by BmNPVs. In the case of AcMNPVs, Dulbecco's Modified Eagle's Medium (DMEM) inhibited the baculovirus-mediated transgene expression in mammalian cells (Hsu et al. 2004, Mähönen et al. 2010). Instead, PBS enhanced its transduction efficiency into mammalian cells (Imai et al. 2016, Salminen et al. 2005). In the presence of DMEM, during its transduction, the genome of BmNPV was detected in HEK293T cells after 1 hour of transduction (Fig. 3). This indicates that BmNPV entered HEK293T cells but GFP was not expressed. The amino acid sequence of GP64 from BmNPV is almost identical to that from AcMNPV (over 95%), but the property of the membrane fusion activity of BmNPV GP64 was lower than that of AcMNPV GP64 (Katou et al. 2010). Especially, over pH 4, the membrane fusion activity of BmNPV was not observed, compared to GP64 from AcMNPV which showed the membrane fusion activity at pH 5. When baculoviruses enter host cells by endocytosis or macropinocytosis, they move through early endosomes and their nucleocapsids are released by the low pH-dependent endosomal fusion (Kataoka et al. 2012, Liu et al. 2014). The replacement of GP64 from

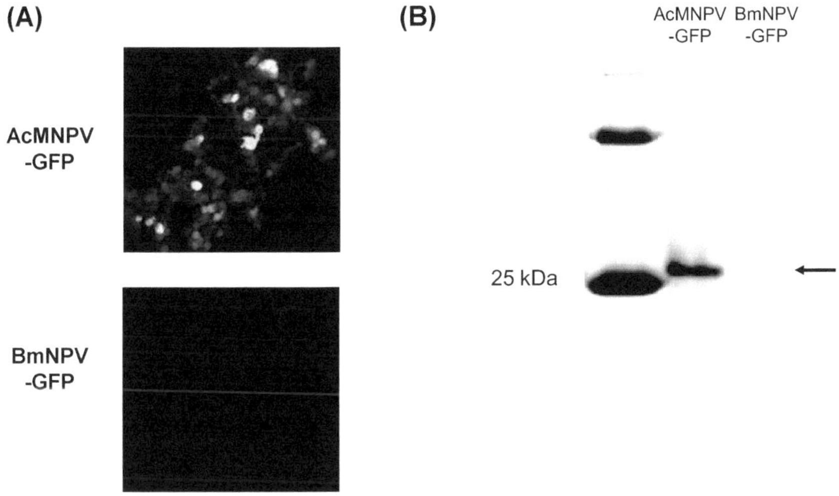

Fig. 2: Expression of GFP in HEK293T cells by the transduction of recombinant AcMNPV or BmNPV. (A) GFP fluorescence in HEK293T cells. (B) Detection of EGFP expressed in HEK293T cells by SDS-PAGE. Each recombinant baculovirus was transduced into HEK293T cells at M.O.I. 300, followed by its cultivation for 2 days.

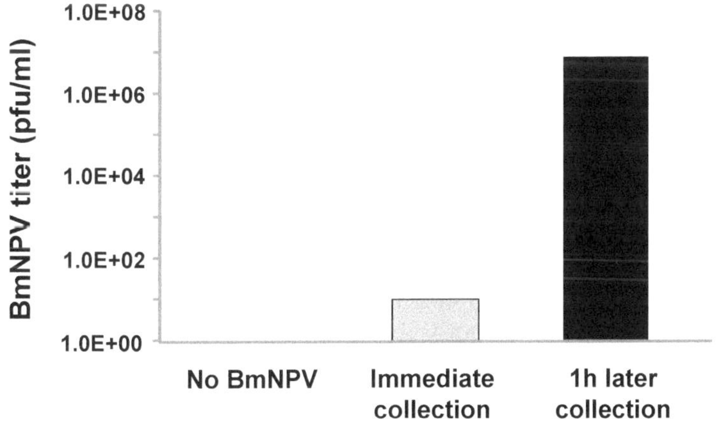

Fig. 3: Quantification of BmNPV genomic DNA in or on HEK293T cells after the transduction of BmNPV-GFP. Each recombinant baculovirus was transduced into HEK293T cells at M.O.I. 300. Cells were washed immediately or after its incubation for 1 hour and total DNA was extracted from cell for the quantification of recombinant BmNPV DNA.

BmNPV with GP64 from AcMNPV led to the success of the BmNPV transduction into mammalian cells (Fig. 4; Kato et al. 2016). These results indicate that GP64 and its transduction condition are important for the BmNPV transduction into mammalian cells.

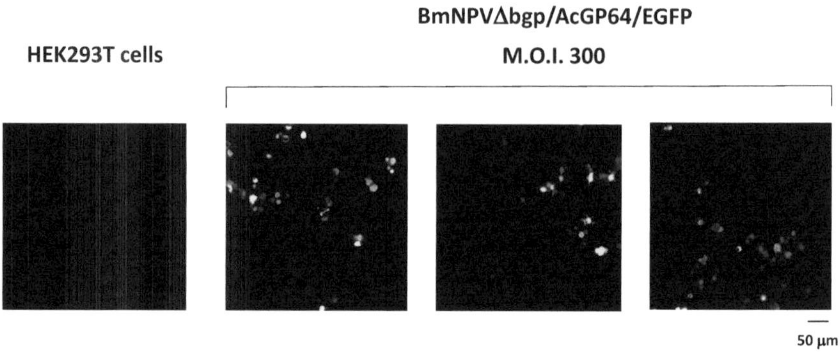

Fig. 4: EGFP expression in HEK293T cells transduced with BmNPVΔbgp/AcGP64/EGFP. Cells were transduced with BmNPVΔbgp/AcGP64/EGFP at M.O.I. 300 and cultivated for 2 days.

3.2 Development of DNA Vaccines-based on BmNPV

Kato et al. (2017) reported that recombinant BmNPV displaying GP64 from AcMNPV was used for the induction of antigen-specific antibody production in mice. The recombinant BmNPV has *N. caninum* SRS2 (NcSRS2) gene expression cassette composed of CMV promoter and SV40 polyadenylation signal and allowed the expression of NcSRS2 in HEK293T cells. In addition, the immunization of this recombinant BmNPV induced the production of NcSRS2-specific antibodies in mice, as well as a DNA vaccine to NcSRS2. This result indicates that BmNPV can also be used as to carry foreign genes into mammalian cells for the induction of gene-specific antibody production by its immunization.

It has been already reported that AcMNPV is a promising tool as a carrier of foreign genes into mammalian cells for vaccine development and gene delivery to targeted tissues. Recently, the gene transduction in mammalian cells by BmNPV has been also developed and applied to vaccine development. BmNPV can be prepared in silkworm larvae at a large scale more easily than AcMNPV and the purification method from haemolymph has already been reported (Kato et al. 2009, 2011). BmNPV is also a promising tool as a gene carrier to mammalian cells in addition to AcMNPV, even though its transduction efficiency should be improved.

4. Future Prospects

Baculoviruses normally infect invertebrates and have been utilized for recombinant protein production in insect cells and insects. Some products produced using baculoviruses have already been approved for commercial use, including on human subjects. In addition, baculoviruses, especially AcMNPV, have been developed through some modifications as vaccines and transgene carriers in vaccinology and gene delivery systems. This indicates that baculoviruses could be used safely on humans. Details of the use of AcMNPV in delivering foreign genes into mammalian cells have already been reported in many papers but it was also shown that BmNPV is equally capable of delivering foreign genes in mammalian cells. BmNPV can be prepared at the higher titer in silkworm haemolymph, compared to that of AcMNPV in cell culture. BmNPV is a promising replacement for AcMNPV in the fields of gene delivery and vaccine development because of its scalability and cost-effectiveness.

References

Abe, T., H. Hemmi, H. Miyamoto, K. Moriishi, S. Tamura, H. Takaku, S. Akira and Y. Matsuura. 2005. Involvement of the Toll-like receptor 9 signaling pathway in the induction of innate immunity by baculovirus. J. Virol. 79: 2847–2858.

Ahi, Y. S., D. S. Bangari and S. K. Mittal. 2011. Adenoviral vector immunity: Its implications and circumvention strategies. Curr. Gene Ther. 11: 307–320.

Bak, X. Y., J. Yang and S. Wang. 2010. Baculovirus-transduced bone marrow mesenchymal stem cells for systemic cancer therapy. Cancer Gene Ther. 17: 721–729.

Balani, P., J. Boulaire, Y. Zhao, J. Zeng, J. Lin and S. Wang. 2009. High mobility group box2 promoter-controlled suicide gene expression enables targeted glioblastoma treatment. Mol. Ther. 17: 1003–1011.

Chen, H. C., H. P. Lee, Y. C. Ho, M. L. Sung and Y. C. Hu. 2006. Combination of baculovirus-mediated gene transfer and rotating-shaft bioreactor for cartilage tissue engineering. Biomaterials. 27: 3154–3162.

Chen, H. C., Y. H. Chang, C. K. Chuang, C. Y. Lin, L. Y. Sung, Y. H. Wang and Y. C. Hu. 2009. The repair of osteochondral defects using baculovirus-mediated gene transfer with de-differentiated chondrocytes in bioreactor culture. Biomaterials. 30: 674–681.

Chen, C. L., J. C. Wu, G. Y. Chen, P. H. Yuan, Y. W. Tseng, K. C. Li, S. M. Hwang and Y. C. Hu. 2015. Baculovirus-mediated miRNA regulation to suppress hepatocellular carcinoma tumorigenicity and metastasis. Mol. Ther. 23: 79–88.

Cho, H., N. H. Park, Y. Jang, Y. D. Gwon, Y. Cho, Y. K. Heo, K. H. Park, H. J. Lee, T. J. Choi and Y. B. Kim. 2017. Fusion of flagellin 2 with bivalent white spot syndrome virus vaccine increases survival in freshwater shrimp. J. Invertebr. Pathol. 144: 97–105.

Choi, H. J., Y. D. Gwon, Y. Jang, Y. Cho, Y. K. Heo, H. J. Lee, K. C. Kim, J. Choi, J. B. Lee and Y. B. Kim. 2015. Effect of AcHERV-GmCSF as an influenza virus vaccine adjuvant. PLoS One. 10: e0129761.

Condreay, J. P., S. M. Witherspoon, W. C. Clay and T. A. Kost. 1999. Transient and stable gene expression in mammalian cells transduced with a recombinant baculovirus vector. Proc. Natl. Acad. Sci. USA. 96: 127–132.

Ge, J., Y. Liu, L. Jin, D. Gao, C. Bai and W. Ping. 2016. Construction of recombinant baculovirus vaccines for Newcastle disease virus and an assessment of their immunogenicity. J. Biotechnol. 231: 201–211.

Grabherr, R. and W. Ernst. 2010. Baculovirus for eukaryotic protein display. Curr. Gene Ther. 10: 195–200.

Gwon, Y. D., S. Kim, Y. Cho, Y. Heo, H. Cho, K. Park, H. J. Lee, J. Choi, H. Poo and Y. D. Kim. 2016. Immunogenicity of virus-like particle forming baculoviral DNA vaccine against pandemic influenza H1N1. PLoS One. 11: e0154824.

Haeseleer, F., Y. Imanishi, D. Saperstein and K. Palczewski. 2001. Gene transfer mediated by recombinant baculovirus into mouse eye. Invest. Ophthalmol. Vis. Sci. 42: 3294–3300.

Hervas-Stubbs, S., P. Rueda, L. Lopez and C. Leclerc. 2007. Insect baculoviruses strongly potentiate adaptive immune responses by inducing type I IFN. J. Immunol. 178: 2361–2369.

Hitchman, R. B., F. Murguía-Meca, E. Locanto, J. Danquah and L. A. King. 2011. Baculovirus as vectors for human cells and applications in organ transplantation. J. Invertebr. Pathol. 107: S49–58.

Ho, Y. C., H. C. Chen, K. C. Wang and Y. C. Hu. 2004. Highly efficient baculovirus-mediated gene transfer into rat chondrocytes. Biotechnol. Bioeng. 88: 643–651.

Hofmann, C., V. Sandig, G. Jennings, M. Rudolph, P. Schlag and M. Strauss. 1995. Efficient gene transfer into human hepatocytes by baculovirus vectors. Proc. Natl. Acad. Sci. USA. 92: 10099–10103.

Hsu, C. S., Y. C. Ho, K. C. Wang and Y. C. Hu. 2004. Investigation of optimal transduction conditions for baculovirus-mediated gene delivery into mammalian cells. Biotechnol. Bioeng. 88: 42–51.

Hüser, A., M. Rudolph and C. Hofmann. 2001. Incorporation of decay-accelerating factor into the baculovirus envelope generates complement-resistant gene transfer vectors. Nat. Biotechnol. 19: 451–455.

Imai, A., T. Tadokoro, S. Kita, M. Horiuchi, H. Fukuhara and K. Maenaka. 2016. Establishment of the BacMam system using silkworm baculovirus. Biochem. Biophys. Res. Commun. 478: 580–585.

Iyori, M., H. Nakaya, K. Inagaki, S. Pichyangkul, D. S. Yamamoto, M. Kawasaki, K. Kwak, M. Mizukoshi, Y. Goto, H. Matsuoka, M. Matsumoto and H. Yoshida. 2013. Protective efficacy of baculovirus dual expression system vaccine expressing *Plasmodium falciparum* circumsporozoite protein. PLoS One. 8: e70819.

Iyori, M., D. S. Yamamoto, M. Sakaguchi, M. Mizutani, S. Ogata, H. Nishiura, T. Tamura, H. Matsuoka and S. Yoshida. 2017. DAF-shielded baculovirus-vectored vaccine enhances protection against malaria sporozoite challenge in mice. Malar. J. 16: 390.

Kaikkonen, M. U., A. I. Maatta, S. Ylä-Herttuala and K. J. Airenne. 2010. Screening of complement inhibitors: Shielded baculoviruses increase the safety and efficacy of gene delivery. Mol. Ther. 18: 987–992.

Kataoka, C., Y. Kaname, S. Taguwa, T. Abe, T. Fukuhara, H. Tani, K. Moriishi and Y. Matsuura. 2012. Baculovirus GP64-mediated entry into mammalian cells. J. Virol. 86: 2610–20.

Kato, T., S. L. Manoha, S. Tanaka and E. Y. Park. 2009. High-titer preparation of *Bombyx mori* nucleopolyhedrovirus (BmNPV) displaying recombinant protein in silkworm larvae by size exclusion chromatography and its characterization. BMC Biotechnol. 9: 55.

Kato, T., F. Suzuki and E. Y. Park. 2011. Purification of functional baculovirus particles from silkworm larval hemolymph and their use as nanoparticles for the detection of human prorenin receptor (PRR) binding. BMC Biotechnol. 11: 60.

Kato, T., S. Sugioka, K. Itagaki and E. Y. Park. 2016. Gene transduction in mammalian cells using *Bombyx mori* nucleopolyhedrovirus assisted by glycoprotein 64 of *Autographa californica* multiple nucleopolyhedrovirus. Sci. Rep. 6: 32283.

Kato, T., K. Itagaki, M. Yoshimoto, R. Hiramatsu, H. Suhaimi, T. Kohsaka and E. Y. Park. 2017. Transduction of a *Neospora caninum* antigen gene into mammalian cells using a modified *Bombyx mori* nucleopolyhedrovirus for antibody production. J. Biosci. Bioeng. 124: 606–610.

Katou, Y., H. Yamada, M. Ikeda and M. Kobayashi. 2010. A single amino acid substitution modulates low-pH-triggered membrane fusion of GP64 protein in *Autographa californica* and *Bombyx mori* nucleopolyhedroviruses. Virology. 404: 204–14.

Kenoutis, C., R. C. Efrose, L. Swevers, A. A. Lavdas, M. Gaitanou, R. Matsas and K. Iatrou. 2006. Baculovirus-mediated gene delivery into mammalian cells does not alter their transcriptional and differentiating potential but is accompanied by early viral gene expression. J. Virol. 80: 4135–4146.

Kinnunen, K., G. Kalesnykas, A. J. Mähönen, S. Laidinen, L. Holma, T. Heikura, K. Airenne, H. Uusitalo and S. Ylä-Herttuala. 2009. Baculovirus is an efficient vector for the transduction of the eye: Comparison of baculovirus- and adenovirus-mediated intravitreal vascular endothelial growth factor D gene transfer in the rabbit eye. J. Gene Med. 11: 382–389.

Kwang, T. W., X. Zeng and S. Wang. 2016. Manufacturing *Ac*MNPV baculovirus vectors to enable gene therapy trials. Mol. Ther. Methods Clin. Dev. 3: 15050.

Lee, H. J., N. Park, H. J. Cho, J. K. Yoon, N. D. Van, Y. K. Oh and Y. B. Kim. 2010. Development of a novel viral DNA vaccine against human papillomavirus: AcHERV-HP16L1. Vaccine. 28: 1613–1619.

Lee, E. X., D. H. Lam, C. Wu, J. Yang, C. K. Tham, W. H. Ng and S. Wang. 2011. Glioma gene therapy using induced pluripotent stem cell derived neural stem cells. Mol. Pharm. 8: 1515–1524.

Liu, Y., K. I. Joo, Y. Lei and P. Wang. 2014. Visualization of intracellular pathways of engineered baculovirus in mammalian cells. Virus Res. 181: 81–91.

Lu, C. H., T. S. Yeh, C. L. Yeh, Y. H. Fang, L. Y. Sung, S. Y. Lin, T. C. Yen, Y. H. Chang and Y. C. Hu. 2014. Regenerating cartilages by engineered ASCs: prolonged TGF-β3/BMP-6 expression improved articular cartilage formation and restored zonal structure. Mol. Ther. 22: 186–195.

Lykhova, A. A., Y. I. Kudryavets, L. I. Strokovska, N. A. Bezdenezhnykh, N. I. Semesiuk, I. N. Adamenko, O. V. Anopriyenko and A. L. Vorontsova. 2015. Suppression of proliferation, tumorigenicity and metastasis of lung cancer cells after their transduction by interferon-beta gene in baculovirus vector. Cytokine. 71: 318–326.

Mähönen, A. J., K. E. Makkonen, J. P. Laakkonen, T. O. Ihalainen, S. P. Kukkonen, M. U. Kaikkonen, M. Vihinen-Ranta, S. Ylä-Herttuala and K. J. Airenne. 2010. Culture medium induced vimentin reorganization associates with enhanced baculovirus-mediated gene delivery. J. Biotechnol. 145: 111–119.

Mäkelä, A. R. and C. Oker-Blom. 2006. Baculovirus display: A multifunctional technology for gene delivery and eukaryotic library development. Adv. Virus Res. 68: 91–112.

Margine, I., L. Martinez-Gil, Y. Y. Chou and F. Krammer. 2012. Residual baculovirus in insect cell-derived influenza virus-like particle preparations enhances immunogenicity. PLoS One. 7: e51559.

Peng, Y., J. Song, J. Lu and X. Chen. 2007. The histone deacetylase inhibitor sodium butyrate inhibits baculovirus-mediated transgene expression in Sf9 cells. J. Biotechnol. 131: 180–187.

Salminen, M., K. J. Airenne, R. Rinnankoski, J. Reimari, O. Välilehto, J. Rinne, S. Suikkanen, S. Kukkonen, S. Ylä-Herttuala, M. S. Kulomaa and M. Vihinen-Ranta. 2005. Improvement in nuclear entry and transgene expression of baculoviruses by disintegration of microtubules in human hepatocytes. J. Virol. 79: 2720–2728.

Sandig, V., C. Hofmann, S. Steinert, G. Jennings, P. Schlag and M. Strauss. 1996. Gene transfer into hepatocytes and human liver tissue by baculovirus vectors. Hum. Gene Ther. 7: 1937–1945.

Sarkis, C., C. Serguera, S. Petres, D. Buchet, J. Ridet, L. Edelman and J. Mallet. 2000. Efficient transduction of neural cells *in vitro* and *in vivo* by a baculovirus-derived vector. Proc. Natl. Acad. Sci. USA. 97: 14638–14643.

Schambach. A., D. Zychlinski, B. Ehrnstroem and C. Baum. 2013. Biosafety features of lentiviral vectors. Hum. Gene Ther. 24: 132–42.

Shan, L., L. Wang, J. Yin, P. Zhong and J. Zhong. 2006. An OriP/EBNA-1-based baculovirus vector with prolonged and enhanced transgene expression. J. Gene Med. 8: 1400–1406.

Strauss, R., A. Huser, S. Ni, S. Tuve, N. Kiviat, P. S. Sow, C. Hofmann and A. Lieber. 2007. Baculovirus-based vaccination vectors allow for efficient induction of immune responses against *Plasmodium falciparum* circumsporozoite protein. Mol. Ther. 15: 193–202.

Suzuki, T., M. O. Chang, M. Kitajima and H. Takaku. 2010. Baculovirus activates murine dendritic cells and induces non-specific NK cell and T cell immune responses. Cell Immunol. 262: 35–43.

Syed, M. S. and J. Kwang. 2011. Oral vaccination of baculovirus-expressed VP28 displays enhanced protection against White Spot Syndrome Virus in *Penaeus monodon*. PLoS One. 6: e26428.

Tani, H., C. K. Limn, C. C. Yap, M. Onishi, M. Nozaki, Y. Nishimune, N. Okahashi, Y. Kitagawa, R. Watanabe, R. Mochizuki, K. Moriishi and Y. Matsuura. 2003. *In vitro* and *in vivo* gene delivery by recombinant baculoviruses. J. Virol. 77: 9799–9808.

Thompson, C. M., E. Petiot, A. Mullick, M. G. Aucoin, O. Henry and A. A. Kamen. 2015. Critical assessment of influenza VLP production in Sf9 and HEK293 expression systems. BMC Biotechnol. 15: 31.

Wang, X., C. Wang, J. Zeng, X. Xu, P. Y. Hwang, W. C. Yee, Y. K. Ng and S. Wang. 2005. Gene transfer to dorsal root ganglia by intrathecal injection: Effects on regeneration of peripheral nerves. Mol. Ther. 12: 314–320.

Wu, Q., L. Fang, X. Wu, B. Li, R. Luo, Z. Yu, M. Jin, H. Chen and S. Xiao. 2009. A pseudotype baculovirus-mediated vaccine confers protective immunity against lethal challenge with H5N1 avian influenza virus in mice and chickens. Mol. Immunol. 46: 2210–2217.

Yoshida, S., M. Kawasaki, N. Hariguchi, K. Hirota and M. Matsumoto. 2009. A baculovirus dual expression system-based malaria vaccine induces strong protection against *Plasmodium berghei* sporozoite challenge in mice. Infect. Immun. 77: 1782–1789.

Part III
Bioproducts from Silkworm

CHAPTER 14

Development of New Biomaterials from Insect and its Virus

Eiji Kotani and *Hajime Mori**

1. Introduction

Some insect viruses, which belong to two families (*Baculoviridae* and *Reoviridae*), encode a protein called polyhedrin that forms protein micro-crystals (polyhedra) in the infected cells (Aruga 1971, Belloncik and Mori 1998). The virus particles, or virions, are protected within these polyhedra and can remain infectious for years outside cells, even in harsh environmental conditions. The polyhedra break down and release the virus only when ingested into the very alkaline environment of the midgut of insect larvae (pH 10–11), resulting in the infection of a new host (Rohrmann 1986). In this chapter we introduce the use of polyhedra as a new biomaterial which has a variety of applications, including development for a stabilization and slow release of growth factors.

We also introduce our new approach for the use of silk proteins as a new biomaterial. Target gene functions in certain organs or tissues in many organisms have been successfully suppressed by gene silencing or editing in the past few years. However, suppression of the target tissue function has not been achieved via simple genetic engineering by using a gene such as cytotoxin family member. We established transgenic silkworms with posterior silk glands (PSGs) that express the enzymatic domain of a

Kyoto Institute of Technology, Matsugasaki, Sakyo-ku, Kyoto 606-8585, Japan.
* Corresponding author: hmori@kit.ac.jp

cytotoxin, named pierisin-1A (P1A). P1A is a recently identified cytotoxic protein from the cabbage butterfly *Pieris rapae* and was found to have a relatively lower DNA ADP-ribosylating activity among the family of pierisin, which had been reported to be an apoptotic inducer for many kinds of mammalian cultured cells. In the later part of this chapter, we describe the biological property of the silkworms with the modified PSGs that produced sericin cocoons with a potential utility in tissue engineering and a new approach through targeted P1A expression, which could be applicable to the development of biologically-useful model organisms with tissue-specific dysfunctions.

2. Polyhedra as a New Biomaterial

2.1 Insect Virus Polyhedra

The genus *Cypovirus* is a member of the family *Reoviridae* that infect insect larvae producing polyhedra in the cytoplasm of mid-gut epithelial cells (Mertens et al. 2005, Mori and Metcalf 2010). The N-terminal sequence of the silkworm *Bombyx mori* cypovirus (BmCPV) turret protein is considered to function as a polyhedrin recognition signal, leading to the occlusion of virus particles in polyhedra (Ikeda et al. 2001). Recently, the atomic structure of BmCPV polyhedra has been determined by using a synchrotron microbeam to collect X-ray diffraction data (Coulibaly et al. 2007). It was elucidated that polyhedra are made from trimeric building blocks of the polyhedron, interlocked into a tight scaffold generated by the amino-terminal α-helix (Coulibaly et al. 2007). These results have been exploited in order to encapsulate a wide variety of foreign proteins into polyhedra (Ijiri et al. 2009).

2.2 Methods for an Encapsulation of Recombinant Proteins

First, we supposed the occlusion of cypovirus particles into polyhedra is due to a specific interaction between CPV polyhedrin and a viral capsid protein. Iodination of BmCPV virion and analysis of the labeled polypeptides by SDS-PAGE first indicated that VP1 and VP3 are outer components of the BmCPV particle (Lewandowski and Traynor 1972). VP3 was then selected in order to investigate the interaction with BmCPV polyhedrin in the occlusion of virus particles into the polyhedra (Ikeda et al. 2001). As incorporation of foreign proteins can be easily assayed through the use of fluorescent proteins, the enhanced green fluorescent protein (EGFP) was fused to the C-terminus of BmCPV VP3 and introduced into the baculovirus expression vector. The fusion protein of VP3 and EGFP was co-expressed with BmCPV polyhedrin in insect cells. The polyhedra were purified and green fluorescence was observed under UV irradiation.

The fluorescence indicated that the VP3-EGFP chimera was incorporated into these polyhedra. In order to identify an essential region of VP3 protein to bind to polyhedrin, either the N-terminal or C-terminal half of VP3 was fused to the N-terminus or C-terminus of EGFP. They were EGFP fused to amino acids 1-448 of VP3 (VP3(N)/EGFP) and EGFP fused to amino acids 428-1057 of VP3 (EGFP/VP3(C)). Each recombinant EGFP was expressed with BmCPV polyhedrin. Green fluorescence was observed with polyhedra by co-infection with VP3(N)/EGFP and polyhedrin. In contrast, no green fluorescence was observed with polyhedra by co-infection with EGFP/VP3(C) and polyhedrin (Ikeda et al. 2006). EGFP was fused with a variety of VP3 regions and the incorporation of EGFP was analyzed in order to identify the minimal region of VP3 required to target the EGFP fusion construct to polyhedra. Finally, it was to be concluded that the encapsulation of EGFP into polyhedra required the VP3 domain between amino acids 1 and 79 and this amino acid sequence was also able to direct the encapsulation of foreign proteins into polyhedra (Ikeda et al. 2006, Mori and Metcalf 2010). The VP3 is the turret protein of BmCPV and the N-terminal sequence forms a separate domain on the outside of the turret (Zhang et al. 2002). This N-terminal domain corresponds to the minimal region of VP3 used to target foreign proteins (fluorescent protein, enzyme, antigen, cytokine, etc.) into BmCPV polyhedra and is named the immobilization signal (VP3 tag) (Ijiri et al. 2009, Ikeda et al. 2006) or polyhedrin-binding domain (PBD) (Yu et al. 2008). It is known that this domain is β-strand rich and located on the outer surface of the CPV turret, thus, it is easily accessible by the polyhedrin trimer. The sheet-like feature is supposed to provide a large interaction surface, which may be the structural basis for its strong and specific binding to the polyhedrin trimer.

The structure of BmCPV polyhedra revealed trimeric building blocks connected by extensive interactions of the α-helix H1, which projects from the main part of the molecule. Each building block consists of a cluster of three identical polyhedrin molecules and the three corresponding H1-helices project outwards from the center of the trimer. The building blocks are interlocked with other identical building blocks to form a tight scaffold, largely stabilized by the H1-helix. Because the H1 helix is at the N-terminus of the molecule and projects outwards from the structure, it may independently form into a helix as the molecule folds. This possibility, together with its role in cross-linking polyhedra, led to the suggestion that the polyhedrin H1-helix might also prove to be a useful tag (H1 tag) for incorporating foreign proteins into polyhedra, like the previously characterized VP3 tag (Ijiri et al. 2009, Mori and Metcalf 2010) (Fig. 1).

The H1-helix sequence was added to either the N-terminus or the C-terminus of EGFP. Each recombinant EGFP (H1/EGFP and EGFP/

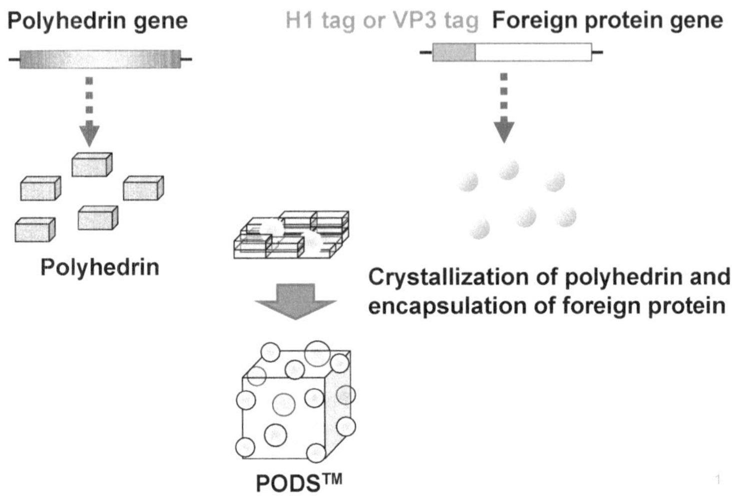

Fig. 1: Schematic diagram of a method for encapsulation of foreign proteins into mico-crystals (polyhedra). We named this method a polyhedra delivery system (PODS). https://www.cellgs.com/.

H1) was co-expressed with BmCPV polyhedrin using the baculovirus expression vector system. Immobilization of EGFP fused with H1-helix was compared with the VP3 tag (Ijiri et al. 2009). The emission of green fluorescence from H1/EGFP polyhedra displayed a greater intensity than either EGFP/H1 or EGFP/VP3 polyhedra. EGFP and *Discosoma* sp. red fluorescent protein (DsRed) were fused with VP3 tag (EGFP/VP3) and H1 tag (H1/DsRed), respectively, and the reverse conformations (H1/EGFP and DsRed/VP3) were also constructed. The double-labeled polyhedra with EGFP and DsRed were isolated and imaged using dual wavelength confocal fluorescence microscopy, showing that multiple proteins can be incorporated into single polyhedra using both VP3 and H1 tags (Ijiri et al. 2009, Mori and Metcalf 2010).

2.3 Encapsulation of Growth Factors

The extracellular matrix (ECM) provides a supra-molecular architecture for the disposition of growth factors which are released after proteolytic cleavage to affect a downstream alteration in the behavior of the responding cells. Conversely, cells that regulate remodeling (or degradation under certain conditions) of the ECM do so via the production and secretion of proteases and protease inhibitors. Each individual cell sends and receives spatio-temporally restricted signals of growth factors (cytokines) from its extracellular environment. Fibroblast growth factor-2 (FGF-2) is one of well-characterized heparin-binding growth factors and regulates cell proliferation, differentiation or migration. The immobilization signal (H1 tag or VP3 tag) was fused to the 18-kDa form of FGF-2 at the N- or C-terminus for the production of polyhedra encapsulating FGF-2 (FGF-2 polyhedra) (Ijiri et al. 2009, Mori et al. 2007). The recombinant FGF-2 proteins were then co-expressed with CPV polyhedrin using a baculovirus expression vector system. Two types of FGF-2 polyhedra induced cell proliferations of mouse fibroblast NIH3T3 cells. Then, we produced other cytokine-encapsulated polyhedra. FGF-7, epidermal growth factor (EGF), leukemia inhibitory factor (LIF), vascular endothelial growth factor (VEGF), endostatin, bone morphogenetic protein-2 (BMP-2), secreted frizzled-related protein 4, etc. were also encapsulated into polyhedra and biological activities were assayed. Human keratinocytes were proliferated by FGF-7 polyhedra and EGF polyhedra (Ijiri et al. 2009). A single addition of LIF-polyhedra to the cell culture medium supported the proliferation of mouse ES continuously for 14 days, suggesting that LIF-polyhedra can be successfully used in the place of a periodic addition of recombinant LIF to the media every 2–3 days. Maintenance of an undifferentiated state of mouse ES cultured with LIF-polyhedra was determined by the detection of pluripotency-related biomarkers Oct3/4 and stage-specific embryonicantigen-1 (SSEA-1) through immunostaining and measurement of alkaline phosphatase activity (Nishishita et al. 2011). Angiogenesis was also controlled by VEGF polyhedra and endostatin polyhedra. The remarkable stability of polyhedra is thought to be applied in slow-release carriers of cytokines and other proteins for tissue engineering or vaccination. Under physiological conditions, polyhedron-derived protein micro-crystals are inert and insoluble. These properties allow us to employ polyhedra as versatile micron-sized carriers. Polyhedra encapsulating BMP-2 enhanced chondrogenic and osteogenic differentiation of progenitor ATDC5 cells (Matsumoto et al. 2012). Absorbable collagen sponge (ACS) impregnated with BMP-2 polyhedra had enough osteogenic activity to promote complete healing in critical-sized bone defects, but ACS with a high dose of rhBMP-2 showed incomplete bone healing, indicating that BMP-2 polyhedra promise to advance state of the art bone healing. Angiogenesis was promoted by polyhedra encapsulating vascular endothelial growth factor. Endostatin-encapsulated polyhedra showed potent anti-endothelial activity, indicating that they may have promise for the treatment of squamous cell carcinoma by inhibiting tumors angiogenesis (Matsumoto et al. 2014). Polyhedra show properties of stabilization, retention and long release, which are very important for drug delivery systems, therefore, we would like to establish optimal protocols, including dosage and scheduling, for tissue engineering or antiangiogenic therapy (Fig. 2).

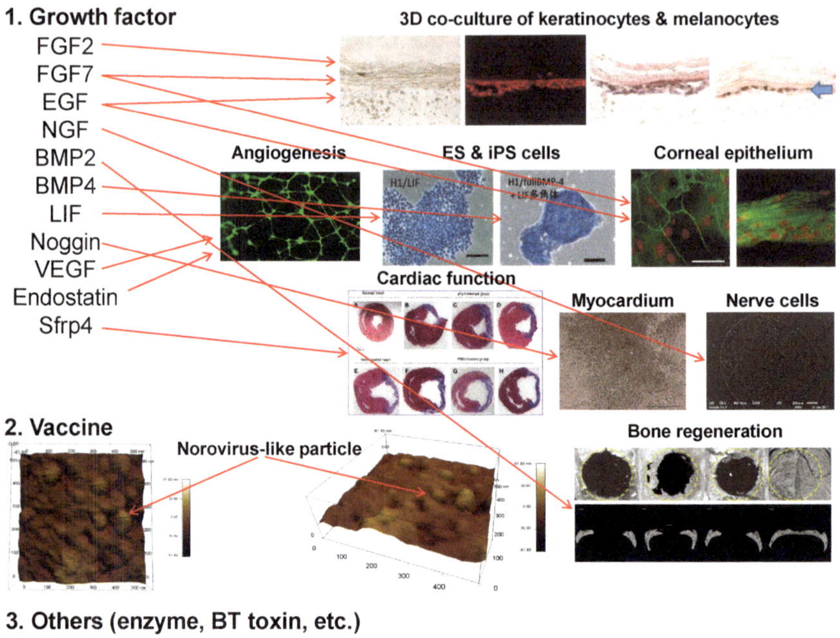

Fig. 2: Applications of PODS on tissue engineering, vaccine, and others. Growth factor-encapsulated polyhedra direct the proliferation and differentiation to cells. Virus-like particles are also occluded into polyhedra.

3. Silk Proteins as New Biomaterials

3.1 Sericin and Fibroin

Silk proteins from silkworm cocoons, mainly composed of fibroin and sericin, are candidate biomaterials with numerous potential biomedical applications or cosmetic uses (Kunz et al. 2016, Omenetto and Kaplan 2010, Thurber et al. 2015). Wild-type silkworm cocoon shells are composed of approximately 70% fibroin protein (fibroin heavy-chain, FibH and fibroin light-chain, FibL), and 25% sericin protein. Fibroin is a raw silk component, specifically produced by the posterior silk glands (PSGs), whereas sericin is a glue-like protein produced mainly in the middle silk glands and promotes cocoon cohesion by surrounding and gluing fibroin threads together (Kunz et al. 2016). In filature, a low silk thread composed of fibroin can be isolated via degumming treatment of the cocoon in hot alkaline water to decompose the sericin; decomposed sericin has been discarded as an industrial waste (Kunz et al. 2016). Genetic manipulations that alter the properties of silk glands would likely help expand the biomedical utilities of silk proteins. Especially the modified silkworm, with loss of fibroin production, would produce a cocoon composed solely of sericin and expand the industrial utilities of sericin.

3.2 Insect Cytotoxic Proteins

Cytotoxic activity for mammalian cells was found in the pupal haemolymph of the cabbage butterfly, *Pieris rapae,* and an identified cytotoxic protein was named pierisin-1 (Kanazawa et al. 2001, Koyama et al. 1996, Takamura-Enya et al. 2001, Watanabe et al. 1999). Research has shown that the addition of purified pierisin-1 to culture media can induce apoptosis of various human cancer cell lines, which can be blocked by Bcl2 (Kanazawa et al. 2001, 2002, Koyama et al. 1996, Nakano et al. 2015, Takamura-Enya et al. 2001, Watanabe et al. 1999). The N-terminal domain of pierisin-1 features an ADP-ribosyltransferase that transfers the ADP-ribose moiety of NAD to the 2′-deoxyguanosine residues of DNA, while the C-terminal region carries a domain that mediates binding to receptors on cell membranes for uptake by target cells, so that it can enter cells with the function of surface receptors from outside cells (Kanazawa et al. 2001). The biological significance of the presence of the pierisin family in the butterfly has been characterized. Its expression exclusively in the pupal haemolymph suggests that pierisin-1 plays an important physiological role in *Pieris* during the developmental stage when the unused larval tissues are discarded. In addition to this, pierisin-1 is thought to function as a biological defense protein that is necessary for the interference with the growth of parasitic wasps, which are not found in the *Pieris* pupa (Takahashi-Nakaguchi et al. 2013). The pierisin family proteins share the structural features with the DNA ADP-ribosyltransferase of different species, such as eubacterium *Streptomyces,* however, they are not fully homologous with CARP-1 found in shellfish, *Meretrixlamaerckii* (Nakano et al. 2013, Nakano et al. 2015). Also, ADP ribosyltransferases, such as insecticidal MTX toxin, were found in the bacterial species, which share common features with pierisin carrying N- and C-terminal domain structure plus 4 ricin-B like domains but recognize non-DNA substrates including proteins (Carpusca et al. 2006). Since pierisin family proteins are found only in the *Pieris* family among insects (Matsumoto et al. 2008, Takamura-Enya et al. 2004), it is likely that the only *Pieris* ancestor had acquired the pierisin ancestral gene for some reason, for example, the horizontal gene transfer from other organisms such as microbes, and evolved to use the pierisin function for its normal development or defense against parasites.

Based on its function, the ectopic expression of pierisin-1 could potentially be used to induce cell or tissue dysfunction for the development of model organisms with modified traits. However, no studies have examined whether cells that are genetically engineered to locally and intracellularly express pierisin-1 *in vivo* undergo apoptosis or have other physiological alterations. This is due to the strong cytotoxicity of the pierisin-1 protein. Even in the ordinary sub-type of *Escherichia coli* cells which proliferate quickly, molecular cloning of pierisin-1 DNA including the enzymatic activity region is found to be quite difficult due to the fact that a slight amount of unexpectedly-translated pierisin-1 protein could be sufficiently harmful for the living cells, although the sequence is designed not to be expressed in the bacterial cells. Effective preparation of pierisin-1 protein by the *in vitro* translation system using the non-living cell extracts is necessary for the investigation of its biological properties (Orth et al. 2011). Thus, the expression of pierisin-1 itself in any kind of living cell has been unsuccessful due to its strong cytotoxicity.

3.3 Utility of Pierisin-1A from Pieris rapae

A genome-wide sequence analysis of chromosomes of the *Pieris* species showed the presence of several homologous genes with pierisin-1 (Shen et al. 2016). Among these, a pierisin-1 homologue named pierisin-1A (P1A) was recently identified (Fig. 3) and its activity of ADP-ribosylation to DNA was compared with that of pierisin-1 (Otsuki et al. 2017). The products obtained by the enzymatic reaction of *in vitro*-translated P1A or pierisin-1 with DNA and β-NAD were examined by HPLC. *In vitro*-translated P1A protein exhibited approximately 5% of DNA ADP-ribosylating activity compared with pierisin-1 (Otsuki et al. 2017). To explore a genetic manipulation strategy for the induction of tissue-specific dysfunction, we

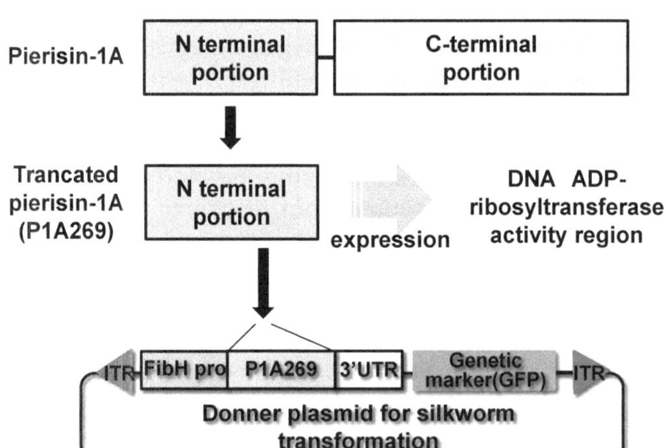

Fig. 3: Schematic diagram of the primary structure of full-length pierisin-1A and truncated pierisin-1A (P1A269) plus diagram of the donner plasmid for silkworm transgenesis in this research. ITR: inverted terminal repeat sequence of the piggyBac transposon; FibH pro: fibroin heavy chain promoter; 3′UTR: 3′ untranslated region.

recently established transgenic silkworms with forced expression of the putative ADP-ribosyltransferase domain from P1A in the PSGs (Otsuki et al. 2017) (Fig. 3).

3.4 Expression of Truncated Pierisin-1A in the Silkworm Posterior Silk Glands

It was necessary to investigate whether P1A can express in the insect cultured cells before consideration of the P1A expression in the organ cells of living silkworm individuals. Recombinant full-length P1A (P1AFull) and the P1AN-terminal portion containing the DNA ADP-ribosyltransferase domain, named P1A269, were confirmed to be transiently expressed in the *Bombyx mori* derived-BmN or *Spodoptera frugiperda* derived-Sf21 cells. Microscopy showed the presence of apoptotic cell fragments in Sf21 cells expressing full-length P1A and P1A269 up to 48 hours post-transfection, and a biochemical analysis revealed an increased level of effector caspase activities of cells, indicating that P1A lacking the C-terminal domain can induce DNA ADP-ribosylation and apoptosis in Sf21 cells. By contrast, the ectopic expression of P1A269 and full-length P1A lead non-apoptotic morphological changes until 96 hours in BmN cells, in which effector caspase was not activated. An investigation of the relationship between P1A function and cellular protein synthesis by measuring reporter luciferase activity from co-transfected plasmid showed that luciferase expression was lowered in BM-N cells (Otsuki et al. 2017), suggesting that P1A could repress protein expression and the morphological change of cell shape resulted from potential cell cycle arrest.

The silkworms carrying a transgene for P1A269 under control of the FibH promoter were generated by the conventional transgenic method using *piggyBac*-expressing helper plasmid and donner plasmid, as described in Fig. 3. The FibH promoter is known to be activated particularly in the PSGs. P1A269 had no potential secretary signal peptide so that it is designed to be expressed as an intracellular protein at the same time as the fibroin expression is activated. P1A269 mRNAs and proteins were confirmed to be expressed in the PSGs of the transgenic silkworms in response to the ecdysteroid secretion. PSGs from the transgenic silkworm larvae at the spinning stage were not ablated by apoptosis, but instead exhibited morphological abnormalities such as a dented appearance as compared with PSGs from non-transformed wild-type larvae (Fig. 4). The transgenic silkworm larvae with the expressed P1A-induced abnormality of PSGs produced thin-layered cocoon shells (Fig. 4) with the 79% weight loss of the cocoons from non-transformed larvae. The lost weight of the obtained thin-layered cocoon shells from the silkworms with PSGs expressing intracellularly P1A269 corresponded well to the potential

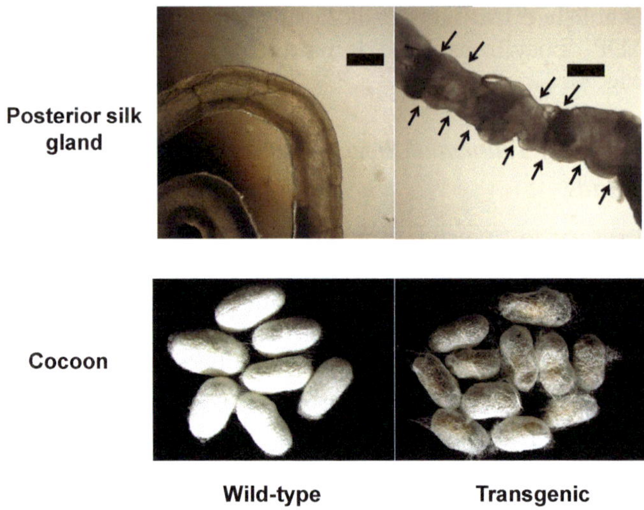

Fig. 4: Microscopic observation (upper) of the partial posterior silk gland of wild-type or the transgenic silkworm with the definite dents on the surface (arrows) and cocoons of wild-type and the transgenic silkworms. Bar, 500 μm.

fibroin weight that could be theoretically predictable in the wild-type cocoon.

3.5 Characteristics of the Bioengineered Silkworms with Loss of the PSG Function

An electrophoresis showed that solely non-fibroin proteins, mainly sericins (Otsuki et al. 2017, Sato et al. 2014, Teramoto et al. 2005) were detectable in the cocoons of the silkworms with PSGs expressing P1A269. RT-qPCR analyses showed that FibH and FibL mRNA levels in the PSGs of the transgenic silkworm larvae were shown to have strongly decreased. These results revealed that intracellularly-expressed P1A represses FibH and FibL protein synthesis in the PSGs of the transgenic silkworms. The messenger RNA level of a housekeeping gene, cytoplasmic actin A3, was not decreased by the P1A269 function, suggesting that P1A269 repressed the transcription of genes, including fibroin genes that are initiated concurrently with the expression of introduced P1A269, and the expression of housekeeping genes are apparently not influenced by P1A269 expression. During whole larval stages, silk gland cells are non-proliferative but grow in size. Thus, it appears that P1A269-induced repression of some gene expression may be responsible for the cellular enlargement of PSG during the larval development and cause the observed abnormalities.

The average weights of the transgenic silkworms after pupation were higher than those of non-transformed wild-type silkworms (Otsuki et al. 2017). This additional weight may be due to the retention of nutritional resources that would otherwise have been devoted to protein production, particularly fibroin, which is needed for cocoon spinning. In classical experiments, the surgical excision of silk glands from 4th and 5th instar larvae caused an accumulation of excess humoral amino acids such as, Gly, Thr, Ser, and Tyr before pupation. Larvae from which the entire silk gland was removed failed to pupate, likely due to the presence of excess amino acids (Fukuda et al. 1955). The transgenic silkworm pupae with excess weight showed no noticeable developmental defects after pupation; these transgenic silkworms were able to mature to adulthood and produce healthy transgenic offspring that can successfully reproduce for many succeeding generations. These results suggest that the retention of nutritional resources for fibroin synthesis does not appear to have negative effects on the growth of these transgenic silkworms.

The pupae with greater weights than wild-type silkworms, which can be obtained through bioengineering using the P1A gene and can be readily taken out alive from sericin cocoon without degumming treatment, could be used as an effective baculovirus host for production of important foreign proteins, such as the BmCPV polyhedra encapsulating cytokines, since the pupae would preserve nutrition that would have been needed for cocoon spinning, so that higher effectivity of the protein synthesis in the transgenic silkworm pupa with greater weights would be expected to be yielded than that in the non-transformed silkworms. Polyhedra are usually produced by the baculovirus system in cultured insect cells, which need the extracellular stimulation from the added mammalian serum for their proliferation (Smith et al. 1983). Preparations of polyhedra in silkworm pupae rather than cultured cells are an ideal system because the possibility of contamination of pathogens such as virus or prions derived from animal serum must be eliminated from the preparation steps of polyhedra, especially for medical use, and no mammal-derived substances such as serum would be contained in the system using silkworm pupa. Also, polyhedra would be readily isolated from pupae with soft tissue and skin by treatment with detergent.

3.6 Cell Proliferation on Sericin from the Transgenic Silkworms

Commercial sericin prepared through a process of degumming the raw silk from regular cocoons is unable to form hydrogels because the high temperature and alkaline pH conditions in degumming results in protein decomposition of sericin. However, intact, soluble, fibroin-free sericin at concentrations detectable by electrophoresis could be readily prepared from the transgenic silkworm sericin cocoons using lithium bromide

(Otsuki et al. 2017, Teramoto et al. 2005). Hydrogel formation of the intact, soluble sericin could then be induced following incubation with ethanol (Fig. 5). The gels can also be fabricated into creams, sheets, sponges or other 3-D shapes that have mechanical strength by air-drying or lyophilization (Fig. 5). It was not possible to use the same chemical method in order to obtain a fibroin-free, intact sericin solution from the wild-type silkworm cocoons due to the high level of fibroin contamination in the extracted low-concentration sericin solution. The fibroin contamination would cause the decreases in yield of the recovered intact, soluble sericin in this preparation (E. Kotani, personal communication). Intact, soluble sericin has moisture-retention properties that are superior to collagen and can gelate in solutions of up to 96% water.

To determine whether sericin hydrogels can act as scaffolds to support cell growth and differentiation, we cultivated the mouse ES cell line EB5 on sericin hydrogels that were overlaid with a medium containing canonical recombinant LIF or LIF-polyhedra (Nishishita et al. 2011, Niwa et al. 2000, Otsuki et al. 2017). Research has shown that the activity of cytokines encapsulated in polyhedra can be stably maintained over long periods (several years) in both *ex vivo* and *in vivo* environments and can be slowly released to continually stimulate the growth and differentiation of many cell types, as we already described (Kotani et al. 2015, Matsumoto et al. 2012, 2014, 2015, Nishishita et al. 2011, Shimabukuro et al. 2014). We selected this culture system using EB5 cells because differentiated EB5 cells

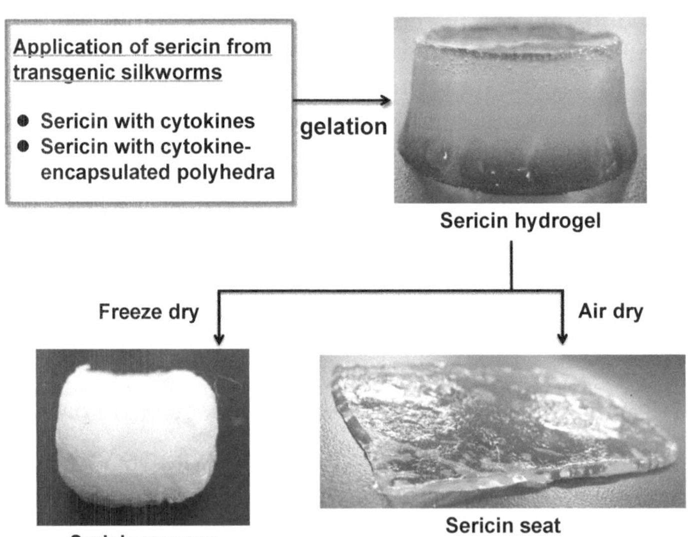

Fig. 5: Envisaged application of sericin from the transgenic silkworms.

are unable to proliferate in media containing blasticidin (Nishishita et al. 2011, Niwa et al. 2000), so that proliferation of only undifferentiated cells can be observed in this system. As a result, EB5 cells grown on sericin hydrogels with LIF-polyhedra and on sericin hydrogels overlaid with a medium containing recombinant LIF were found to form characteristic dome-shaped colonies while expressing alkaline phosphatase as a marker enzyme of the undifferentiated cells. Our results suggested that sericin hydrogels can serve as a scaffold that supports EB5 cell growth, provided that LIF is available and that the LIF could be released from the polyhedron micro-crystals, diffuse through the hydrogel in an active state and be released into the medium, continually stimulating the proliferation of undifferentiated EB5 cells.

4. Concluding Remark

In this chapter, we proposed the effective use of the BmCPV polyhedron micro-crystals, which can stabilize and slowly release the encapsulated cytokines, and can be applicable for cell growth control *in vivo* and *ex vivo*. Also, we demonstrated the potential utilities of the intact, soluble sericin from the transgenic silkworms expressing the truncated P1A protein in the PSGs. We previously generated transgenic silkworms that produced FGF-2-polyhedra inside silkworm PSGs, allowing for the prevention of loss of FGF-2 activity in the larval organ (Kotani et al. 2015). These PSGs could be fabricated into fibrous materials that stimulated NIH3T3 cell proliferation, suggesting that polyhedron-encapsulated cytokines produced in silkworm silk glands can be useful for producing pathogen-free artificial extracellular matrices (Kotani et al. 2015). In the future, further bioengineering could be performed in order to generate a transgenic silkworm with a middle silk gland that produces sericin incorporating polyhedron-encapsulated active cytokines, from which pathogen-free artificial extracellular matrices could be fabricated for use in tissue engineering. The use of bioengineered silkworms with the important trait can be practical by the application of the technology of BmCPV polyhedra for the protection of cytokine activity in the living silkworm organs, or in the fabricated sericin materials (Fig. 5). The production of these transgenic silkworms can be easily and economically scaled, particularly given the strong industrial foundation of the silk industry in Japan where technology to effectively rear silkworms on aseptic artificial diets is available in mass scale. Thus, from the above evidence and observations, integration among the distinct technologies of production of the cytokine-encapsulated polyhedron micro-crystals and bioengineering of silkworms to modify the properties of cocoons can promote development of silkworms, which have useful properties that are applicable in biomedical fields.

Acknowledgments

This work was supported by Grant-in-Aid for Scientific Research (Grant Nos. 15K07794 and 26450467) from The Ministry of Education, Culture, Sports, Science and Technology, Japan. This study was partly supported by The Japan Society for the Promotion of Science (JSPS) Program for Advancing Strategic International Networks to Accelerate the Circulation of Talented Researchers (Grant No. S2802) and Core-to-Core Program, B Asia-Africa Science Platforms.

References

Aruga, H. 1971. Cytoplasmic polyhedrosis of the silkworm—Historical, economical and epizootiological aspects. pp. 3–57. *In*: H. Aruga and Y. Tanada (eds.). The Cytoplasmic Polyhedrosis Virus of the Silkworm. University of Tokyo Press, Tokyo, Japan.

Belloncik, S. and H. Mori. 1998. Cypoviruses. *In*: L. K. Miller and L. A. Ball (eds.). The Insect Viruses. Plenum Press, New York, UK.

Carpusca, I., T. Jank and K. Aktories. 2006. *Bacillus sphaericus* mosquitocidal toxin (MTX) and pierisin: The enigmatic offspring from the family of ADP-ribosyltransferases. Mol. Microbiol. 62: 621–630.

Coulibaly, F., E. Chiu, K. Ikeda, S. Gutmann, P. W. Haebel, C. Schulze-Briese, H. Mori and P. Metcalf. 2007. The molecular organization of cypovirus polyhedra. Nature. 446: 97–101.

Fukuda, T., J. Kirimura, M. Matsuda and T. Suzuki. 1955. Biochemical studies on the formation of the silkprotein. I. The kinds of free amino acids concerned in the biosynthesis of the silkprotein. J. Biochem. 42: 341–346.

Ijiri, H., F. Coulibaly, G. Nishimura, D. Nakai, E. Chiu, C. Takenaka, K. Ikeda, H. Nakazawa, N. Hamada, E. Kotani, P. Metcalf, S. Kawamata and H. Mori. 2009. Structure-based targeting of bioactive proteins into cypovirus polyhedra and application to immobilized cytokines for mammalian cell culture. Biomaterials. 30: 4297–4308.

Ikeda, K., S. Nagaoka, S. Winkler, K. Kotani, H. Yagi, K. Nakanishi, S. Miyajima, J. Kobayashi and H. Mori. 2001. Molecular characterization of *Bombyx mori* cytoplasmic polyhedrosis virus genome segment 4. J. Virol. 75: 988–995.

Ikeda, K., H. Nakazawa, A. Shimo-Oka, K. Ishio, S. Miyata, Y. Hosokawa, S. Matsumura, H. Masuhara, S. Belloncik, R. Alain, N. Goshima, N. Nomura, K. Morigaki, A. Kawai, T. Kuroita, B. Kawakami, Y. Endo and H. Mori. 2006. Immobilization of diverse foreign proteins in viral polyhedra and potential application for protein microarrays. Proteomics. 6: 54–66.

Kanazawa, T., M. Watanabe, Y. Matsushima-Hibiya, T. Kono, N. Tanaka, K. Koyama, T. Sugimura and K. Wakabayashi. 2001. Distinct roles for the N- and C-terminal regions in the cytotoxicity of pierisin-1, a putative ADP-ribosylating toxin from cabbage butterfly, against mammalian cells. Proc. Natl. Acad. Sci. U S A. 98: 2226–2231.

Kanazawa, T., T. Kono, M. Watanabe, Y. Matsushima-Hibiya, T. Nakano, K. Koyama, N. Tanaka, T. Sugimura and K. Wakabayashi. 2002. Bcl-2 blocks apoptosis caused by pierisin-1, a guanine-specific ADP-ribosylating toxin from the cabbage butterfly. Biochem. Biophys. Res. Commun. 296: 20–25.

Kotani, E., N. Yamamoto, I. Kobayashi, K. Uchino, S. Muto, H. Ijiri, J. Shimabukuro, T. Tamura, H. Sezutsu and H. Mori. 2015. Cell proliferation by silk gut incorporating FGF-2 protein microcrystals. Sci. Rep. 5: 11051.

Koyama, K., K. Wakabayashi, M. Masutani, K. Koiwai, M. Watanabe, S. Yamazaki, T. Kono, K. Miki and T. Sugimura. 1996. Presence in *Pieris rapae* of cytotoxic activity against human carcinoma cells. Jpn. J. Cancer Res. 87: 1259–1262.

Kunz, R. I., R. M. Brancalhao, L. F. Ribeiro and M. R. Natali. 2016. Silkworm sericin: Properties and biomedical applications. Biomed. Res. Int. 2016: 8175701.

Lewandowski, L. J. and B. L. Traynor. 1972. Comparison of the structure and polypeptide composition of three double-stranded ribonucleic acid-containing viruses (diplornaviruses): Cytoplasmic polyhedrosis virus, wound tumor virus and reovirus. J. Virol. 10: 1053–1070.

Matsumoto, Y., T. Nakano, M. Yamamoto, Y. Matsushima-Hibiya, K. Odagiri, O. Yata, K. Koyama, T. Sugimura and K. Wakabayashi. 2008. Distribution of cytotoxic and DNA ADP-ribosylating activity in crude extracts from butterflies among the family *Pieridae*. Proc. Natl. Acad. Sci. U S A. 105: 2516–2520.

Matsumoto, G., T. Ueda, J. Shimoyama, H. Ijiri, Y. Omi, H. Yube, Y. Sugita, K. Kubo, H. Maeda, Y. Kinoshita, D. G. Arias, J. Shimabukuro, E. Kotani, S. Kawamata and H. Mori. 2012. Bone regeneration by polyhedral microcrystals from silkworm virus. Sci. Rep. 2: 935.

Matsumoto, G., R. Hirohata, K. Hayashi, Y. Sugimoto, E. Kotani, J. Shimabukuro, T. Hirano, Y. Nakajima, S. Kawamata and H. Mori. 2014. Control of angiogenesis by VEGF and endostatin-encapsulated protein microcrystals and inhibition of tumor angiogenesis. Biomaterials. 35: 1326–1333.

Matsumoto, G., T. Ueda, Y. Sugita, K. Kubo, M. Mizoguchi, E. Kotani, N. Oda, S. Kawamata, N. Segami and H. Mori. 2015. Polyhedral microcrystals encapsulating bone morphogenetic protein 2 improve healing in the alveolar ridge. J. Biomater. Appl. 30: 193–200.

Mertens, P. P. C., S. Rao and H. Zhou. 2005. Cypovirus. pp. 522–533. *In*: C. M. Fauquet, M. A. Mayo, J. Maniloff, U. Desselberger and L. A. Ball (eds.). Virus Taxonomy: Eighth Report of the International Committee on Taxonomy of Viruses. Elsevier Academic Press, London UK.

Mori, H., C. Shukunami, A. Furuyama, H. Notsu, Y. Nishizaki and Y. Hiraki. 2007. Immobilization of bioactive fibroblast growth factor-2 into cubic proteinous microcrystals (*Bombyx mori* cypovirus polyhedra) that are insoluble in a physiological cellular environment. J. Biol. Chem. 282: 17289–17296.

Mori, H. and P. Metcalf. 2010. Cypoviruses. pp. 307–324. *In*: S. Asgari and K. N. Johnson (eds.). Insect Virology. Caister Academic Press, Norfolk, UK.

Nakano, T., Y. Matsushima-Hibiya, M. Yamamoto, A. Takahashi-Nakaguchi, H. Fukuda, M. Ono, T. Takamura-Enya, H. Kinashi and Y. Totsuka. 2013. ADP-ribosylation of guanosine by SCO5461 protein secreted from Streptomyces coelicolor. Toxicon. 63: 55–63.

Nakano, T., A. Takahashi-Nakaguchi, M. Yamamoto and M. Watanabe. 2015. Pierisins and CARP-1: ADP-ribosylation of DNA by ARTCs in butterflies and shellfish. Curr. Top. Microbiol. Immunol. 384: 127–149.

Nishishita, N., H. Ijiri, C. Takenaka, K. Kobayashi, K. Goto, E. Kotani, T. Itoh, H. Mori and S. Kawamata. 2011. The use of leukemia inhibitory factor immobilized on virus-derived polyhedra to support the proliferation of mouse embryonic and induced pluripotent stem cells. Biomaterials. 32: 3555–3563.

Niwa, H., J. Miyazaki and A. G. Smith. 2000. Quantitative expression of Oct-3/4 defines differentiation, dedifferentiation or self-renewal of ES cells. Nat. Genet. 24: 372–376.

Omenetto, F. G. and D. L. Kaplan. 2010. New opportunities for an ancient material. Science. 329: 528–531.

Orth, J. H., B. Schorch, S. Boundy, R. Ffrench-Constant, S. Kubick and K. Aktories. 2011. Cell-free synthesis and characterization of a novel cytotoxic pierisin-like protein from the cabbage butterfly *Pieris rapae*. Toxicon. 57: 199–207.

Otsuki, R., M. Yamamoto, E. Matsumoto, S. I. Iwamoto, H. Sezutsu, M. Suzui, K. Takaki, K. Wakabayashi, H. Mori and E. Kotani. 2017. Bioengineered silkworms with butterfly cytotoxin-modified silk glands produce sericin cocoons with a utility for a new biomaterial. Proc. Natl. Acad. Sci. U S A. 114: 6740–6745.

Rohrmann, G. F. 1986. Polyhedrin structure. J. Gen. Virol. 67(Pt 8): 1499–1513.

Sato, M., K. Kojima, C. Sakuma, M. Murakami, Y. Tamada and H. Kitani. 2014. Production of scFv-conjugated affinity silk film and its application to a novel enzyme-linked immunosorbent assay. Sci. Rep. 4: 4080.

Shen, J., Q. Cong, L. N. Kinch, D. Borek, Z. Otwinowski and N. V. Grishin. 2016. Complete genome of *Pieris rapae*, a resilient alien, a cabbage pest and a source of anti-cancer proteins. F1000Res. 5: 2631.

Shimabukuro, J., A. Yamaoka, K. Murata, E. Kotani, T. Hirano, Y. Nakajima, G. Matsumoto and H. Mori. 2014. 3D co-cultures of keratinocytes and melanocytes and cytoprotective effects on keratinocytes against reactive oxygen species by insect virus-derived protein microcrystals. Mater. Sci. Eng. C Mater. Biol. Appl. 42: 64–69.

Smith, G. E., M. D. Summers and M. J. Fraser. 1983. Production of human beta interferon in insect cells infected with a baculovirus expression vector. Mol. Cell. Biol. 3: 2156–2165.

Takahashi-Nakaguchi, A., Y. Matsumoto, M. Yamamoto, K. Iwabuchi, Y. Totsuka, T. Sugimura and K. Wakabayashi. 2013. Demonstration of cytotoxicity against wasps by pierisin-1: A possible defense factor in the cabbage white butterfly. PLoS One. 8: e60539.

Takamura-Enya, T., M. Watanabe, Y. Totsuka, T. Kanazawa, Y. Matsushima-Hibiya, K. Koyama, T. Sugimura and K. Wakabayashi. 2001. Mono(ADP-ribosyl)ation of 2′-deoxyguanosine residue in DNA by an apoptosis-inducing protein, pierisin-1, from cabbage butterfly. Proc. Natl. Acad. Sci. U S A. 98: 12414–12419.

Takamura-Enya, T., M. Watanabe, K. Koyama, T. Sugimura and K. Wakabayashi. 2004. Mono(ADP-ribosyl)ation of the N2 amino groups of guanine residues in DNA by pierisin-2, from the cabbage butterfly, *Pieris brassicae*. Biochem. Biophys. Res. Commun. 323: 579–582.

Teramoto, H., K. Nakajima and C. Takabayashi. 2005. Preparation of elastic silk sericin hydrogel. Biosci. Biotechnol. Biochem. 69: 845–847.

Thurber, A. E., F. G. Omenetto and D. L. Kaplan. 2015. *In vivo* bioresponses to silk proteins. Biomaterials. 71: 145–157.

Watanabe, M., T. Kono, Y. Matsushima-Hibiya, T. Kanazawa, N. Nishisaka, T. Kishimoto, K. Koyama, T. Sugimura and K. Wakabayashi. 1999. Molecular cloning of an apoptosis-inducing protein, pierisin, from cabbage butterfly: Possible involvement of ADP-ribosylation in its activity. Proc. Natl. Acad. Sci. U S A. 96: 10608–10613.

Yu, X., L. Jin and Z. H. Zhou. 2008. 3.88 A structure of cytoplasmic polyhedrosis virus by cryo-electron microscopy. Nature. 453: 415–419.

Zhang, H., X. K. Yu, X. Y. Lu, J. Q. Zhang and Z. H. Zhou. 2002. Molecular interactions and viral stability revealed by structural analyses of chemically treated cypovirus capsids. Virology. 298: 45–52.

Chapter 15

Production of Medical and Cosmetic Products using Transgenic Silkworms

Masahiro Tomita

1. Introduction

Recombinant proteins have diverse medical and industrial applications. The sources of the biologics used in medicine have changed from animal tissues or human blood to recombinant human proteins because of the high safety and reliability requirements. A variety of recombinant protein production systems have been developed; these utilize bacteria (Cederbaum et al. 1984), yeasts (Cregg et al. 1993), mammalian cells (Lalonde and Durocher 2017) or insect cells (Cox 2012) as hosts. The systems based on microorganisms such as bacteria and yeasts produce proteins at relatively low cost, but these host organisms cannot synthesize high-molecular-weight proteins with mammalian-type post-translational modifications. Cultured Chinese hamster ovary (CHO) cells are frequently used to produce biologics (including antibodies). CHO cells can synthesize proteins with complex structures and post-translational modifications that also meet regulatory requirements. However, enormous investment is required for the construction and running of the bioreactors needed to culture CHO cells. Therefore, most CHO-cell-produced biologics are expensive, and the resulting high medical costs represent a considerable socioeconomic burden.

Immuno-Biological Laboratories Co., Ltd. 1091-1 Naka, Fujioka-Shi, Gunma 375-0005, Japan.
E-mail: do-tomita@ibl-japan.co.jp

The silkworm, *Bombyx mori*, spins silk threads composed of silk proteins in order to build a cocoon. During the 4,000–5,000-year history of sericulture, the silk productivity of the silkworm has been improved by selective breeding. As a result, the silkworm can synthesize a vast amount of silk proteins in a short period of time. Because it is a bulk product, the technology exists to handle silk on a very large scale. Therefore, the silkworm is a promising candidate as a host organism for production of recombinant proteins on a large scale.

To generate silkworms that produce recombinant proteins continuously, a transgenic silkworm system was developed. This system involves the introduction of foreign genes into silkworms using the *piggyBac* transposon to establish transgenic silkworm lines into the genomes, to which the foreign genes are integrated (Tamura et al. 2000). Recombinant proteins are expressed in silk glands, specialized organs in which silk proteins are synthesized, and are secreted as a constituent of silk threads.

2. Expression of Recombinant Proteins in the Silk Gland

The two main protein components of silk are fibroin and sericin (Grzelak 1995). Fibroin is the constituent of the silk core and is synthesized in the posterior silk gland (PSG). The fibroin synthesized in PSG cells is secreted into the PSG lumen and transported into the middle silk gland (MSG) (Fig. 1a), where it is coated with sericin. These silk proteins are transported to the anterior silk gland, from which they are pulled by a figure-eight movement of the worm's head. Sericin binds the two silk threads together as they emerge from a pair of glands. Thus, silk threads are composed of an

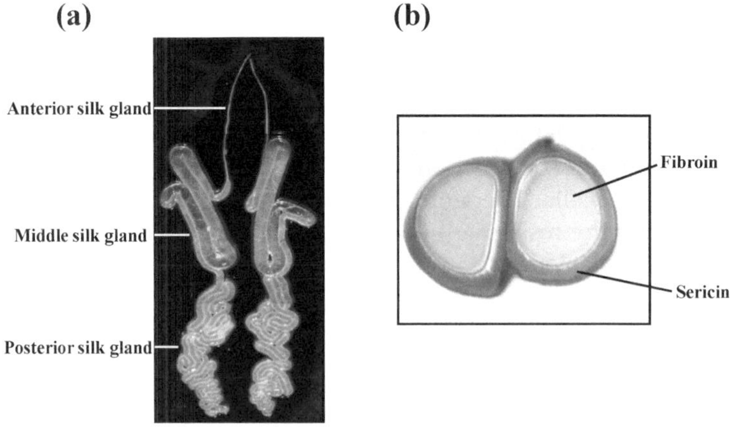

Fig. 1: Structures of the silk gland and silk thread. (a) Gross appearance of dissected silk glands. (b) Cross section of a silk thread stained with Azure B.

inner fibroin core and outer sericin layer (Fig. 1b). The fibroin and sericin weight contents of silk are approximately 75% and 25%, respectively.

In the transgenic silkworm system, the localization of recombinant proteins in silk threads can be controlled by modulating the location in which they are expressed: expression in the PSG leads to localization of the proteins in the inner fibroin core, and expression in the MSG leads to their localization in the outer sericin layer. Recombinant proteins in the insoluble fibroin threads, i.e., those produced in the PSG, must be extracted using strong chaotropic salts, such as guanidine thiocyanate or lithium thiocyanate (Kurihara et al. 2007, Tomita et al. 2003). Therefore, the biological activities of recombinant proteins are mostly lost during the extraction process. By contrast, expression of the proteins in the MSG leads to their localization in the outer sericin layer of silk threads (Ogawa et al. 2007, Tatematsu et al. 2010, Tomita et al. 2007). Because the outer sericin layer is more hydrophilic than the fibroin thread, recombinant proteins in the sericin layer can be recovered from cocoons by immersion in mild aqueous solutions, such as buffered saline, containing neutral detergents. Sericin forms insoluble meshwork structures composed of hydrophilic random coils and locally formed antiparallel β-sheets. The recombinant proteins are released from this sericin meshwork upon immersion. To date, various proteins have been produced using this MSG-expression method (Tomita 2011), several of which are used in commercial products.

3. Production of a Laminin 511-E8 Fragment

Laminin 511 is a member of the laminin family, distributed as extracellular proteins in the basement membranes of animal tissues. The receptor for laminin 511 is α6β1 integrin, which is abundantly expressed in embryonic stem (ES) cells and induced pluripotent stem (iPS) cells. ES and iPS cells have considerable potential for regenerative medicine, human disease modelling and drug discovery. Therefore, laminin 511 could be an important substrate molecule for the culture of ES and iPS cells. Indeed, recombinant laminin 511 synthesized by mammalian cells is a useful substrate for the culture of ES and iPS cells (Domogatskaya et al. 2008, Rodin et al. 2010). However, full-length laminin 511 is a large heterotrimeric protein that is difficult to produce on a large scale. Miyazaki et al. (2012) developed a recombinant laminin 511-E8 fragment that contained the α6β1 integrin-recognition site. The fragment promoted the growth of ES and iPS cells and sustained their self-renewal over long-term dissociated cell passaging. Laminin 511-E8 was produced using a CHO cell-expression system and launched by Nippi, Inc. as a commercial product. This product has a good reputation but is expensive because of the inadequate protein yield of CHO cell-based systems.

Production of laminin 511-E8 using a transgenic silkworm system has been investigated. Laminin 511-E8 is composed of three subunits, α5, β1 and γ1. A vector carrying the genes encoding the three subunits was constructed and injected into silkworm eggs in order to generate transgenic silkworms. The resultant silkworms synthesized laminin 511-E8 in the MSG and secreted it into the sericin layers of silk threads. Laminin 511-E8 was extracted from the silk threads by immersing cocoons in a neutral pH buffer containing a mild detergent.

The extracted laminin 511-E8 was purified by two column-chromatography steps, followed by sodium dodecyl sulfate-polyacrylamide gel electrophoresis (SDS-PAGE). Under reducing conditions, bands corresponding to the α5, β1 and γ1 subunits were evident on the gel. The electrophoretic mobility of each subunit was slightly greater than those of CHO-produced laminin 511-E8 (Fig. 2a). As explained below, the *N*-glycan chains on silkworm-produced proteins are shorter than those on mammalian proteins. Thus, the greater mobilities of the silkworm-produced subunits indicate smaller *N*-glycan structures. To investigate subunit assembly, purified laminin 511-E8 was analyzed under non-reducing conditions. The silkworm-produced laminin 511-E8 was detected as the α5 monomer and the β1-γ1 dimer linked by disulfide bonds (Fig. 2b). CHO-produced laminin 511-E8 yielded similar results. Thus, the silkworm-produced laminin 511-E8 had the same structure as that produced in CHO cells.

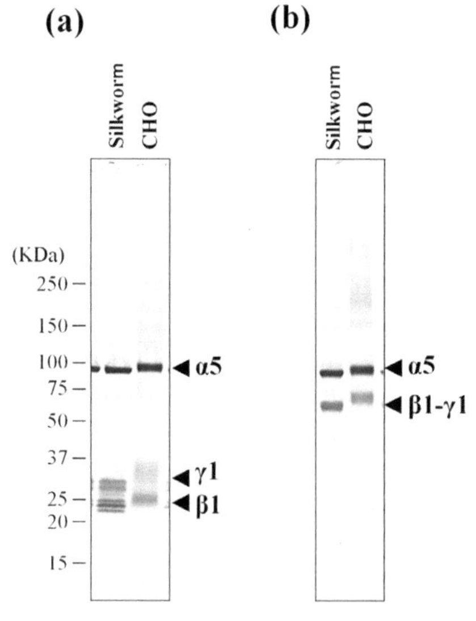

Fig. 2: SDS-PAGE analysis of purified laminin 511-E8. Silkworm-derived (Silkworm) and CHO-derived (CHO) laminin 511-E8 fragments were electrophoresed under reducing (a) and non-reducing (b) conditions, and silver stained.

The affinity of silkworm-produced laminin 511-E8 for α6β1 integrin was similar to that of CHO-produced laminin (Fig. 3a). iPS cells were cultured on dishes coated with silkworm-produced laminin 511-E8 in order to evaluate its suitability as a substrate. iPS cells on silkworm-produced laminin 511-E8-coated dishes formed tightly packed and flattened colonies (Fig. 3b), similar to those formed by iPS cells on dishes coated with CHO-produced laminin 511-E8. Fluorescence-activated cell sorting (FACS) analysis after three passages on silkworm-produced laminin 511-E8 revealed that the cells expressed, SSEA4, Tra1–60 and OCT3/4, markers of undifferentiated cells. Thus, the silkworm-produced laminin 511-E8 facilitated maintenance of iPS cells.

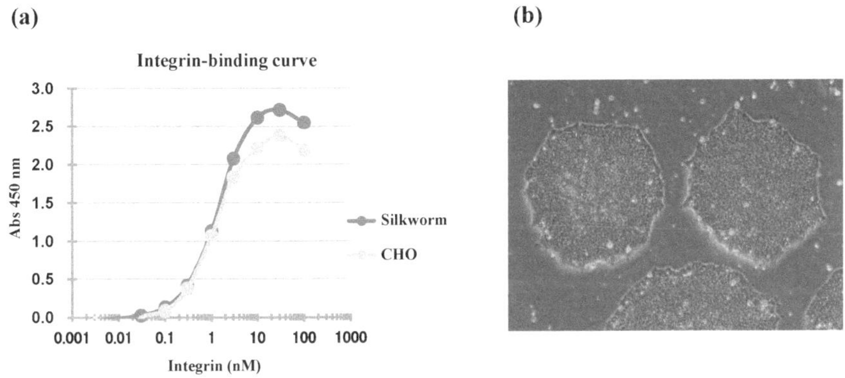

Fig. 3: Characterization of silkworm-produced laminin 511-E8. (a) Binding of silkworm-produced (Silkworm) and CHO-produced laminin 511 E8 (CHO) to α6β1 integrin. Data from Nippi, Inc. (b) iPS cells were cultured on dishes coated with silkworm-produced laminin 511-E8. Photograph provided by Dr. K. Sekiguchi of Osaka University.

As described above, the structure and function of silkworm-produced laminin 511-E8 are almost identical to those of CHO-produced laminin 511-E8. In addition, the production cost of the transgenic silkworm system is markedly lower than that of CHO cell-based systems. Immuno-Biological Laboratories Co., Ltd. (IBL) commercially produced laminin 511-E8 at an industrial scale using transgenic silkworms, and the product was marketed as "iMatrix-511 silk" through Nippi, Inc. and Matrixome, Inc. iMatrix-511 silk is widely used as a low-cost xeno-free substrate for culture of ES and iPS cells.

4. Production of Human Type I Collagen Chains for Cosmetics

Type I collagen is a component of the mammalian extracellular matrix, and is present in the dermis, tendon and bone at high rates, where it plays a structural role. Collagen is used as a biomedical material because it is strong, stable and compatible with living tissue. Moreover, collagen is used in cosmetics as a moisturizer. Traditionally, the major source of collagen was cow skin; however, this is associated with a risk for transmission of the abnormal prion protein that causes bovine spongiform encephalopathy (BSE).

Adachi et al. (2010) reported the production of a human type I collagen subunit, α1(I)-chain, using transgenic silkworms. The gene encoding the full-length triple-helix region of the human α1(I)-chain was introduced to silkworms, resulting in high-level expression in the MSG. Because the silk glands lack a prolyl hydroxylase (Adachi et al. 2005), the synthesized α1(I)-chain contained no hydroxyprolines, which are required for formation of the triple-helix structure of collagen. Therefore, the biochemical features of the α1(I)-chain were similar to those of gelatin rather than collagen. The suitability of the α1(I)-chain as a cell culture substrate was also determined. Human skin fibroblasts seeded on α1(I)-chain-coated dishes attached and spread, but at low chain concentrations, and the rate of spread was lower than that on collagen.

In the cosmetics industry, fish-derived collagens or their hydrolysis products are used in moisturizers. These materials have a lower risk for pathogen contamination but a higher risk for allergic reactions than animal-derived collagens because of the low sequence similarity between human and fish collagens. In fact, about one-third of allergic reactions to eating fish are reportedly caused by fish collagen, as evidenced by detection of IgE against the α1(I)- and α2(I)-chains of fish type I collagen (Sakaguchi et al. 2000).

In recent years, reports of anaphylactic reactions following percutaneous sensitization by proteins or their hydrolysis products in cosmetic products are accumulating. For example, in Japan, percutaneous sensitization to hydrolyzed wheat proteins in facial soap led to allergic reactions (Hiragun et al. 2013). Percutaneous sensitization by fish collagen in a skin-care product was also reported. The sensitized patient experienced anaphylaxis after ingesting a dietary supplement or gummy candy, both of which contained hydrolyzed fish collagen (Fujimoto et al. 2016).

To use the recombinant human α1(I)-chain produced from silkworms as a skin moisturizer with low allergenicity, the reactivity of human α1(I)-chain was surveyed to IgE from patients with fish collagen allergy via an ELISA-based method. The IgE reacted strongly with salmon collagen, but weakly to the human α1(I)-chain (Fig. 4a). Because the triple-helix structure is not required in cosmetics, an α1(I)-chain without triple helices can be used. IBL launched the α1(I)-chain as a novel cosmetic ingredient, Neosilk-Human Collagen I. This ingredient was used as a component of skin-care products under the brand name "frais vent" (Fig. 4b).

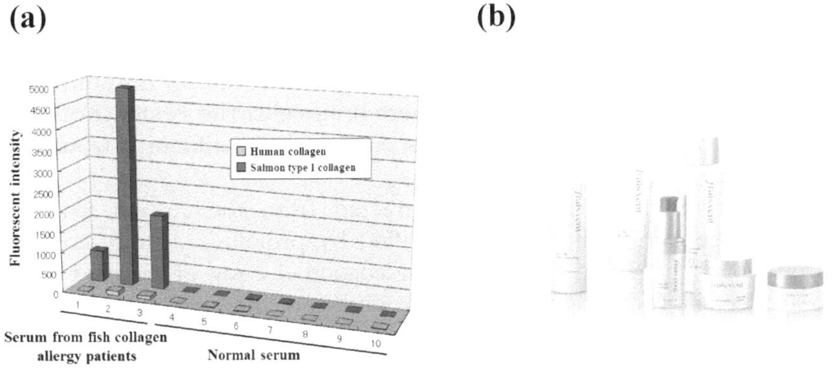

Fig. 4: Recombinant human collagen α1(I)-chain produced by transgenic silkworms. (a) Reactivity of IgE from patients with allergy to fish collagen to recombinant human collagen. Serum from three collagen-allergic and seven normal subjects was reacted with recombinant human collagen α1(I)-chain or salmon type I collagen. Collagen-bound IgE in serum was detected by an ELISA-based method. Data are derived from contract research to Dr. M. Sakaguchi of Azabu University. (b) Series of skin-care products containing recombinant human collagen α1(I)-chain.

5. Production of Monoclonal Antibodies for Diagnostic Use

Monoclonal antibodies (mAbs) are used to detect analytes in diagnostic specimens, such as blood or urine. Antibodies play a central role in the immune system, and recombinant mAbs comprise the fastest growing class of therapeutic proteins. Thus, mAbs are used in various medical applications, so there is an increasing need for their cost-effective production. In addition to the CHO-system, numerous mAb production systems using plants (Yusibov et al. 2016), filamentous fungi (Ward et al. 2004), chickens (Zhu et al. 2005) and insect cells (Hsu et al. 1997, Johansson et al. 2007) have been developed.

The transgenic silkworm system has also been investigated for cost-effective production of mAbs. Iizuka et al. (2009) reported production of a mouse mAb (IgG) in a transgenic silkworm system. They generated three transgenic lines, L-, H- and L/H-, which synthesized the mouse IgG L-chain, H-chain, and both the L- and H-chains, respectively. The L-line silkworm secreted the L-chain as a monomer into the cocoon, whereas the H-line silkworm secreted the H-chain as a dimer and higher-molecular-weight aggregates. In the L/H-line, the co-expressed L- and H-chains formed fully assembled L_2H_2, which was secreted as a major product into the cocoon. The L-chain monomer and H-chain dimer were not detected in the L/H-line cocoon. The amount of H-chain in the L/H-line cocoon was 2.3-fold higher than that in the H-cocoon. Thus, a recombinant mAb with an L_2H_2 structure was synthesized in MSG cells and preferentially secreted into the cocoon. In vertebrate antibody-producing cells, the quality-control mechanism for synthesis and secretion of IgG is present, by which the L_2H_2 molecule is efficiently synthesized in, and secreted from, the cells (Bole et al. 1986, Hendershot et al. 1987). The MSG cells of the silkworm may possess a similar mechanism to vertebrate cells. The existence of such a mechanism for IgG synthesis and secretion in the silkworm would be unexpected.

The recombinant mAb was extracted from the cocoon using a 3 M urea-containing buffer and purified by protein-G affinity column chromatography. From 500 mg of the cocoon, 1.2 mg of purified mAb was collected. The antigen-binding properties of the purified mAb were identical to those of a natural mAb from the hybridoma from which the introduced IgG genes were obtained.

Based on the above, IBL investigated the suitability of the silkworm-produced mAb for use in ELISA kits. The genes encoding the anti-amyloid-β mAb were cloned from hybridomas and introduced into silkworms in order to generate transgenic lines. The performance of the silkworm recombinant mAb as a capture antibody in an ELISA was comparable to that of the hybridoma-derived original mAb using serum and cerebral fluid samples (Fig. 5a). The silkworm-derived mAb exhibited higher lot-to-lot consistency. IBL now uses the silkworm-derived mAb in an ELISA kit for detecting amyloid-β (Fig. 5b).

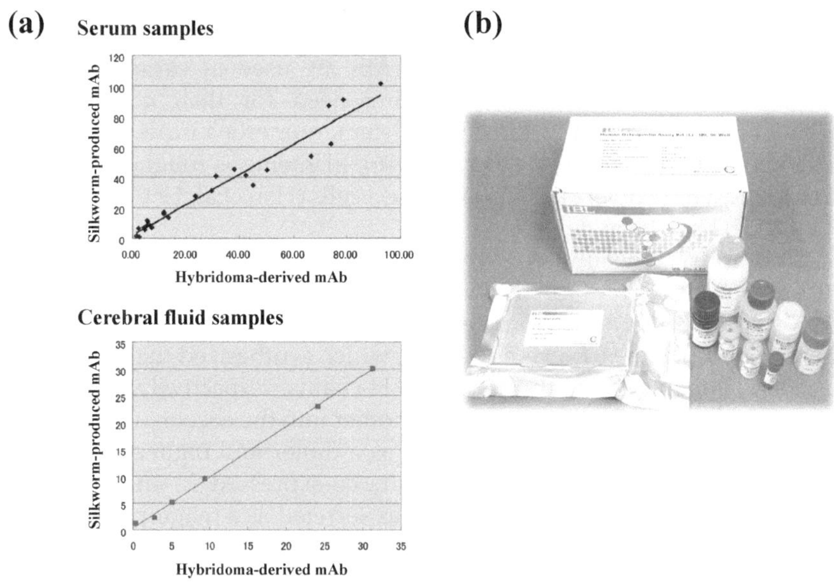

Fig. 5: ELISA kit using a silkworm-produced mAb. (a) Reactivity of the silkworm- and hybridoma-derived mAbs to serum (upper) and cerebral fluid (lower). (b) ELISA kit containing a silkworm-produced anti-amyloid-β mAb as a capture antibody.

IBL also supplies the silkworm-produced mAbs to several manufacturers of diagnostic products. mAbs are produced by injecting hybridomas into the peritoneal cavity of mice; the mAb is collected from their ascitic fluid. However, mass production of mAbs by this method is prohibited in many European countries for reasons of ethics. Silkworms are not subject to animal protection regulations and, thus, are a viable alternative to live mice for mAb production.

6. *N*-Glycan Structures of Silkworm-produced mAbs and their Therapeutic Potential

The *N*-glycans attached to glycoproteins produced in mammalian cells are of a complex-type terminally galactosylated and/or sialylated. By contrast, the major *N*-glycans in insects have paucimannose and high-mannose structures (Kubelka et al. 1994, Kulakosky et al. 1998) (Fig. 6). Paucimannose-type *N*-glycans are characteristic of insects and are not found in mammals. This type of *N*-glycan is formed from a hybrid-type *N*-glycan by removal of a GlcNAc residue by the Golgi membrane-associated enzyme β-*N*-acetylglucosaminidase (Altmann et al. 1995). Insect *N*-glycans also have considerable numbers of fucose residues α-1,3- and/or α-1,6-linked to the core GlcNAc residue (Kubelka et al. 1994, Staudacher et al. 1992). For example, the ratios of *N*-glycans with α-1,3-fucose and α-1,6-fucose, and both α-1,3- and α-1,6-fucoses, to the total amounts of *N*-glycans in membrane glycoproteins in Sf-21 cells are 1.8, 15, and 8.8%, respectively (Kubelka et al. 1994).

The *N*-glycan structure of the silkworm-produced mAb was different from those of insect proteins (Iizuka et al. 2009) (Fig. 6), e.g., paucimannose-type glycans were not detected. The major *N*-glycan types were the oligomannose-type (five mannose residues) and the terminally *N*-acetylglucosaminylated complex-type *N*-glycans. Moreover, the mAb *N*-glycans lacked fucoses linked to the core GlcNAc residue, and neither α-1,3- nor α-1,6-fucoses were detected. To determine the reason for these differences, the *N*-glycan structures of proteins in the cocoon and two larval tissues (MSGs and fat bodies) of wild-type silkworm were analyzed. The *N*-glycan structures of endogenous proteins in the cocoon and MSGs were similar to those of the recombinant mAb. In contrast, most fat-body *N*-glycans were of the fucosylated paucimannose type and the high-mannose type with more than seven mannose residues (Iizuka et al. 2009). Thus, the unique structural features of the *N*-glycans in the recombinant mAb are attributable to the tissue specificity of the silk glands. The amounts and/or kinds of *N*-glycan–processing enzymes in the MSGs might differ from those in other silkworm tissues.

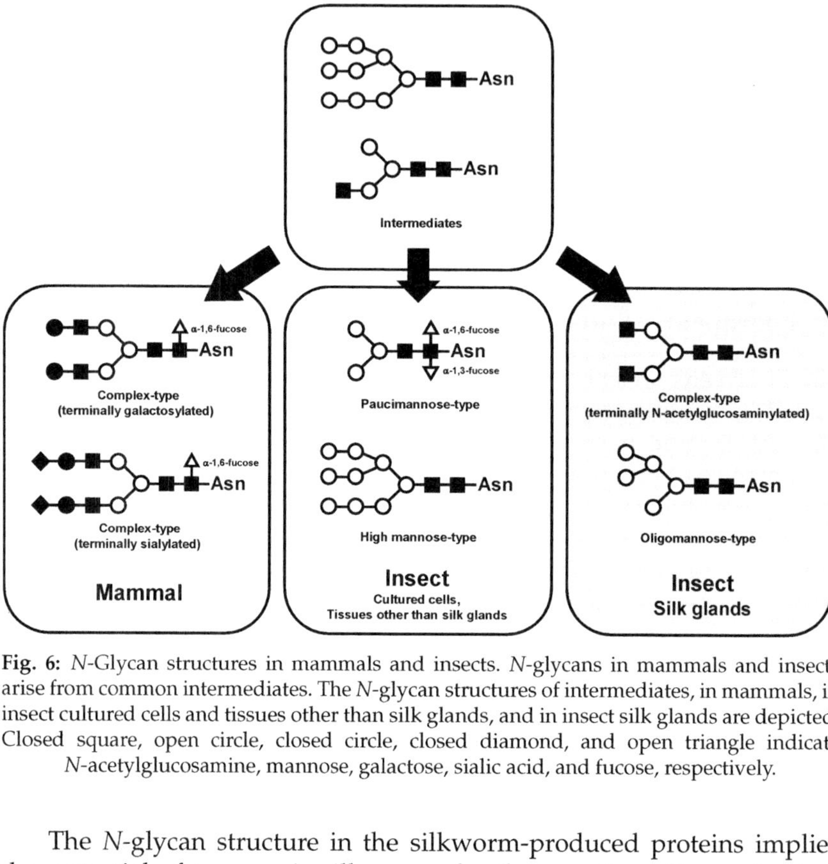

Fig. 6: *N*-Glycan structures in mammals and insects. *N*-glycans in mammals and insects arise from common intermediates. The *N*-glycan structures of intermediates, in mammals, in insect cultured cells and tissues other than silk glands, and in insect silk glands are depicted. Closed square, open circle, closed circle, closed diamond, and open triangle indicate *N*-acetylglucosamine, mannose, galactose, sialic acid, and fucose, respectively.

The *N*-glycan structure in the silkworm-produced proteins implies the potential of transgenic silkworms for therapeutic mAb production. Although the presence of oligomannose *N*-glycans is not always favorable for therapeutic use of mAbs, the absence of core fucosylation is beneficial. Fucose residues α-1,3-linked to GlcNAc show high antigenicity in humans (Bencúrová et al. 2004). Therefore, the presence of α-1,3-fucose in recombinant glycoproteins produced by insect cells is an important issue for their use in therapeutic applications. This may not be an issue with the transgenic silkworm system because no α-1,3-fucose residues were detected in the recombinant proteins. The absence of α-1,6-fucose also supports the use of this system in the production of mAbs. The absence of α-1,6-fucose enhances the antibody-dependent cellular cytotoxicity (ADCC) activity of IgG (Shields et al. 2002, Shinkawa et al. 2003). Thus, the transgenic silkworm system may enable production of mAbs with high ADCC activity.

Tada et al. (2015) reported that a silkworm-produced anti-CD20 mAb with the same amino acid sequence as rituximab had similar antigen-binding properties, but stronger ADCC activity than CHO-produced rituximab. The *N*-glycan structures of the silkworm-produced anti-CD20 mAb were similar to those of the aforementioned mouse mAb, i.e., no α-1,3- and α-1,6-fucoses in the core. Kurogochi et al. (2015) reported production of an anti-Her2 mAb known as trastuzumab using a transgenic silkworm system. The *N*-glycans of the mAb lacked core fucoses, and the mAb exhibited higher ADCC activity than CHO-produced trastuzumab (Fig. 7). Furthermore, the silkworm-produced mAb was used as an acceptor for transglycosylation, resulting in generation of glycoengineered mAbs with homogenous glycans that had various ADCC activities. Thus, the transgenic silkworm may facilitate production of mAbs with enhanced ADCC activity for treatment of cancer and infectious diseases.

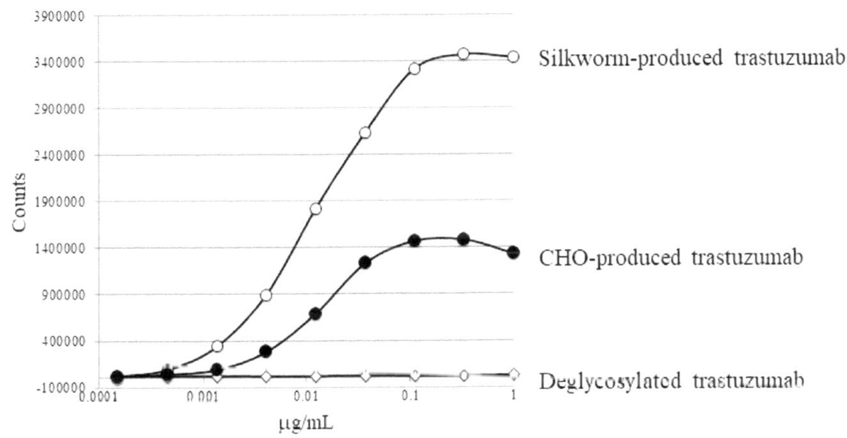

Fig. 7: ADCC activity of silkworm-produced and CHO-produced trastuzumab. Deglycosylated CHO-produced trastuzumab was used as the negative control.

7. Challenges in Production of Biologics using Transgenic Silkworms

Various recombinant proteins have been produced using transgenic silkworm systems, several of which have been launched as diagnostic and cosmetic products by IBL. However, IBL considers the production of therapeutic biologics using transgenic silkworms to be a principal goal. A promising candidate for silkworm-produced biologics is mAbs with enhanced ADCC activities. Another candidate is fibrinogen. Fibrinogen is used as a hemostatic agent, but because it is sourced from human blood there is a risk for virus transmission. IBL has produced human fibrinogen with normal hemostatic activity in transgenic silkworms, and is, in collaboration with a pharmaceutical company, developing the protein as a novel hemostatic agent.

To produce biologics, transgenic silkworms must be reared under good manufacturing practice (GMP)-compliant conditions. Therefore, IBL invested about 1 billion yen to construct a GMP-compliant pilot plant for producing biologics using transgenic silkworms. The plant is equipped for R&D and GMP production for clinical trials. The GMP areas consists of silkworm-rearing and purification rooms. The silkworm-rearing rooms are grade C clean rooms with four semi-automated silkworm-rearing devices (Fig. 8). The rearing device consists of automatically operated rearing-shelfs with controlled temperature, humidity and hygiene. Silkworms are reared on a heat-sterilized artificial diet, which facilitates bioburden control in the rearing devices. Using the four devices, about 90,000 silkworms can be reared simultaneously. IBL is currently engaged in developing several biologics in this pilot facility.

(a) **(b)**

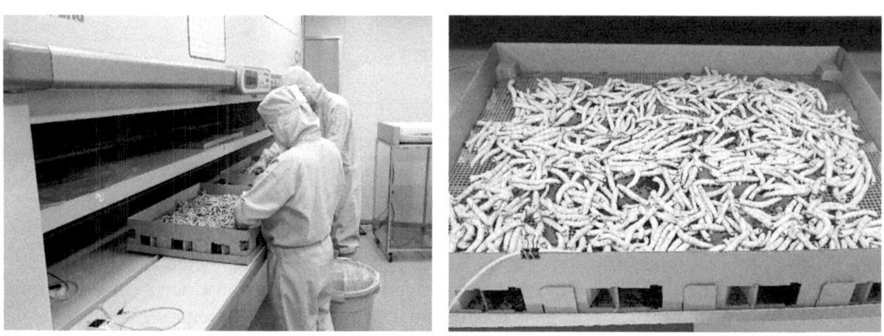

Fig. 8: GMP-compliant silkworm rearing. (a) A semi-automated silkworm-rearing device. (b) Silkworms being reared in the device.

8. Conclusions

Transgenic silkworms show potential for the cost-effective production of recombinant proteins at an industrial scale. Unlike production using microorganisms, silkworms can synthesize multi-subunit high-molecular-weight proteins such as laminin and fibrinogen. Furthermore, silkworms enable post-translational modifications of proteins, although the glycosylation pattern is different from that of mammalian cells. Because it is a bulk product, the technology exists to handle silk on a very large scale. Therefore, the silkworm shows promise for production of recombinant proteins on a large scale. Using the silkworm system, the yields of purified recombinant proteins are 2–5 mg per cocoon. Therefore, several hundred thousand cocoons are required in order to produce 1 kg recombinant protein. This corresponds to one to several lots of rearing in a sericultural farmhouse, i.e., the minimum scale used in sericulture. Thus, mass production of recombinant protein-containing cocoons is feasible. The knowledge accumulated over thousands of years of sericulture will facilitate cost-effective protein production using transgenic silkworms.

References

Adachi, T., M. Tomita and K. Yoshizato. 2005. Synthesis of prolyl 4-hydroxylase alpha subunit and type IV collagen in hemocytic granular cells of silkworm, *Bombyx mori*: Involvement of type IV collagen in self-defense reaction and metamorphosis. Matrix Biol. 24: 136–154.

Adachi, T., X. Wang, T. Murata, M. Obara, H. Akutsu, M. Machida, A. Umezawa and M. Tomita. 2010. Production of a non-triple helical collagen alpha chain in transgenic silkworms and its evaluation as a gelatin substitute for cell culture. Biotechnol. Bioeng. 106: 860–870.

Altmann, F., H. Schwihla, E. Staudacher, J. Glössl and L. März. 1995. Insect cells contain an unusual, membrane-bound b-N-acetylglucosaminidase probably involved in the processing of protein N-glycans. J. Biol. Chem. 270: 17344–17349.

Bencúrová, M., W. Hemmer and M. Focke-Tejkl. 2004. Specificity of IgG and IgE antibodies against plant and insect glycoprotein glycans determined with artificial glycoforms of human transferrin. Glycobiology. 14: 457–466.

Bole, D. G., L. M. Hendershot and J. F. Kearney. 1986. Post-translational association of immunoglobulin heavy chain binding protein with nascent heavy chains in non-secreting and secreting hybridomas. J. Cell Biol. 102: 1558–1566.

Cederbaum, S. D., G. C. Fareed, M. A. Lovett and L. J. Shapiro. 1984. Recombinant DNA in medicine. West J. Med. 141: 210–222.

Cox, M. M. 2012. Recombinant protein vaccines produced in insect cells. Vaccine. 30: 1759–1766.

Cregg, J. M., T. S. Vedvick and W. C. Raschke. 1993. Recent advances in the expression of foreign genes in *Pichia pastoris*. Biotechnology (N Y). 11: 905–910.

Domogatskaya, A., S. Rodin, A. Boutaud and K. Tryggvason. 2008. Laminin-511 but not -332, 111, or -411 enables mouse embryonic stem cell self-renewal *in vitro*. 2008. Stem Cells. 26: 2800–2809.

Fujimoto, W., M. Fukuda, T. Yokooji, T. Yamamoto, A. Tanaka and H. Matsuo. 2016. Anaphylaxis provoked by ingestion of hydrolyzed fish collagen probably induced by epicutaneous sensitization. Allergol. Int. 65: 474–476.

Grzelak, K. 1995. Control of expression of silk protein genes. Comp. Biochem. Physiol. B. Biochem. Mol. Biol. 110: 671–681.

Hendershot, L., D. Bole, G. Köhler and J. F. Kearney. 1987. Assembly and secretion of heavy chains that do not associate post-translationally with immunoglobulin heavy chain-binding protein. J. Cell Biol. 104: 761–767.

Hiragun, M., K. Ishii, T. Hiragun, H. Shindo, S. Hihara, H. Matsuo and M. Hide. 2013. The sensitivity and clinical course of patients with wheat-dependent exercise-induced anaphylaxis sensitized to hydrolyzed wheat protein in facial soap—secondary publication. Allergol. Int. 62: 351–358.

Hsu, T. A., N. Takahashi, Y. Tsukamoto, K. Kato, I. Shimada, K. Masuda, E. M. Whiteley, J. Q. Fan, Y. C. Lee and M. J. Betenbaugh. 1997. Differential N-glycan patterns of secreted and intracellular IgG produced in *Trichoplusia ni* cells. J. Biol. Chem. 272: 9062–9070.

Iizuka, M., S. Ogawa, A. Takeuchi, S. Nakakita, Y. Kubo, Y. Miyawaki, J. Hirabayashi and M. Tomita. 2009. Production of a recombinant mouse monoclonal antibody in transgenic silkworm cocoons. FEBS J. 276: 5806–5820.

Johansson, D. X., K. Drakenberg, K. H. Hopmann, A. Schmidt, F. Yari, J. Hinkula and M. A. Persson. 2007. Efficient expression of recombinant human monoclonal antibodies in *Drosophila* S2 cells. J. Immunol. Methods. 318: 37–46.

Kubelka, V., F. Altmann, G. Kornfeld and L. März. 1994. Structures of the N-linked oligosaccharides of the membrane glycoproteins from three lepidopteran cell lines (Sf-21, IZDMb-0503, Bm-N). Arch. Biochem. Biophys. 308: 148–157.

Kulakosky, P. C., P. R. Hughes and H. A. Wood. 1998. N-linked glycosylation of a baculovirus-expressed recombinant glycoprotein in insect larvae and tissue culture cells. Glycobiology. 8: 741–745.

Kurihara, H., H. Sezutsu, T. Tamura and K. Yamada. 2007. Production of an active feline interferon in the cocoon of transgenic silkworms using the fibroin H-chain expression system. Biochem. Biophys. Res. Commun. 20: 976–980.

Kurogochi, M., M. Mori, K. Osumi, M. Tojino, S. Sugawara, S. Takashima, Y. Hirose, W. Tsukimura, M. Mizuno, J. Amano, A. Matsuda, M. Tomita, A. Takayanagi, S. Shoda and T. Shirai. 2015. Glycoengineered monoclonal antibodies with homogeneous glycan (M3, G0, G2, and A2) using a chemoenzymatic approach have different affinities for FcγRIIIa and variable antibody-dependent cellular cytotoxicity activities. PLoS One. 10(7): e0132848. Doi: 10.1371/journal.pone.0132848. eCollection 2015.

Lalonde, M. E. and Y. Durocher. 2017. Therapeutic glycoprotein production in mammalian cells. J. Biotechnol. 251: 128–140.

Miyazaki, T., S. Futaki, H. Suemori, Y. Taniguchi, M. Yamada, M. Kawasaki, M. Hayashi, H. Kumagai, N. Nakatsuji, K. Sekiguchi and E. Kawase. 2012. Laminin E8 fragments support efficient adhesion and expansion of dissociated human pluripotent stem cells. Nat. Commun. 3: 1236.

Ogawa, S., M. Tomita, K. Shimizu and K. Yoshizato. 2007. Generation of a transgenic silkworm that secretes recombinant proteins in the sericin layer of cocoon: Production of recombinant human serum albumin. J. Biotechnol. 128: 531–544.

Rodin, S., A. Domogatskaya, S. Ström, E. M. Hansson, K. R. Chien, J. Inzunza, O. Hovatta and K. Tryggvason. 2010. Long-term self-renewal of human pluripotent stem cells on human recombinant laminin-511. Nat. Biotechnol. 28: 611–615.

Sakaguchi, M., M. Toda, T. Ebihara, S. Irie, H. Hori, A. Imai, M. Yanagida, H. Miyazawa, H. Ohsuna, Z. Ikezawa and S. Inouye. 2000. IgE antibody to fish gelatin (type I collagen) in patients with fish allergy. J. Allergy Clin. Immunol. 106: 579–584.

Shields, R. L., J. Lai, R. Keck, L. Y. O'Connell, K. Hong, Y. G. Meng, S. H. Weikert and L. G. Presta. 2002. Lack of fucose on human IgG1 N-linked oligosaccharide improves binding to human FccRIII and antibody-dependent cellular toxicity. J. Biol. Chem. 277: 26733–26740.

Shinkawa, T., K. Nakamura, N. Yamane, E. Shoji-Hosaka, Y. Kanda, M. Sakurada, K. Uchida, H. Anazawa, M. Satoh, M. Yamasaki, N. Hanai and K. Shitara. 2003. The absence of fucose but not the presence of galactose or bisecting N-acetylglucosamine of human IgG1 complex-type oligosaccharides shows the critical role of enhancing antibody-dependent cellular cytotoxicity. J. Biol. Chem. 278: 3466–3473.

Staudacher, E., V. Kubelka and L. März. 1992. Distinct N-glycan fucosylation potentials of three lepidopteran cell lines. Eur. J. Biochem. 207: 987–993.

Tada, M., K. Tatematsu, A. Ishii-Watabe, A. Harazono, D. Takakura, N. Hashii, H. Sezutsu and N. Kawasaki. 2015. Characterization of anti-CD20 monoclonal antibody produced by transgenic silkworms (*Bombyx mori*). MAbs. 7: 1138–1150.

Tamura, T., C. Thibert, C. Royer, T. Kanda, E. Abraham, M. Kamba, N. Komoto, J. L. Thomas, B. Mauchamp, G. Chavancy, P. Shirk, M. Fraser, J. C. Prudhomme and P. Couble. 2000. Germline transformation of the silkworm *Bombyx mori* L. using a piggyBac transposon-derived vector. Nat. Biotechnol. 18: 81–84.

Tatematsu, K., I. Kobayashi, K. Uchino, H. Sezutsu, T. Iizuka, N. Yonemura and T. Tamura. 2010. Construction of a binary transgenic gene expression system for recombinant protein production in the middle silk gland of the silkworm *Bombyx mori*. Transgenic Res. 19: 473–87.

Tomita, M., H. Munetsuna, T. Sato, T. Adachi, R. Hino, M. Hayashi, K. Shimizu, N. Nakamura, T. Tamura and K. Yoshizato. 2003. Transgenic silkworms produce recombinant human type III procollagen in cocoons. Nat. Biotechnol. 21: 52–56.

Tomita, M., R. Hino, S. Ogawa, M. Iizuka, T. Adachi, K. Shimizu, H. Sotoshiro and K. Yoshizato. 2007. A germline transgenic silkworm that secretes recombinant proteins in the sericin layer of cocoon. Transgenic. Res. 16: 449–465.

Tomita, M. 2011. Transgenic silkworms that weave recombinant proteins into silk cocoons. Biotechnol. Lett. 33: 645–654.

Ward, M., C. Lin, D. C. Victoria, B. P. Fox, J. A. Fox, D. L. Wong, H. J. Meerman, J. P. Pucci, R. B. Fong, M. H. Heng, N. Tsurushita, C. Gieswein, M. Park and H. Wang. 2004. Characterization of humanized antibodies secreted by *Aspergillus niger*. Appl. Environ. Microbiol. 70: 2567–2576.

Yusibov, V., N. Kushnir and S. J. Streatfield. 2016. Antibody production in plants and green algae. Annu. Rev. Plant Biol. 67: 669–701.

Zhu, L., M. C. van de Lavoir, J. Albanese, D. O. Beenhouwer, P. M. Cardarelli, S. Cuison, D. F. Deng, S. Deshpande, J. H. Diamond, L. Green, E. L. Halk, B. S. Heyer, R. M. Kay, A. Kerchner, P. A. Leighton, C. M. Mather, S. L. Morrison, Z. L. Nikolov, D. B. Passmore, A. Pradas-Monne, B. T. Preston, V. S. Rangan, M. Shi, M. Srinivasan, S. G. White, P. Winters-Digiacinto, S. Wong, W. Zhou and R. J. Etches. 2005. Production of human monoclonal antibody in eggs of chimeric chickens. Nat. Biotechnol. 23: 1159–1169.

CHAPTER 16

Comparative Investigation of Influenza Virus-like Particles from Synthetic Chemical Substance and Silkworm

Kuniaki Nerome,[1,*] *Kazumichi Kuroda*[2] and *Shigeo Sugita*[3]

1. Introduction

Modern civilization has undoubtedly resulted in the destruction of nature and has caused changes in biological ecosystems. A consequence of the changing environment is the occurrence of a variety of genetic mutations in animals and plants. The emergence of new viral pathogens and cancers may be associated with genetic evolution and the incidence of the related diseases is increasing day-by-day worldwide (Katanoda et al. 2015) (International agency for research on cancer, WHO, World cancer report 2014, http://publications.iarc.fr/Non-Series-Publications/World-Cancer-Report 2014). In Fig. 1b, we can see chickens in green grasses and shrubs. In fact, all vertebrate RNA viruses may originate from small vertebrates such as reptile amphibian, fish and bird. For example, Picornaviridae, Filoviridae (Ebola), Astroviridae, Caliciviridae, Paramyxoviridae, Orthomyxoviridae and other RNA virus family have all come from amphibian, fish and mammal (Shi et al. 2018).

[1] The Institute of Biological Resources, 893-2, Nakayama, Nago, Okinawa 905-0004, Japan.
[2] Nihon University School of Medicine, 30-1 Oyaguchi-kamicho, Itabashi-ku, Tokyo 173-8610, Japan.
[3] Equine Research Institute, Japan Racing Association, 321-4 Tokami-cho, Utsunomiya-shi, Tochigi 320-0856, Japan.
* Corresponding author: rnerome_ibr@train.ocn.ne.jp

On the basis of the above speculation, is it an unreasonable speculation that these vertebrate RNA viruses may pass through plant species? In the case of influenza virus, the virus evolution route can be followed from plant to human via birds. Figure 1a presents a typical peaceful scene of a river in the deep forest of a Japanese farming village in Hokkaido, showing water, trees and even fish under the water. The well-known Japanese archipelago lies off the east coast of the Asian continent, which extends from northeast to southwest. Most of the country is covered with deep forest and contains natural water sources in abundance. Furthermore, a large number of wild birds migrate to Japan from northern countries in autumn.

As shown in the right part of Fig. 1a, all sixteen sub-types of influenza A virus are distributed throughout several bird species, including ducks, swans and wild geese, and they cause unprecedented damage via highly

a

b

Fig. 1: Bird migration scenery in the winter of Hokkaido (a) and chickens and virus families in green grasses (b).

pathogenic avian influenza viruses (HPAI). Over two million chickens were killed in relevant farms in Japan between 2004 and 2015 in order to control the outbreak (Fig. 2b). Furthermore, 2,730,000 chickens were killed in Japan over this 11-year period. Undoubtedly, healthy chickens were included in this culling. Culling of livestock is the principal strategy used to control communicable viral diseases. It might be more appropriate to place emphasis on antiviral drugs or disinfection of the livestock husbandry area. In addition, increased priority should be given to the use of veterinary vaccination in controlling viral outbreaks.

This chapter deals with a virus-like particle (VLP) vaccine and its usefulness in controlling current and emerging viral threats. Along with three chickens around a shrub, a number of RNA virus family members and influenza virus are shown in Fig. 1b. Moreover, the fact that the ancestors of all RNA viruses have originated from plants, suggests that future emerging viruses will have a similar origin. Considering the increased rate of virus emergence, the development of VLP vaccines has become much more important.

2. Origin of VLP Vaccine

The development of improved influenza vaccines, with high immunogenicity and minimal adverse reactivity is important for the successful use of all inactivated vaccines. The use of modern technologies for purification or disruption of viral particles has resulted in safer inactivated vaccines, including sub-unit vaccines. Drs. Hiromich Mizutani and Kuniaki Nerome first developed an experimental ether-split, or sub-unit, vaccine in Japan in 1969 and this test vaccine was subsequently approved by the Ministry of Health and Welfare of Japanese Government in 1972 and is still used in Japan. In a mass immunization study of schoolchildren in Japan, which has been carried out since 1972, the split vaccine, which had the advantage of being safe in young children, appeared to be less immunogenic than the whole virus particle vaccine. An analysis of excess mortality over a period of 40 years in Japan showed that vaccines did not prevent the occurrence of high rates of excess mortality caused by influenza and pneumonia in six influenza seasons (1965, 1968, 1976, 1983, 1991, 1993) (Fig. 2a).

Japan was obliged to develop a more potent and effective influenza vaccine in order to reduce the excess mortality figures. However, this vaccine still contained unnecessary structural proteins such as polymerases, ribonucleoproteins and matrix proteins of influenza virus, which could hamper the attempt to increase the antigenic content of the vaccine. In order to solve this problem, K. Nerome invented a new type of potent influenza vaccine, which is described below.

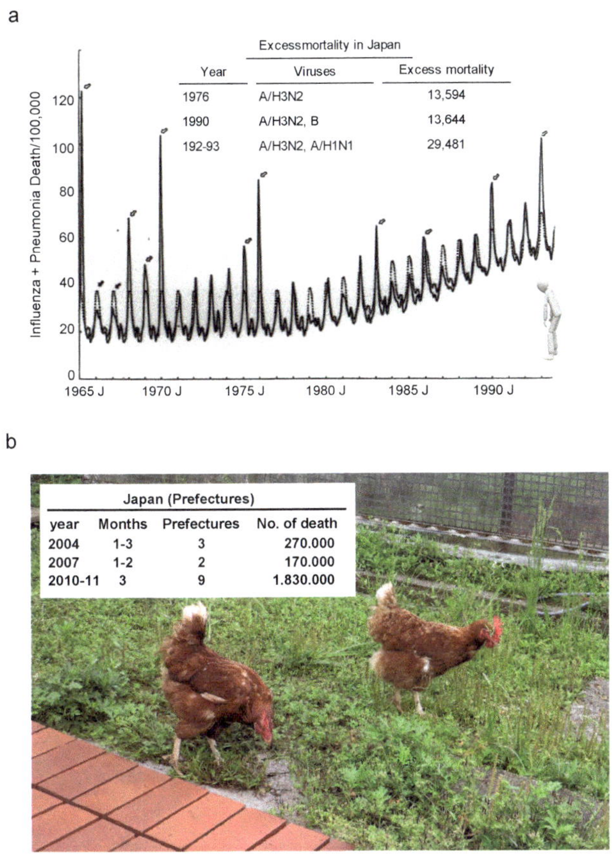

Fig. 2: Excess mortality in Japan (a) and numbers of chickens killed in Japan (b).

2.1 Preparation of the First VLP (Virosome) Vaccine

First, Nerome selected and used three substances and two subunit components of influenza virus. Muramyldipeptide (MDP), [6-O-(2-tetradecyl)-N-acetyl muramyl-L-isoglutamine], cholesterol, n-octylglucoside, neuraminidase (Arunachalam) and hemagglutinin (HA) were mixed and the mixture was stirred overnight after sonication. The resultant mixture was dialyzed against serial dilutions of 1,000–2,000 volumes using Mg^{++} and Ca^{++} ion-free phosphate buffered saline (PBS (–)) with stirring to remove n-octylglucoside. During dialysis, a change from clarity to opacity was observed, which indicated the formation of virosome (VLP). A series of technical procedures is shown in Fig. 3a–c. A model of liposome (virosome, VLP) is shown in Fig. 3e. The initial name given to this vaccine was liposome as artificial membrane vaccine. Later, on the basis of the structure observed under electron microscope, the

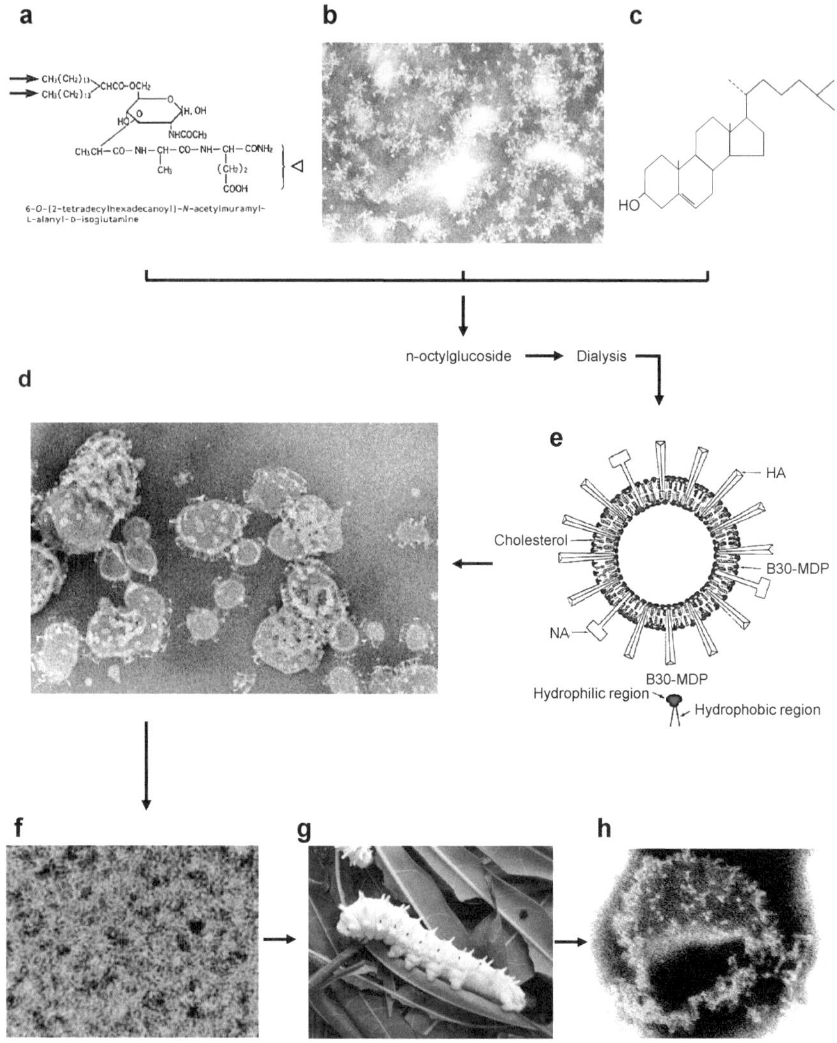

Fig. 3: Chemical structure of B30-muramyldipeptide (B30-MDP) (a), purified hemagglutinin (HA) from egg grown influenza virus (b), chemical structure of cholesterol (c), MDP-virosome (VLP) (d), design of virosome (e) and purified HA from silkworm grown VLP (f), silkworm larvae (g) and the first VLP from silkworm (h).

vaccine was named virus-like artificial liposome (virosome) vaccine. As seen in the electron micrograph (Fig. 3d) and the model structure (Fig. 3e), recent VLP vaccine was very similar to the virosome vaccine. Therefore, we assumed the virosome vaccine as the first VLP vaccine prepared using synthesized chemical lipids like B30-MDP (Fig. 3a) (Bright et al. 2007, Nerome et al. 1990).

2.2 Immune Responses against the First VLP (Virosome) Vaccine

The humoral immune responses in mice, elicited by virosome vaccine prepared using MDP, appeared to be higher than those induced by HANA proteins (HA and neuraminidase (NA)) (data not shown). For example, antibody titers against the Bangkok/1/79 (H3N2) and A/Philippine/2/82 (H3N2) were determined. As a result, the sub-unit vaccine without B30-MDP appeared to elicit low levels of antibody production throughout the entire immunization period. In contrast, the antibody titer in B30-MDP virosome vaccinated mice was more than 3-fold higher than that measured in the sub-unit vaccinated mice. In addition, the immune responses elicited by the trivalent B30-MDP vaccine in humans were much higher than those elicited in mice (Fig. 4a). The responses against B30-MDP vaccine in humans, were much higher compared to the responses against sub-unit vaccine. The HI titers in the humans vaccinated with sub-unit vaccine were estimated to be 14, 14, and 20 against A/Yamagata/120/86 (H1N1), A/Fukushima/K29/85 (H3N2), and B/Nagasaki 1/87, respectively. In contrast, the titers

a

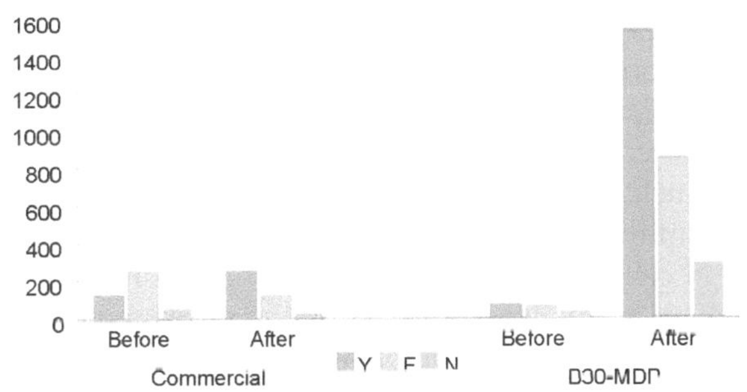

b

Challenge viruses	Spleen cells primed	Pulmonary virus titers (log_{10} PFU) on Day 5
A/Bangkok/1/79 (H3N2)	MDP Virosome	3.3 ± 0.4
	f-HANA	4.9 ± 0.6
	PBS	5.1 ± 0.5
A/Philippines/2/82 (H1N1)	MDP Virosome	3.9 ± 0.6
	f-HANA	5.0 ± 0.5
	PBS	5.2 ± 0.2
A/Yamagata/120/86 (H1N1)	MDP Virosome	5.2 ± 0.5
	f-HANA	5.1 ± 0.7
	PBS	5.3 ± 0.8

Fig. 4: Comparison of antibody rise in humans after immunization with commercial seasonal vaccines and experimental B30-MDP vaccine (a) and cellular immunity based on the killer cell activity of B30-MDP vaccine (b).

in humans vaccinated with B30-MDP vaccine containing the same antigens were significantly higher and were 1,536 (H1N1), 853 (H3N2) and 298 (B), respectively. Interestingly, the above result strongly suggested that B30-MDP vaccine should be very useful in humans as seasonal influenza vaccines. Figure 4b also shows the characteristics of the cellular immune responses induced by the B30-MDP vaccine. In addition to specific stimulation of lymphocytes based on the delayed cellular immunity (data not shown), killer cell activity was also induced in mice immunized with the B30-MDP-virosome vaccine (Fig. 4b). Additionally, enhancement of the cellular immune response, measured by delayed type hypersensitivity reactions, was also observed in the guinea pigs immunized with MDP-virosome vaccine (data not shown). The development of a series of virosome-VLP vaccines and their application has been a success story. A future challenge is to produce the hemagglutinin sub-unit, a crucial component in the preparation of the VLP.

3. Mass Production of VLP Vaccines in Silkworm

The urgent push to develop high-quality human H3N2, swine H1N2, and avian H5 vaccines was to save humans, swine and also a large number of chickens from a considerable amount of damage. First, we cloned HA genes of Hong KongH3N2, A/PR/8/34 (H1N1) and A/swine/Ehime/1/80 (H1N2). They were inserted into a transfer vector in order to produce recombinant *Bombyx mori* nuclearpolyhedrovirus (BmNPV) and their expressions were confirmed by immunofluorescence method using silk worm (*Bombyx mori*)-derived Bm-N cells, which were infected with recombinant BmNPV. The expression levels were also examined by the measurement of HA titers of homogenate prepared from recombinant BmNPV infected Bm-N cells. This recombinant BmNPV could also produce HA proteins in silkworm larvae. When larvae were infected with the recombinant virus, the homogenates of the larvae showed high expression levels with 8192-16384 HA titers. Figure 3h is the first electron micrograph of VLP vaccine prepared with HA proteins produced in recombinant virus-infected larvae. Although this premature VLP did not show typical VLP structure, it was covered with HA-like projections. Mean expression level of three recombinant BmNPVs expressing HA was estimated to be 11,000 HA-titer and this expression level was approximately 10 times higher than that of allantoic fluid of chicken eggs inoculated with high yielding strains of H3N2 and H1N1 viruses. In May 1997, a 5-year-old boy in Hong Kong died from an infection of avian H5N1 virus. In the WHO informal meeting, the development of a vaccine against this virus was thoroughly discussed and Nerome developed avian H5 silkworm VLP vaccine using cloned nonoptimized HA gene. Its expression level was almost similar to that of above three strains. In Fig. 3f and g, HA protein produced in silkworm larvae were shown for preparation of first silkworm derived VLP.

4. Establishment of Large Scale Production of Avian Influenza VLP Vaccine

Our research group had successfully produced large amount of HA proteins using cloned single HA gene in silkworm larvae (Fig. 3g). In order to enhance the production in silkworms we designed target HA gene optimized to silkworm specific codons. To improve the immunogenicity of influenza sub-unit vaccines, a number of VLPs have been employed in insect cells (Bright et al. 2007, Cox 2008, Galarza et al. 2005, Gavrilov et al. 2011, Latham and Galarza 2001, Pan et al. 2010) and in this system M1 (matrix 1) and NA genes, other than the HA gene, were co-expressed for production of HA sub-unit vaccine. In contrast, our system produced large amounts of VLPs through the expression of a single synthetic and silk worm codon optimized gene.

4.1 The Mass Production of an Artificial Influenza H5 VLP Vaccine in Silkworm Pupae

As discussed before, in 1997, avian H5N1 influenza virus was isolated from a boy in Hong Kong. The 1997 H5N1 outbreak in Hong Kong was the first known-case of H5N1 infecting humans (Hiromoto et al. 2000, Hiromoto et al. 2000, Shortridge et al. 1998) (WHO. H5N1 highly pathogenic avian influenza: timeline of major events. WHO: 2014, at www.who.int/influenza/human_animal_interface/h5na1_avian_influenza_update20140714.pdf?ua=1&=1). This outbreak in Hong Kong was not restricted to chickens; 18 humans were also infected, 6 of whom died (Hiromoto et al. 2000). Because the World Health Organization has highlighted the risk of a human pandemic resulting from the avian HPAI virus (WHO. Unprecedented spread of avian influenza requires broad collaboration. WHO:2014 at http://www.who.int/mediacentre/news/releases/2004/pr7/en/), avian influenza vaccines for chickens and humans have been developed. Further research on developing vaccines with more efficacious properties is underway (Baz et al. 2013, Steel 2011).

Although studies to develop more effective and safer vaccines are ongoing worldwide (Arunachalam 2014, Baz et al. 2013, Karron et al. 2009, Steel 2011, Suguitan et al. 2009, Suguitan et al. 2006, Treanor et al. 2006), the development of a vaccine against HPAI H5 viruses is associated with intrinsic problems such as low immunogenicity of the vaccines (Treanor et al. 2006) and the biohazard risk of using an infectious HPAI virus for the production of the vaccine. To solve these problems, we modified H5 HA gene that lack four basic amino acids located between HA1 and HA2 subunits and associated with pathogenicity of HPAI viruses (Fig. 6a). Prior to the above experiment, we selected the H5 vaccine viruses because a lot of H5 antigenic variants have circulated in wild ducks.

As shown in Fig. 5a, a large number of H5N1 viruses were present in wild ducks (Nerome et al. 2015), some of which caused severe outbreaks in many parts of the world. As seen in Fig. 5a, they were evolutionarily divided into 3 branches (subclades such as 2.3.4: 2.3.2: 2.3.2.1). Most of these viruses appeared to show a similar evolutionary rate. However, a small number of H5N1 virus (clade 2.3) evolved slowly; these were frequently transmitted between wild ducks and chickens (Fig. 5b). Therefore, we selected a vaccine strain (A/tufted duck/Fukushima/16/2011 (H5N1)) belonging to clade 2.3 and synthesized the HA gene of this virus. We showed that the chimeric HA DNA gene contained 77.5% of HA gene and approximately 22% of silkworm codon nucleotides. The chimeric gene was inserted into a baculovirus transfer vector, pBM-8, in order to prepare H5-HA recombinant baculovirus. The resultant recombinant baculovirus, H5-HA-BmNPV, was inserted into Bm-N cells and the expression of H5-HA gene was confirmed by immunofluorescence method and western-blot analysis (Nerome et al. 2015).

After the confirmation of the expression H5-HA gene, H5HA-BmNPV was inoculated into silkworm pupae. As shown in Fig. 8b, apparent HA production in the pupae was detected in the first day after inoculation and HA titers increased gradually over 4 days. Interestingly, the codon optimization effect indicated by the final HA titers was more than 500,000 (Fig. 8b). To characterize the structure, the homogenates of the pupae were centrifuged through a 10–50% (w/w) sucrose density gradient at 25,000 rpm for 120 minutes using sw28 swing rotor (Hitachi). After centrifugation, the gradient was fractionated, and each fraction was examined for protein, sucrose concentration and HA activity. As shown in Fig. 8a, the peaks of HA activity and protein concentration were observed in fraction 25 and six fractions (fraction 21–26) were collected. The collected fractions were further centrifuged at 25,000 rpm for 4 h and the pellet was examined under H-7600 electron microscope (Hitachi). As shown in Fig. 6b–d, the VLP (approximately 100 nm diameter) was covered with HA projections. They were also similar to the original H5N1 influenza viral particles (Fig. 7a) and H3N2 Hong Kong (Fig. 7d, f). The length of the projections was estimated to be 140 Å. Additionally, a large number of triangle-shaped head structures were observed in Fig. 6d. The long, filamentous, authentic H5 influenza virus structures were observed in Fig. 7a and their approximate diameter was one third of the large VLP structure (Fig. 7b). Calculation of surface projections on the authentic virion and the VLP revealed that thirteen projections are arranged on the 100 nm envelope of virus and VLP structures (Fig. 7c–f).

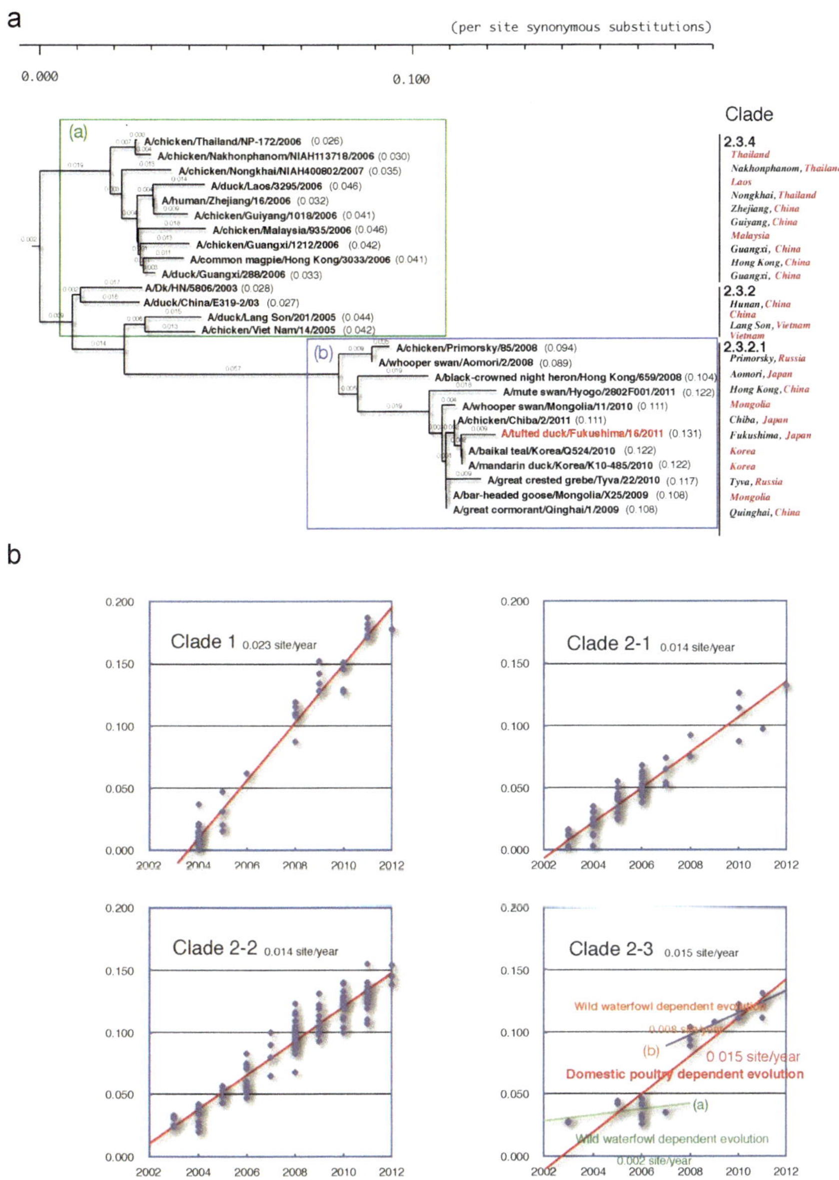

Fig. 5: Evolutionary pathways of HA gene of avian H5N1 viruses (a) and evolutionary rate of future evolutionary branches for selection of vaccine strain (b).

In order to examine the immune responses of the H5HA VLP antigens produced in silkworm pupae, we prepared the pupae homogenates with HA titer of 16,000 and 0.5 ml of the homogenates were injected

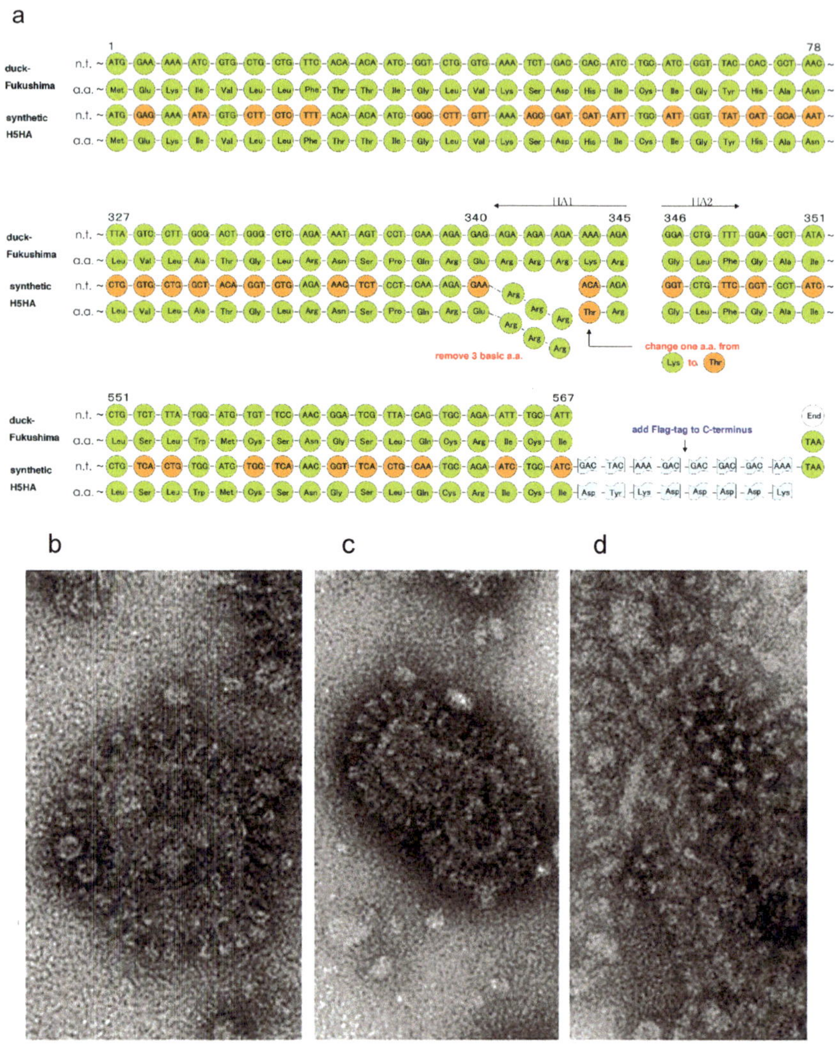

Fig. 6: Model sequence arrangement of chimera DNA consisting of authentic avian H5HA codons (green) and silkworm codons (orange) (a) and electron microscopic photographs (b–d).

intradermally into 1-month-old female chickens, subsequent injections were performed on days 16 and 33 post first injection (Fig. 8c). The HI titers were examined for their sera collected on days 17, 31, 45 and 55 post first injection. As a result, a higher HI antibody production in chickens was confirmed, shown by nearly 4,000 HI titer at 30 days after immunization. Simultaneously, the plaque-inhibition titer of the sera collected at 30 days was determined against A/duck/Singapore/F119-3/97 (H5N3), which

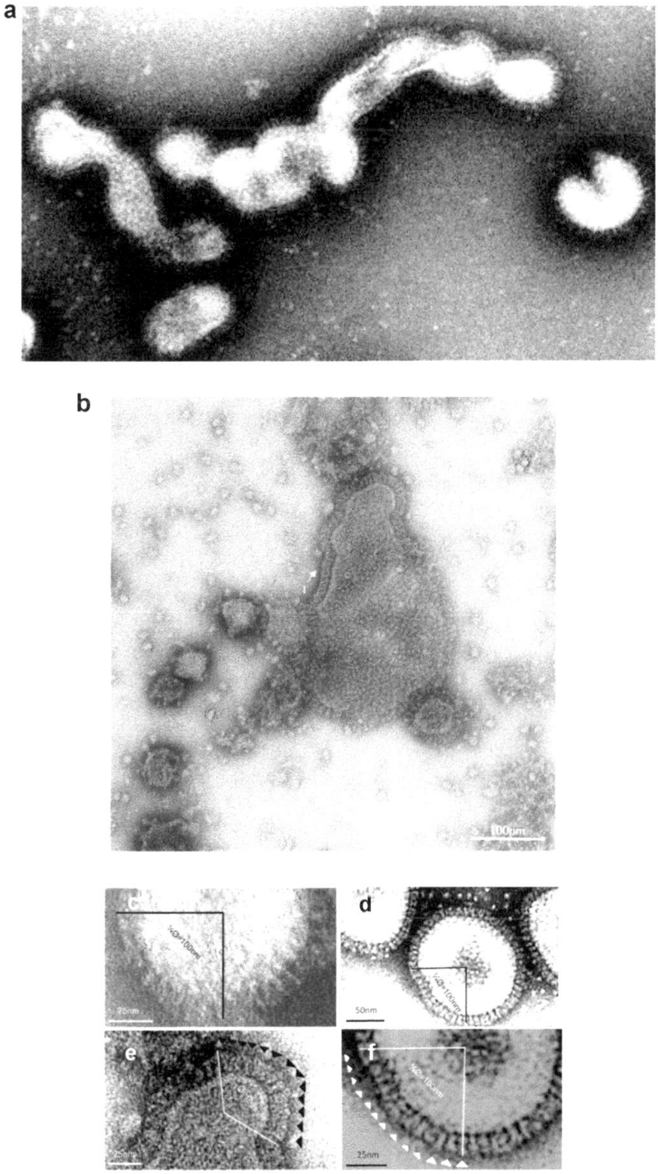

Fig. 7: Electron micrographs of authentic H5N1 virus (a), H5HA VLP produced in silkworms (b), HA projection arrangement of VLP and authentic H3N2 viruses (c–f).

was antigenically distantly related to H5 A/Fukushima virus. As shown in Fig. 8d, despite antigenic differences between both viruses, antiserum to H5 Fukushima VLP antigen efficiently inhibited plaque formation by A/duck/Singapore/F119-3/97 virus.

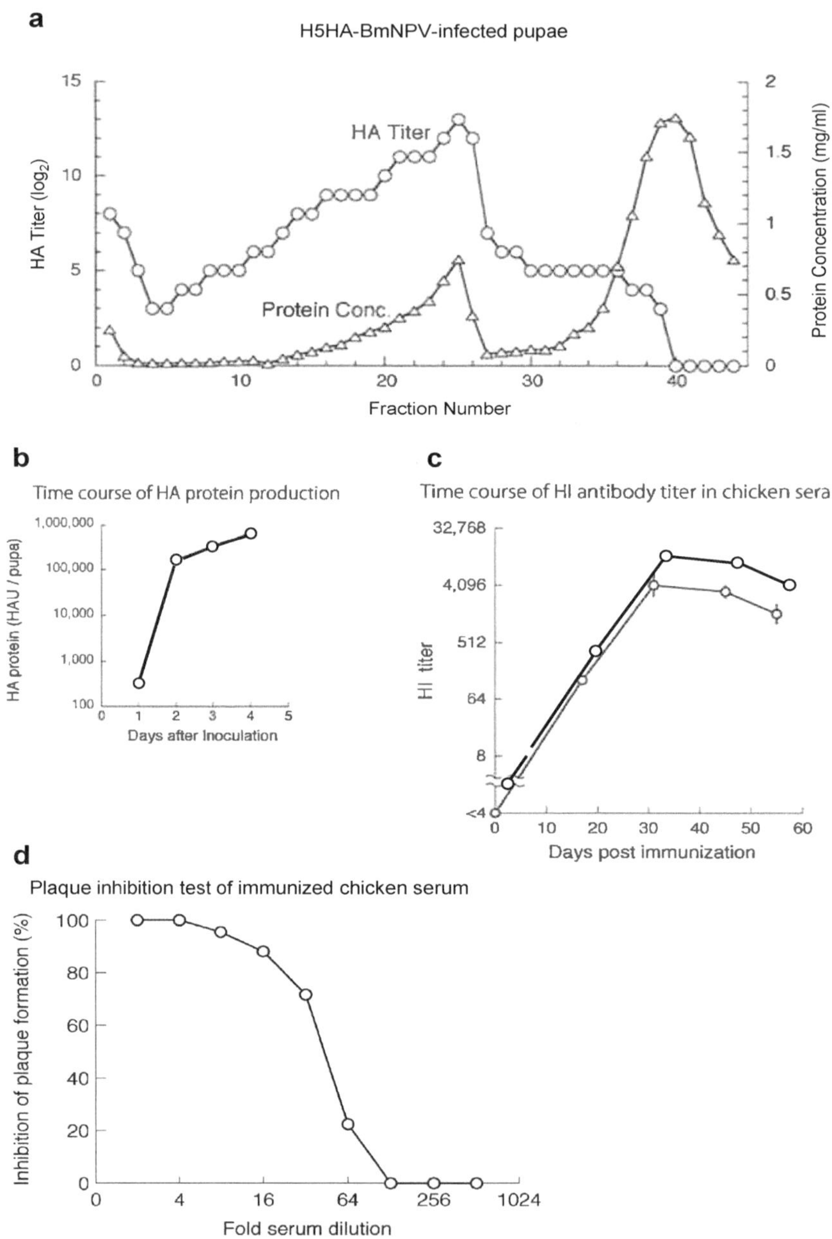

Fig. 8: Purification of H5VLP grown in silkworm pupae (a), its time course production in pupae (b), its antibody production in chicken (c), and plaque inhibition carve of A/duck/ Singapore (H5N3) virus by antibody against H5 VLP antigen (d).

4.2 Quantitative Analysis of Avian H7 Influenza VLP Vaccine Produced in Silkworm Pupae

Since the first isolation of avian H5N1 HPAI influenza virus in Hong Kong, avian virus has posed a threat to society, suggesting the advent of a possible pandemic strain. The isolation of the H7N9 in birds in 2013 and subsequent human cases was more concerning and the scientific community feared that it would be the next major pandemic strain (Arunachalam 2014, Chen et al. 2014, Ke et al. 2014, Tan et al. 2015, Wang et al. 2014). During this outbreak, 440 human infections were confirmed, including 122 fatalities. We have targeted a virus isolated from wild duck in Korea in 2010 (Kim et al. 2012) for vaccine development. In this study, we first synthesized codon optimized HA gene of A/duck/Korea/A76/2010 (H7N7) (H7-Korea). The synthesized gene was inserted into pBm-8 plasmid using the In-Fusion technique to produce H7-KoreaHA-BmNPV recombinant. Codon-optimization was accomplished by introduction of frequently used silkworm codons into HA gene of A/duck/Korea (H7N7) virus (Fig. 9a). As a result, a total of 384 silkworm codons (22.9%) were introduced and 1295 (77.1%) viral codons remained. According to these procedures, recombinant baculovirus containing HA genes of Shanghai/1/20/2013 (H7N9) (H7-Shanghai) and A/Anhui/1/2013 (H7N9) (H7-Anhui) were also prepared. Expression of these H7 HA genes were confirmed in Bm-N cells using immunofluorescent method and western-blot analysis.

The expression was also analyzed at transcription level. Quantity of mRNA was determined by real-time RT-PCR (Nerome et al. 2017). As shown in Fig. 9b, mRNA of HA genes was detected at 24 h after recombinant virus infection, however, the expression levels were much higher at 48 h after infection. In agreement with exprcssion of m-RNA, HA production was confirmed (Fig. 9c) in H5 and H7 recombinant, but wild recombinant strain did not show HA production.

H7-KoreaHA-BmNPV was also inoculated into silkworm larvae and pupae and they were homogenized in three cycle sonification as described previously (Nerome et al. 2017). In Fig. 10a, fractionation profiles of the homogenates on sucrose density gradient (10–50%) centrifugation at 27,000 rpm for 120 minutes were shown. Hemagglutination activity was separated into heavy (HF) and light (Treanor et al. 2006) fractions (Fig. 10a). It appeared that HF fraction contained complete HA VLP (Fig. 10b, c) and LF fraction contained small hemagglutinin particles (Fig. 10d).

Antigenic analysis of H7 HA and H5 HA VLP showed that VLP belonging to H5 and H7 sub-types characteristically reacted with each homologous antiserum (data not shown). Oral inoculation into mice and chickens with H5-Fukushima and H7-Korea HA VLP resulted in higher HI antibody production (Fig. 11a, b). Also, to investigate antibody production

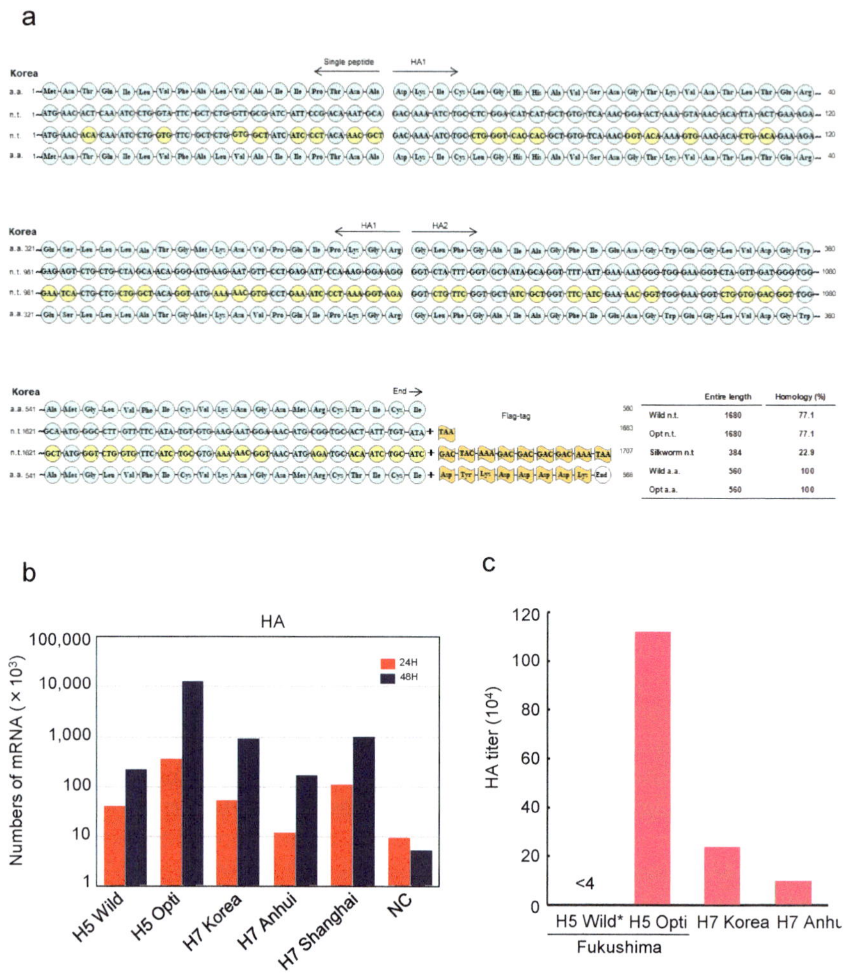

Fig. 9: Design of H7HA DNA consisting of authentic H7HA sequence (blue) and silkworm specific codon (yellow) (a). Comparison of m-RNA synthesis in Bm-N cells infected with H5 and H7 Bm-NPV recombinants (b) and comparison about HA activity of HA VLP produced in pupae (d).

against H5 and H7 VLP antigens which orally inoculated in mice, their immune reaction were confirmed by Fluorescent antibody test (Fig. 11c–f). Furthermore, similar immune responses were determined in western blot analysis (Fig. 11g, h).

These results suggested that the approximate 100 nm VLP structure enhanced the immunogenicity of HA antigen, because they showed good antibody responses without any adjuvant materials. Particularly, it should be emphasized that the production of antibodies was confirmed in mice

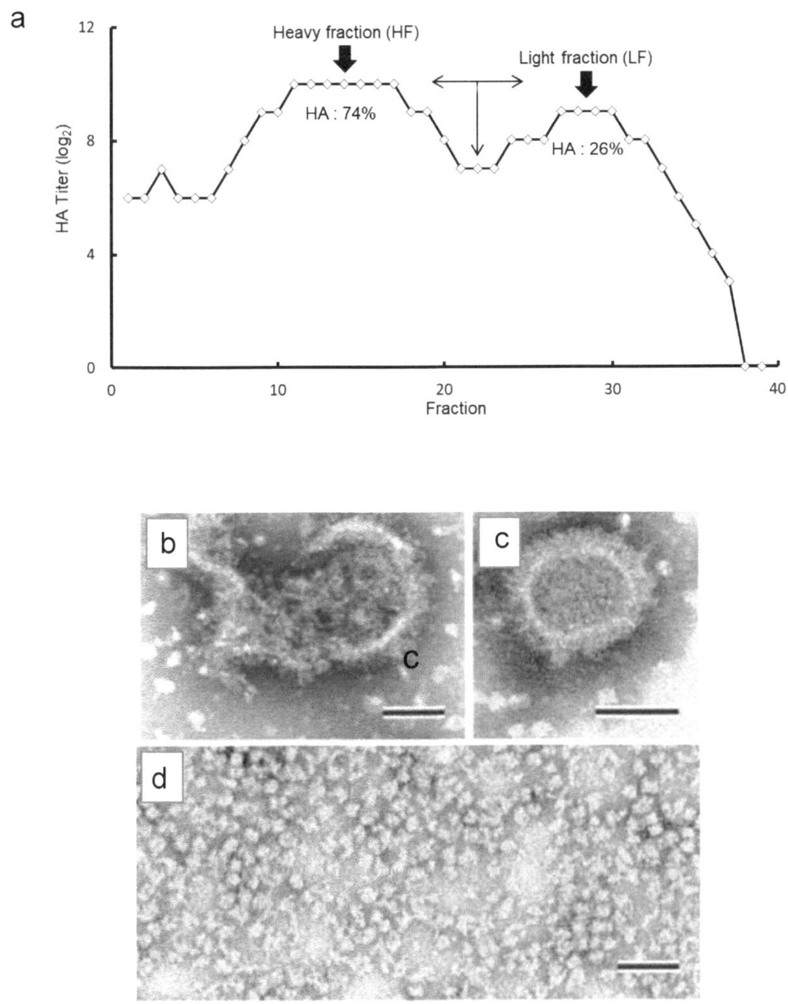

Fig. 10: Purification of H7 Korea VLP antigen grown in silkworm pupae (a) and electron-microscopic observation of H7 VLP structure (b–d).

immunized through oral route (Fig. 11c, d). Negative control did not show remarkable fluorescent antibody reaction (Fig. 11e, f). Western blot analysis also confirmed induction antibody production in mice immunized oral with H5 and H7 VLP (Fig. 11g, h).

5. Experimental Design of Divalent Influenza VLP Vaccines

Finally, we investigated future multi-valent influenza vaccine using baculovirus silkworm system. In the first stage, we produced a recombinant

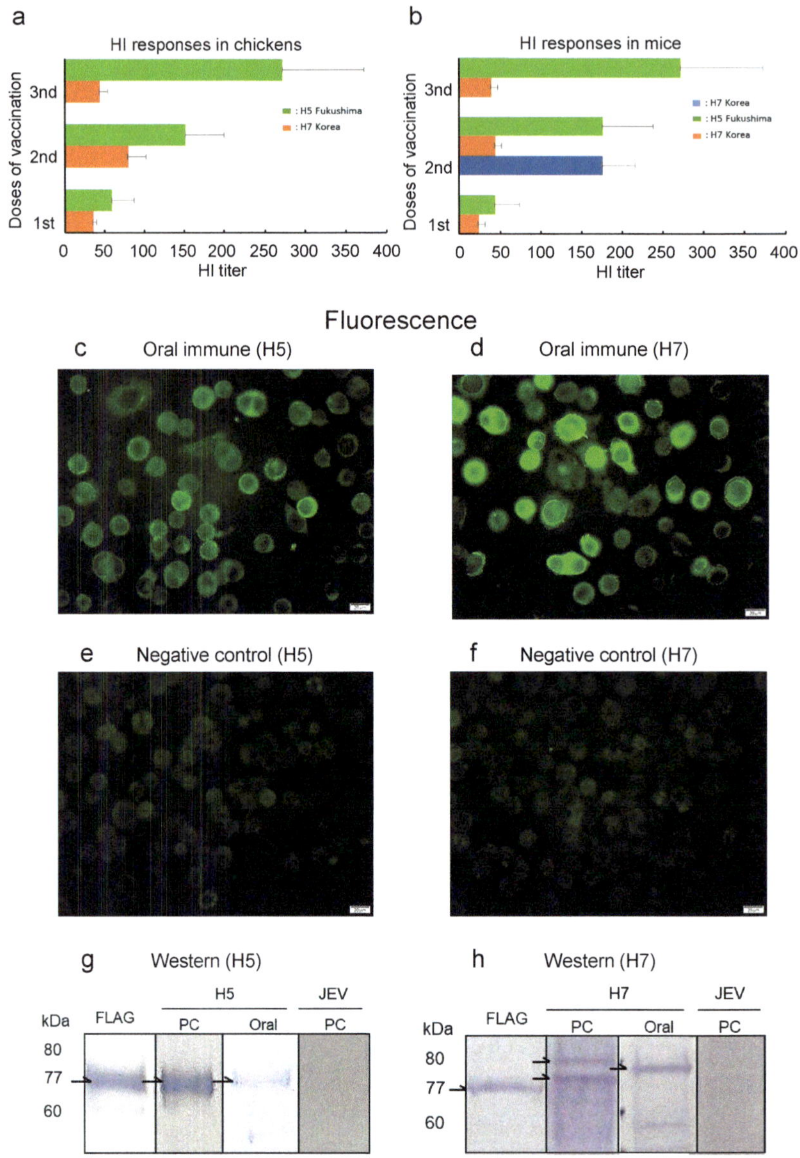

Fig. 11: Immune response in chickens (a) and mice (b). Fluorescence antibody test of oral immunized with H5 Fukushima (c, e) and H7 Korea (d, f) influenza VLP antigens produced in silkworms, and their western blot analysis (g, h).

virus containing codon-optimized H5 and H7 genes. The two target HA genes were connected to T2A sequence and were expected to self-cleave into two HA proteins post-transcriptionally (Fig. 12a, b). This recombinant

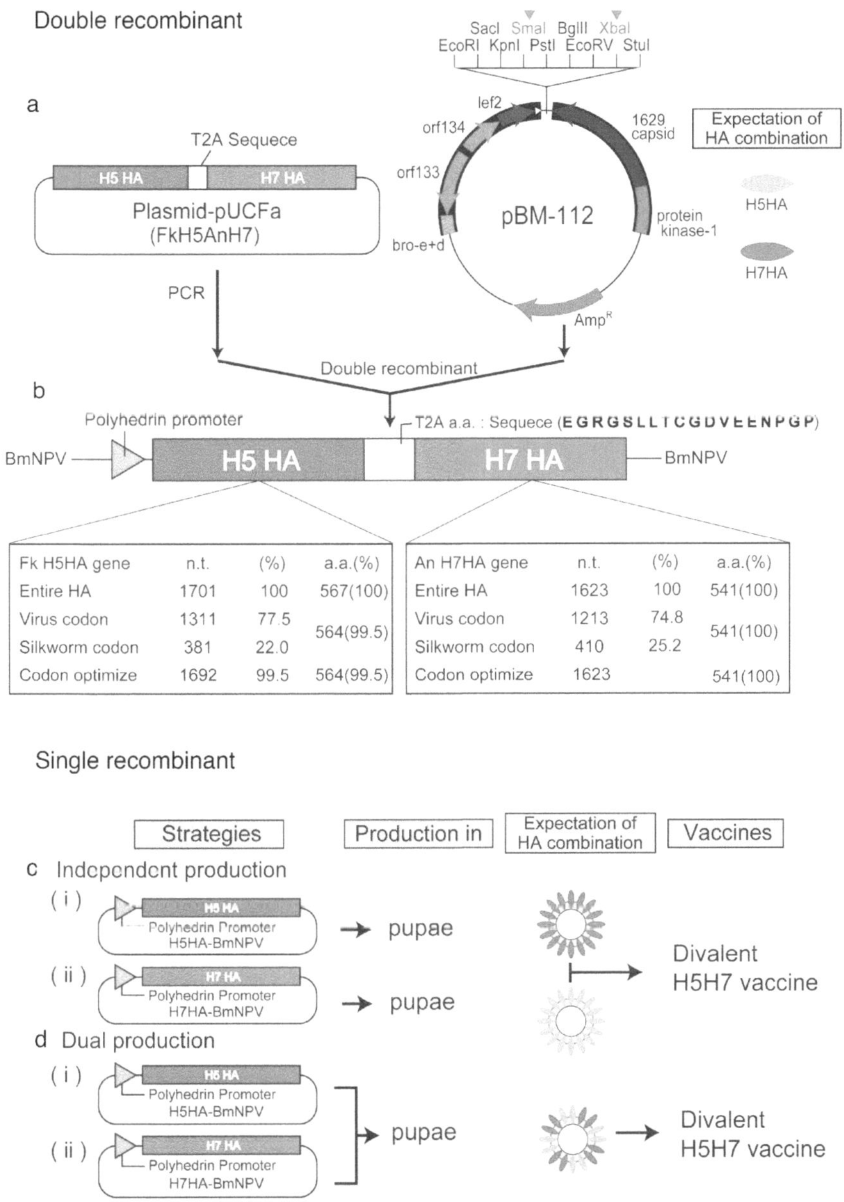

Fig. 12: Strategic design of preparation of divalent H5 and H7 influenza vaccines based on the double recombinant (a, b), mixed infection (c) and dual infection (d).

appeared to be expressed highly on the basis of mRNA. However, HA gene expression was poor on the protein level. In contrast, simultaneous infection of silkworm pupae with H5 and H7 recombinants showed high mRNA expression (Fig. 9b) and also produced large amounts of both HA antigens and the level of HA antigen production was 200,000 HA titer. For this reason, using a combination of recombinants has an advantage while preparing divalent vaccines. Simultaneously, it was noteworthy that, despite the differences in multiplicity of infection between H5 and H7 recombinants in silkworm pupae, an almost identical HA protein production was observed under the condition of multiplicity of injection (MOI), suggesting possible equivalent production of H5 and H7 vaccines. As a result, a large number of divalent influenza VLP vaccines may be able to be produced in silkworm using dual or multiple infection of two or more recombinant viruses (Fig. 12a–d).

6. Conclusion

Recently, a variety of vaccines in different virus fields have been developed using useful modern technologies such as cell culture-derived vaccine, DNA-based vaccine, cold adapted live vaccine and reverse genetics-derived vaccine. When considering the advent of a variety of emerging viruses, we have to develop new and economical vaccines with high efficacy and low side-effects in a short-period of time. The development of VLP vaccines may satisfy the above conditions; especially the time required from gene design to chemical synthesis to production process, which is only a few months. In this period, we can rapidly respond to the requirements of every situation such as the emergence of viruses from the pandemic avian influenza viruses. Furthermore, as described earlier, VLP structure consists of a cell membrane which is approximately 100 nm in diameter and this structural unit plays an important role in immune responses like adjuvant reaction. Therefore, we can prepare potent and safer vaccines in a shorter period, possibly corresponding to all emerging viruses. The last merit of VLP vaccine relates to their low production cost. For example, the use of high yielding baculovirus recombinant system results in approximately more than 400,000 HA titer. As a result, one silkworm pupa can produce influenza VLP vaccines sufficient for approximately 50 people, based on the estimation of the unpurified HA titer. In addition, including new functions in the VLP vaccine might increase their usefulness.

Acknowledgement

We are very grateful to Dr. Reiko Nerome (The Institute of Biological Resources, Okinawa) for her helpful preparation of the manuscript and figures.

References

Arunachalam, R. 2014. Adaptive evolution of a novel avian-origin influenza A/H7N9 virus. Genomics. 104(6 Pt B): 545–553. Doi: 10.1016/j.ygeno.2014.10.012.

Baz, M., C. J. Luke, X. Cheng, H. Jin and K. Subbarao. 2013. H5N1 vaccines in humans. Virus Res. 178(1): 78–98. Doi: 10.1016/j.virusres.2013.05.006.

Bright, R. A., D. M. Carter, S. Daniluk, F. R. Toapanta, A. Ahmad, V. Gavrilov, M. Massare, P. Pushko, N. Mytle, T. Rowe, G. Smith and T. M. Ross. 2007. Influenza virus-like particles elicit broader immune responses than whole virion inactivated influenza virus or recombinant hemagglutinin. Vaccine. 25(19): 3871–3878. Doi: 10.1016/j.vaccine.2007.01.106.

Chen, Z., K. Li, L. Luo, E. Lu, J. Yuan, H. Liu, J. Lu, B. Di, X. Xiao and Z. Yang. 2014. Detection of avian influenza A(H7N9) virus from live poultry markets in Guangzhou, China: A surveillance report. PLoS One. 9(9): e107266. Doi: 10.1371/journal.pone.0107266.

Cox, M. M. 2008. Progress on baculovirus-derived influenza vaccines. Curr. Opin. Mol. Ther. 10(1): 56–61.

Galarza, J. M., T. Latham and A. Cupo. 2005. Virus-like particle vaccine conferred complete protection against a lethal influenza virus challenge. Viral Immunol. 18(2): 365–372. Doi: 10.1089/vim.2005.18.365.

Gavrilov, V., T. Orekov, C. Alabanza, U. Porika, H. Jiang, K. Connolly and S. Pincus. 2011. Influenza virus-like particles as a new tool for vaccine immunogenicity testing: Validation of a neuraminidase neutralizing antibody assay. J. Virol Methods. 173(2): 364–373. Doi: 10.1016/j.jviromet.2011.03.011.

Hiromoto, Y., T. Saito, S. Lindstrom and K. Nerome. 2000. Characterization of low virulent strains of highly pathogenic A/Hong Kong/156/97 (H5N1) virus in mice after passage in embryonated hens' eggs. Virology. 272(2): 429–437. Doi: 10.1006/viro.2000.0371.

Hiromoto, Y., Y. Yamazaki, T. Fukushima, T. Saito, S. E. Lindstrom, K. Omoe, R. Nerome, W. Lim, S. Sugita and K. Nerome. 2000. Evolutionary characterization of the six internal genes of H5N1 human influenza A virus. J. Gen. Virol. 81(Pt 5): 1293–1303. Doi: 10.1099/0022-1317-81-5-1293.

Karron, R. A., K. Talaat, C. Luke, K. Callahan, B. Thumar, S. Dilorenzo, J. McAuliffe, E. Schappell, A. Suguitan, K. Mills, G. Chen, E. Lamirande, K. Coelingh, H. Jin, B. R. Murphy, G. Kemble and K. Subbarao. 2009. Evaluation of two live attenuated cold-adapted H5N1 influenza virus vaccines in healthy adults. Vaccine. 27(36): 4953–4960. Doi: 10.1016/j.vaccine.2009.05.099.

Katanoda, K., M. Hori, T. Matsuda, A. Shibata, Y. Nishino, M. Hattori, M. Soda, A. Ioka, T. Sobue and H. Nishimoto. 2015. An updated report on the trends in cancer incidence and mortality in Japan, 1958–2013. Jpn. J. Clin. Oncol. 45(4): 390–401. Doi: 10.1093/jjco/hyv002.

Ke, C., J. Lu, J. Wu, D. Guan, L. Zou, T. Song, L. Yi, X. Zeng, L. Liang, H. Ni, M Kang, X. Zhang, H. Zhong, J. He, J. Lin, D. Smith, D. Burke, R. A. Fouchier, M. Koopmans and Y. Zhang. 2014. Circulation of reassortant influenza A(H7N9) viruses in poultry and humans, Guangdong Province, China, 2013. Emerg. Infect. Dis. 20(12): 2034–2040. Doi: 10.3201/eid2012.140765.

Kim, H. R., C. K. Park, Y. J. Lee, J. K. Oem, H. M. Kang, J. G. Choi, O. S. Lee and Y. C. Bae. 2012. Low pathogenic H7 sub-type avian influenza viruses isolated from domestic ducks in South Korea and the close association with isolates of wild birds. J. Gen. Virol. 93(Pt 6): 1278–1287. Doi: 10.1099/vir.0.041269-0.

Latham, T. and J. M. Galarza. 2001. Formation of wild-type and chimeric influenza virus-like particles following simultaneous expression of only four structural proteins. J. Virol. 75(13): 6154–6165. Doi: 10.1128/JVI.75.13.6154-6165.2001.

Nerome, K., Y. Yoshioka, M. Ishida, K. Okuma, T. Oka, T. Kataoka, A. Inoue and A. Oya. 1990. Development of a new type of influenza sub-unit vaccine made by muramyldipeptide-liposome: Enhancement of humoral and cellular immune responses. Vaccine. 8(5): 503–509.

Nerome, K., S. Sugita, K. Kuroda, T. Hirose, S. Matsuda, K. Majima, K. Kawasaki, T. Shibata, O. N. Poetri, R. D. Soejoedono, N. L. Mayasari, S. Aqungpriyono and R. Nerome. 2015. The large-scale production of an artificial influenza virus-like particle vaccine in silkworm pupae. Vaccine. 33(1): 117–125. Doi: 10.1016/j.vaccine.2014.11.009.

Nerome, K., S. Matsuda, K. Maegawa, S. Sugita, K. Kuroda, K. Kawasaki and R. Nerome. 2017. Quantitative analysis of the yield of avian H7 influenza virus hemagglutinin protein produced in silkworm pupae with the use of the codon-optimized DNA: A possible oral vaccine. Vaccine. 35(5): 738–746. Doi: 10.1016/j.vaccine.2016.12.058.

Pan, Y. S., H. J. Wei, C. C. Chang, C. H. Lin, T. S. Wei, S. C. Wu and D. K. Chang. 2010. Construction and characterization of insect cell-derived influenza VLP: Cell binding, fusion, and EGFP incorporation. J. Biomed. Biotechnol. 2010: 506363. Doi: 10.1155/2010/506363.

Shi, M., X. D. Lin, X. Chen, J. H. Tian, L. J. Chen, K. Li, W. Wang, J. -S.Eden, J. -J. Shen, L. Liu, E. C. Holmes and Y. Z. Zhang. 2018. The evolutionary history of vertebrate RNA viruses. Nature. 556(7700): 197–202. Doi: 10.1038/s41586-018-0012-7.

Shortridge, K. F., N. N. Zhou, Y. Guan, P. Gao, T. Ito, Y. Kawaoka, S. Kodihalli, S. Krauss, D. Markwell, K. G. Murti, M. Norwood, D. Senne, L. Sims, A. Takada and R. G. Webster. 1998. Characterization of avian H5N1 influenza viruses from poultry in Hong Kong. Virology. 252(2): 331–342. Doi: 10.1006/viro.1998.9488.

Steel, J. 2011. New strategies for the development of H5N1 subtype influenza vaccines: Progress and challenges. BioDrugs. 25(5): 285–298. Doi: 10.2165/11593870-000000000-00000.

Suguitan, A. L. Jr., J. McAuliffe, K. L. Mills, H. Jin, G. Duke, B. Lu, C. J. Luke, B. Murphy, D. E. Swayne, G. Kemble and K. Subbarao. 2006. Live, attenuated influenza A H5N1 candidate vaccines provide broad cross-protection in mice and ferrets. PLoS Med. 3(9): e360. Doi: 10.1371/journal.pmed.0030360.

Suguitan, A. L. Jr., M. P. Marino, P. D. Desai, L. M. Chen, Y. Matsuoka, R. O. Donis, H. Jin, D. E. Swayne, G. Kemble and K. Subbarao. 2009. The influence of the multi-basic cleavage site of the H5 hemagglutinin on the attenuation, immunogenicity and efficacy of a live attenuated influenza A H5N1 cold-adapted vaccine virus. Virology. 395(2): 280–288. Doi: 10.1016/j.virol.2009.09.017.

Tan, K. X., S. A. Jacob, K. G. Chan and L. H. Lee. 2015. An overview of the characteristics of the novel avian influenza A H7N9 virus in humans. Front Microbiol. 6: 140. Doi: 10.3389/fmicb.2015.00140.

Treanor, J. J., J. D. Campbell, K. M. Zangwill, T. Rowe and M. Wolff. 2006. Safety and immunogenicity of an inactivated subvirion influenza A (H5N1) vaccine. N. Engl. J. Med. 354(13): 1343–1351. Doi: 10.1056/NEJMoa055778.

Wang, X., S. Fang, X. Lu, C. Xu, B. J. Cowling, X. Tang, B. Peng, W. Wu, J. He, Y. Tang, X. Xie, S. Mei, D. Kong, R. Zhang, H. Ma and J. Cheng. 2014. Seroprevalence to avian influenza A(H7N9) virus among poultry workers and the general population in southern China: A longitudinal study. Clin. Infect. Dis. 59(6): e76–83. Doi: 10.1093/cid/ciu399.

CHAPTER 17

Production of Feline and Canine Interferon from Silkworm

Tsuyoshi Tanaka and *Takashi Tanaka**

1. Introduction

Interferon (IFN) is a kind of cytokine that is included in biodefense and immunological function. IFN is involved in biological defense and immune functions. Our company initiated the development of IFN preparations for companion animals and commercialized the world's first feline interferon (FeIFN) preparation indicated for feline Calicivirus infection (i.e., the common cold in cats) in 1994. This preparation is commercially available in 45 countries worldwide at present. Furthermore, the canine interferon γ (CaIFN) preparation, Type II IFN preparation, was approved as a drug for treating atopic dermatitis in dogs.

Both products are manufactured using a silkworm-baculovirus expression system, in which silkworms infected with gene recombinant baculovirus express and accumulate the protein in the silkworm larval haemolymph. In this chapter, the manufacturing process of animal IFN using the silkworm-baculovirus expression system is introduced.

Toray Industries, Inc. Ehime Plant, Veterinary Medicine Production Sect., 1515, Tsutsui Masaki-cho Iyogun, Ehime 791-3193, Japan.
E-mail: Tsuyoshi_Tanaka@nts.toray.co.jp
* Corresponding author: Takashi_Tanaka2@nts.toray.co.jp

2. Feline Interferon

IFNs are bioactive substances that were discovered as a viral inhibitory factor by Nagano and Kojima (1954), and later named "interferons" by Isaacs and Lindenman (1957). Type I IFNs including IFN-α, IFN-β and IFN-ω have been used in the treatment of chronic hepatitis B and C as anti-viral drugs, as well as for treating brain tumors and melanomas and other cancers as anti-tumor drugs. IFNs are drugs with species specificity, therefore, administration of human IFN (HuIFN) to animals results in the production of anti-HuIFN antibody. Nowadays, dogs and cats are positioned as family members or companions of their owners rather than merely as pets and are, thus, called companion animals. From a healthcare perspective, drug preparations for dogs and cats have not been adequately developed. In dogs and cats, infections with parasites, bacteria and viruses often occur, and viral infections are frequently fatal. Because of this, development of fundamental treatment methods was eagerly awaited. In 1993, the FeIFN preparation indicated for feline Calicivirus infection was commercialized in Japan, and in 1997, an additional indication for canine parvovirus infection was approved. The feline Calicivirus infection is a disease that produces common cold symptoms such as oral and lingual ulcers, coughing, sneezing, decreased appetite and fever, and may be fatal in physically weak kittens. The canine parvovirus infection, on the other hand, is a viral disease with high mortality that affects intestinal mucosal cells and thus causes severe bloody diarrhea. Although symptomatic treatment had been provided for these infections, the development of the FeIFN preparation allowed radical treatment of these viral diseases. In addition, this preparation was approved in the EU in 2001 and is currently available in 45 countries worldwide.

2.1 Development of Feline Interferon Production

To develop a production method for FeIFN using gene recombination techniques, various production systems were compared in terms of their productivity. The results showed that the productivity of the FeIFN using silkworms was higher than that of any of the conventional ones using *Escherichia coli*, yeast, hamster cells (Chinese hamster ovary cells or CHO cells) and monkey cells (COS cells). Most notably, the productivity within silkworm bodies was found to be higher than that in silkworm cell culture (Fig. 1).

The silkworm production system has been established as follows: silkworms are infected with the recombinant baculovirus, which is infectious in silkworms and prepared by integrating the gene of interest into the viral genome, to produce and accumulate the substance of interest

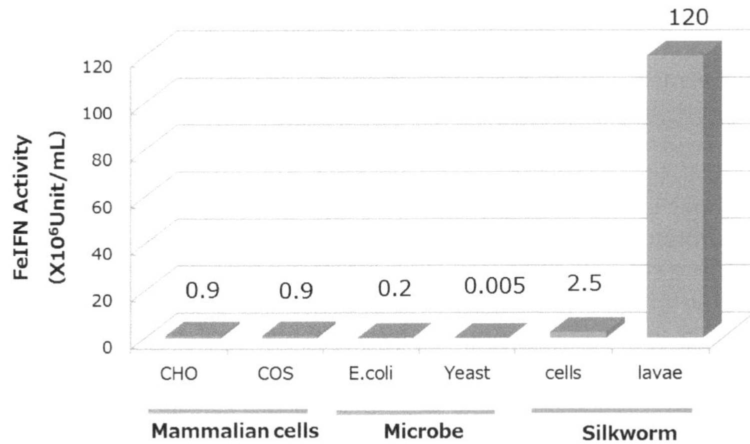

Fig. 1: Comparison of feline interferon productivity in various expression systems.

in silkworm larval haemolymph. Baculoviruses that are infectious in insects such as silkworms are covered with a crystalline protein, called polyhedra. The polyhedra proteins are formed by polyhedrin, reflecting the high expression level of the polyhedrin gene. With the focus on this high productivity, a production system using a polyhedrin gene promoter was devised. Smith et al. (1983) reported production of HuIFN-β employing the above system, and Maeda et al. (1985) reported production of HuIFN-α with this system.

2.2 Preparation of Recombinant Virus

The gene encoding FeIFN (FeIFN gene) was cloned from a feline thymocyte (LSA-1 cell) cDNA library, based on the antiviral activity of the transient expression product in monkey COS-1 cells (Nakamura et al. 1992). Then, a plasmid with the FeIFN gene inserted downstream from the polyhedrin promoter in the transfer vector (pBM030) was prepared, and homologous recombination between the plasmid and *Bombyx mori* (silkworm) nuclear polyhedrovirus (BmNPV) DNA in silkworm BM-N cells was carried out in order to prepare recombinant BmNPV coding the FeIFN gene. Screening for the recombinant virus was carried out by limiting dilution and plaque assay to obtain recombinant virus strains capable of producing FeIFN (Sakurai et al. 1992).

2.3 Manufacturing Method of Feline Interferon

Silkworm larvae at the 5th instar were inoculated with the recombinant BmNPV. This was followed by feeding of the silkworms with artificial diet for approximately 4 days. During this period, FeIFN is actively expressed and accumulated in haemolymph in substantial amounts. This larval haemolymph is recovered from the silkworms and then treated with hydrochloric acid in order to inactivate the recombinant BmNPV and silkworm-derived proteinases. Then, the treated haemolymph is neutralized and subjected to centrifugation, blue affinity chromatography and copper chelate affinity chromatography, in this order, to purify the FeIFN protein. The FeIFN protein fraction is then desalted to prepare the FeIFN drug substance.

To the drug substance, purified gelatin and D-sorbitol are added as a stabilizer and a filler, respectively, and the formulated mixture is then lyophilized in order to obtain the product.

2.4 Unification of Feline Interferon Form

At the early development stage, the purified FeIFN protein was found to be a mixture of two variants by reverse-phase high-performance liquid chromatography (RP-HPLC) (Fig. 2A). Because two forms of the gene product were produced from one gene, cleavage to remove the signal sequence was considered to take place at two positions in the precursor FeIFN protein during the processing step. N-terminal amino acid sequence analysis of these two forms revealed that one form had Cys-Asp-Leu-Pro- at the N-terminal end, which is the same as that of IFN-α, while the other form had Leu-Gly-Cys-Asp-Leu-Pro at the N-terminal end. These results suggested that the cleavage took place at two positions, between Gly and Cys and between Ser and Leu. Bioassay of these two forms further revealed that the form with Cys-Asp-Leu-Pro at the N-terminal end exerted activity, while the form with Leu-Gly-Cys-Asp-Leu-Pro at the N-terminal end did not.

Investigations were carried out in order to control the cleavage site of the signal sequence so that the cleavage would take place only between Gly and Cys to produce the form with Cys-Asp-Leu-Pro at the N-terminal end. Detailed analyses focusing on the appearance frequencies of amino acid residues around the cleavage site of the signal sequence were reported (von Heijne et al. 1986). The FeIFN signal sequence according to the above report indicated that cleavage was likely to take place between Ser at the position of -3 and Leu at the position of -2. Substitution from Ser to Val at the position of -3 would presumably protect the site between positions -3 and -2 from cleavage. The recombinant BmNPV designed to substitute Ser with Val in the FeIFN signal sequence by modifying the DNA sequence was prepared and silkworms were inoculated with it. The resultant FeIFN protein was then analyzed for the N-terminal amino acid sequence. The result demonstrated production of a single form of FeIFN protein with Cys at the N-terminal end, as expected (Fig. 2B) (Ueda et al. 1993).

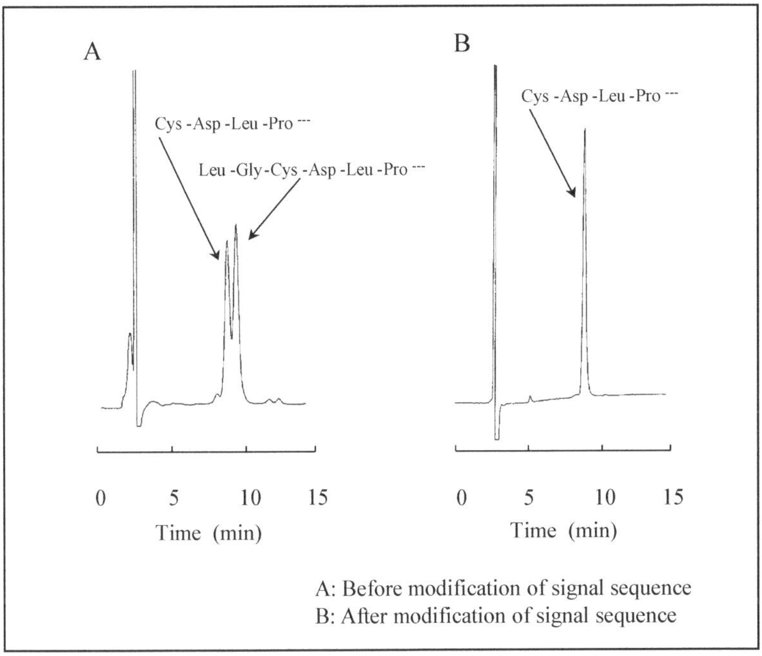

Fig. 2: Reverse phase HPLC analysis of purified feline interferon. (A) In original sequence, the cleavage of signal sequence occurred at two different sites. (B) Changing Ser at position -3 into Val leads to homogeneous production of feline interferon.

2.5 Glycosylation

Recombinant proteins expressed while employing a silkworm-baculovirus expression system undergo post-translational modifications, such as glycosylation, unlike those expressed in *E. coli*. Carbohydrate chains have many functions, serving as protein stabilizers or markers in a cell recognition process. *N*-linked oligosaccharides are well known to be major carbohydrate chains. Although N-linked oligosaccharide is also attached to proteins produced in the silkworm-baculovirus expression system, its structure is slightly different from that in mammals, including humans. *N*-linked oligosaccharides in mammals undergo further sialylation at their non-reducing terminal, but those in silkworms are rather primitive in comparison to those in mammalian species; *N*-linked oligosaccharides in silkworms are characterized by a low-mannose type structure and shorter carbohydrate chains (Misaki et al. 2003).

Analysis of the carbohydrate chain structure in FeIFN proteins expressed in the baculovirus-silkworm expression system indicated the presence of two types of *N*-linked oligosaccharides, which were shorter than those in mammalian species (Fig. 3).

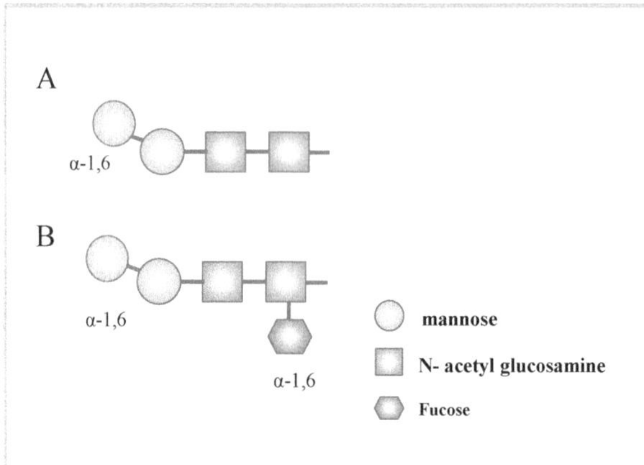

Fig. 3: FeIFN proteins expressed in the baculovirus-silkworm expression system indicated the presence of two types of *N*-linked oligosaccharides.

3. Canine Interferon-γ

IFN-γ is a type II IFN that mainly plays active roles in immuno-stimulation and immune regulation, rather than functioning as an anti-viral factor. In humans, type II IFN has been used for the treatment of renal cancer, chronic granulomatous disease, mycosis fungoides and adult T-cell leukemia, etc.

One report (Hanifin et al. 1993) showed that IFN-γ was effective for treating atopic dermatitis (AD), thus, its use as a new therapeutic drug for AD was initially expected. Various investigations were carried out. In patients with AD, high total IgE levels in serum and eosinophilia in blood are observed, suggesting enhanced activity of Th2 cells, one of the helper T cell types. According to the Th1/Th2 concept proposed by Mosmann et al. (1986), helper T cells are classified into two types, Th1 cells and Th2 cells, based on the cytokines produced in these cells. Th1 cells produce IFN-γ and interleukin (IL)-2, and Th2 cells produce IL-4, IL-5 and IL-10. This Th1/Th2 balance is said to be involved in the development of various diseases as well as in defense systems. In patients with AD, profound Th2 dominance was reported (Mosmann et al. 1986). The high total IgE level in serum from patients with AD is considered to result from a class switch in B cells, in which IL-4 produced by Th2 cells induces differentiation into IgE producing cells. Th1 cells and Th2 cells are mutually exclusive; IFN-γ produced by Th1 cells suppresses Th2 cell proliferation and inhibits their IL-4 production.

In other words, administration of IFN-γ is expected to normalize the Th1/Th2 balance, which is profoundly Th2 dominant in patients with AD, and thereby reduce the total IgE level in the serum, leading to alleviation of AD symptoms.

Research and development of canine IFN-γ was initiated with the proposed indication for AD, because IFN-γ was reported to be effective in human patients with AD; no serious adverse effects were reported; AD is relatively common in dogs, but no effective curative treatments are available except for symptomatic treatments. A positive effect is expected based on the onset mechanism of AD and the mechanism underlying the actions of IFN.

3.1 Virus Preparation

Since the cDNA of the CaIFN-γ gene was cloned by Devos et al. (1992), the primer was designed based on the reported sequence. The cDNA of the CaIFN-γ gene was cloned from a canine lymphocyte cDNA library employing the reverse transcriptase-polymerase chain reaction (RT-PCR). Using the obtained cDNA, the recombinant virus was prepared using the same strategy as that for expression of FeIFN.

3.2 Development of Canine Interferon-γ

Silkworm larvae at the 5th instar are inoculated with the recombinant virus and fed an artificial diet, designed specifically for silkworms, at 25°C for approximately 4 days. The silkworm larval haemolymph containing the produced CaIFN-γ is treated with 0.01% benzalkonium chloride in order to inactivate the recombinant BmNPV. Then, the treated haemolymph is subjected to centrifugation, copper chelate affinity chromatography, anion exchange chromatography and blue affinity chromatography, in this order, so as to purify the CaIFN-γ protein. The CaIFN-γ protein fraction is then applied to gel filtration chromatography in order to prepare the CaIFN-γ drug substance. To this drug substance, stabilizers including Acacia and L-cysteine are added, and the formulated mixture is then lyophilized in order to obtain the product.

3.3 Variations in Carbohydrate Chain Binding

The CaIFN-γ productivity in various expression systems was investigated and silkworms showed the most favorable productivity among those examined. Although the productivity was high, CaIFN-γ products expressed in silkworms were found to exist in complex and varied forms. Western blotting of CaIFN-γ-expressed silkworm larval haemolymph using anti-CaIFN-γ serum showed at least four main bands (Okano et al. 2000). This variability was considered to mainly be attributable to the use of different *N*-linked oligosaccharide binding, because the CaIFN-γ gene sequence suggests that the product protein has two *N*-linked oligosaccharide binding sites, and Western blotting after *N*-glycosidase treatment showed only one main band. Because CaIFN-γ with and without the carbohydrate chain showed the same activity, recombinant BmNPV, in which both Asn16 and Asn83 of *N*-linked oligosaccharide binding sites were designed to be Gln, was prepared, and Western blotting of the larval haemolymph of silkworms infected with the recombinant baculovirus showed a single band of CaIFN-γ.

3.4 Variations in C-terminal Structure and Unification

The crude purified fraction from the first copper chelate affinity chromatography was applied to anion exchange chromatography resin and CaIFN-γ was eluted by increasing the NaCl concentrations stepwise. The eluted fractions were analyzed by RP-HPLC (Fig. 4). The results revealed the presence of further multiple CaIFN-γ protein forms. Structural analysis of the main peak from each fraction showed that the

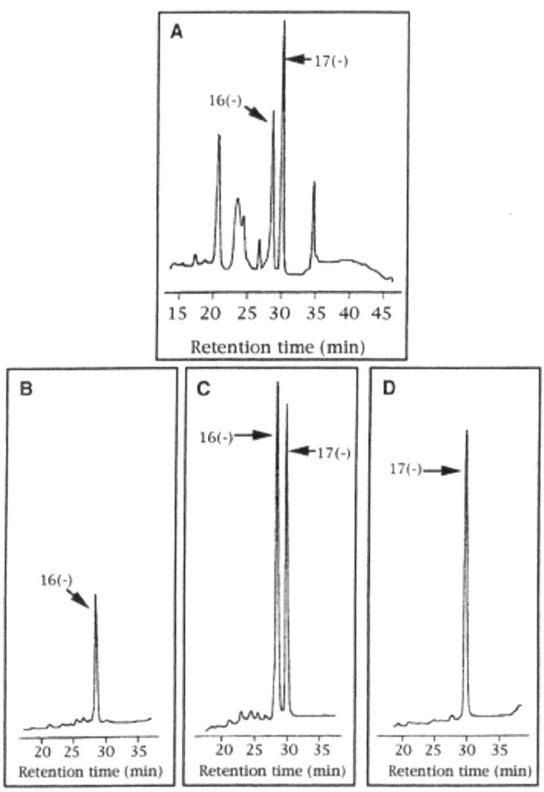

Fig. 4: HPLC analysis of rCaIFN-γ separated by anion exchange chromatography. (A) Chelating Sepharose passing through fraction. (B) Q Sepharose eluted fraction (25 mM NaCl). (C) Q Sepharose eluted fraction (35 mM NaCl). (D) Q Sepharose eluted fraction (50 mM NaCl).

peak at the retention time of about 28 min was derived from the CaIFN-γ protein with the C-terminal 16 amino acid residues deleted; and the peak at the retention time of about 29 minutes was derived from that with the C-terminal 17 amino acid residues deleted (these forms are referred to as C16(–) and C17(–), respectively). Furthermore, molecular weights of proteins in fractions B to D were analyzed by gel filtration chromatography. The analysis indicated that proteins in these fractions formed dimers. Accordingly, of the three elution fractions of CaIFN-γ protein from the anion exchange chromatography, fraction B consisted mainly of homodimers of C16(–); fraction C consisted mainly of heterodimers of C16(–) and C17(–); fraction D consisted mainly of homodimers of C17(–). The other investigations revealed that addition of EDTA to the treatment procedure for the silkworm larval haemolymph suspension mostly suppressed the generation of C17(–), and that the remaining C17(–) was removed by anion exchange chromatography. Consequently, highly purified C16(–) was obtained as the product of interest (Okano et al. 2000).

4. Securing the Quality of Drugs Manufactured using Silkworms

Conventionally, *E. coli*, yeasts and animal cells have been used in the manufacture of biopharmaceuticals. Considerations to ensure the quality of drugs manufactured using silkworms for the production of biopharmaceuticals are discussed.

4.1 Bank Preparation

In the manufacture of biopharmaceuticals, cells, regarded as a substance required for production, are controlled in the cell bank system, because the properties such a substance affect the uniformity of the product of interest and the impurity profile. The bank system ensures that the manufacturing process is always initiated with cells derived from a single cell pool.

In the silkworm-baculovirus expression system, recombinant BmNPV and silkworms correspond to substances for production, if the conventional definition of substances for production is applied. The bank system might be applied to the control of baculovirus but a bank system for control of silkworms has not as yet been put into practice. It is therefore important to use silkworms of consistent quality for production by maintaining the silkworm strain for use and strictly controlling the rearing method and conditions.

4.2 Raw Materials

The major raw materials are silkworms and the feed given to them. For silkworms, the rearing environment should be monitored, and individual growth conditions should be controlled and assessed. The artificial diet is mainly made from plant materials such as mulberry leaves and defatted soybeans; it is regarded as being reasonably safe because it is manufactured through heat treatment, which substantially reduces the risk of introducing contaminants such as microorganisms.

4.3 Containment of Recombinant Virus

Containment of recombinant BmNPV is a key subject to be addressed for practical implementation of a manufacturing process using such substances. Silkworms, which have been reared by humans for several thousand years, are a completely domesticated insect, and thus cannot survive in any field environment. Rooms for rearing silkworms are controlled under negative pressure relative to the atmospheric pressure, and the air is exhausted through a high-efficiency-particulate-air (HEPA) filter to the outside of the building. Accordingly, the recombinant BmNPV has no chance to spread outside.

Because the recombinant BmNPV does not have the polyhedrin gene, it is not capable of forming the polyhedra which protect the viruses themselves from the external environment. Even in the event of leakage, the recombinant BmNPV cannot survive outside. Furthermore, the silkworm larval haemolymph containing FeIFN is treated with hydrochloric acid. These treatment steps completely inactivate the recombinant BmNPV. As described above, the production process using silkworms is strictly controlled, making it a safe production process.

4.4 Process-related Impurities

Process-related impurities that should be controlled include silkworm-derived proteins, silkworm-derived DNA and baculovirus. For each of these potential impurities, appropriate acceptable limits should be set to implement both evaluation and control.

4.5 Contaminants Such as Adventitious Viruses

Generally, silkworms do not serve as hosts for viruses infecting humans, therefore, the risk of contamination with such viruses is low. Even for manufacturing processes using insects as substrates. However, measures to ensure antiviral safety are considered to be necessary, because the guideline (ICH Q5A guideline), in which requirements for ensuring the antiviral safety of cell substrates are specified, contains the following statement: "This document is concerned with testing and evaluation of the antiviral safety of biotechnology products derived from characterized cell lines of human or animal origin (i.e., mammalian, avian, insect)."

5. Conclusions

In the production of biopharmaceuticals, needs and required levels of technologies for manufacturing recombinant proteins have been increasing. We adopted a production method using the silkworm-baculovirus expression system and achieved commercialization of FeIFN and CaIFN-γ by carrying out the entire development process from research to production (Fig. 5). Our investigations demonstrated quite a high productivity of the expression system using silkworms, although it may depend on the type and properties of the protein to be expressed. The silkworm-baculovirus expression system does not require a large capital investment, and its running costs are cheap. The system is also scalable simply by changing the number of silkworms for use. In addition, as the history of sericulture is long, the accumulation of advanced technologies and knowledge about the rearing and breeding of silkworms has been substantial. We aspire to contribute to advancement in a medical and veterinary field by employing the silkworm-baculovirus expression system suitable for production of biopharmaceuticals.

Fig. 5: Veterinary medicines produced in silkworm-baculovirus expression system. Feline interferon is approved as a treatment of feline and canine viral diseases in 45 countries. Canine interferon-γ is approved as a treatment of canine atopic dermatitis in Japan.

References

Devos, K., F. Duerinck, K. van Audenhove and W. Fiers. 1992. Cloning and expression of the canine interferon-gamma gene. J. Interferon. Res. 12(2): 95–102.

Hanifin, J. M., L. C. Schneider, D. Y. Leung, C. N. Ellis, H. S. Jaffe, A. E. Izu, L. R. Bucalo, S. E. Hirabayashi, S. J. Tofte and G. Cantu-Gonzales. 1993. Recombinant interferon gamma therapy for atopic dermatitis. J. Am. Acad. Dermatol. 28(2 Pt 1): 189–197.

ICH harmonized tripartite guideline viral safety evaluation of biotechnology products derived from cell lines of human or animal origin Q5A(R1). 1990.

Isaacs, A. and J. Lindenmann. 1957. Virus interference. I. The interferon. Proc. R. Soc. Lond. B Biol. Sci. 147: 258–267.

Maeda, S., T. Kawai, M. Obinata, H. Fujiwara, T. Horiuchi, Y. Saeki, Y. Sato and M. Furusawa. 1985. Production of human alpha-interferon in silkworm using a baculovirus vector. Nature. 315(6020): 592–594.
Misaki, R., H. Nagaya, K. Fujiyama, I. Yanagihara, T. Honda and T. Seki. 2003. N-linked glycan structures of mouse interferon-beta produced by *Bombyx mori* larvae. Biochem. Biophys. Res. Commun. 311(4): 979–986.
Mosmann, T. R., H. Cherwinski, M. W. Bond, M. A. Giedlin and R. L. Coffman. 1986. Two types of murine helper T cell clone. I. Definition according to profiles of lymphokine activities and secreted proteins. J. Immunol. 136(7): 2348–2357.
Nagano, Y. and Y. Kojima. 1954. Immunizing property of vaccinia virus inactivated by ultraviolet rays. C. R. Seances Soc. Biol. Fil. 148: 1700–1702.
Nakamura, N., T. Sudo, S. Matsuda and A. Yanai. 1992. Molecular cloning of feline interferon cDNA by direct expression. Biosci. Biotech. Biochem. 56(2): 211–214.
Okano, F., M. Satoh, T. Ido, N. Okamoto and K. Yamada. 2000. Production of canine IFN-γ in silkworm by recombinant baculovirus and characterization of the product. J. Interferon Cytokine Res. 20(11): 1015–1022.
Sakurai, T., Y. Ueda, M. Sato and A. Yanai. 1992. Feline interferon production in silkworm by recombinant baculovirus. J. Vet. Med. Sci. 54(3): 563–565.
Smith, G. E., M. D. Summers and M. J. Fraser. 1983. Production of human beta interferon in insect cells infected with a baculovirus expression vector. Mol. Cell Biol. 3(12): 2156–2165.
Ueda, Y., T. Sakurai and A. Yanai. 1993. Homogeneous production of feline interferon in silkworm by replacing single amino acid code in signal peptide region in recombinant baculovirus and characterization of the product. J. Vet. Med. Sci. 55(2): 251–258.
von Heijne, G. 1986. A new method for predicting signal sequence cleavage sites. Nucleic Acids Res. 14(11): 4683–4690.

Index

A

AcMNPV 220–229
Apolipophorin 39, 51
artificial diet adaptability 23, 35, 36
Arylphorin 39, 40
atomic force microscopy 95, 96
Autographa californica 130

B

BacMam system 129–131, 133, 137–139
bacmid 208–211
Baculovirus 108, 110, 117–120, 122–124
beneficial traits 4, 8
bilayer 205, 207, 208, 210–212, 214, 215
Biofactory 124
biomaterial 235, 236, 240
biosynthesis 39
BmNPV 220, 225–229
Bombyrin 52–56
Bombyx mori 18, 20, 38–41, 43–45, 49, 52, 54, 57
Bombyx mori nucleopolyhedrovirus 130
brain-corpus allatum-prothoracic gland system 85

C

Calicivirus infection 287, 288
canine interferon γ 287, 292, 293, 297
Carbohydrate 163, 166, 171, 178
Chemo-enzymatic synthesis 164, 167–169, 172, 173
Cluster effect 175, 176
collagen 255–257
cross breeding 18, 21, 24, 28

D

development and metamorphosis 62

E

evolution 38, 39, 48, 51, 53, 56

F

ferritin 39, 46, 48–51, 56

G

Gene delivery 220, 223, 224, 228, 229
gene expression 90
Genome editing 144–150, 152
genome information 3, 4, 8, 13, 14
Germline transformation 143, 144
germplasm resources 19–22, 25
glycans 107–113, 117–120, 122–124
glyco-engineering 116
Glycopolymer 166, 170
glycosylation 107, 108, 111, 112, 114, 116, 118, 120, 124
Glycosyltransferase 163–175, 177
GMP 262
GTPase protein 194, 201

H

high-throughput cell based assay system 138
human killer cell immunoglobulin-like receptor 2DL1 181

I

Influenza virus 266–270, 273, 274, 279, 284
insulin-like 90, 93, 97–99, 101, 103
interferon 287–293, 297

J

juvenile hormone 62, 70, 78–80, 82, 85, 86

L

laminin 253–255, 262
binding site 89
lipid metabolism 93, 94, 96
Lipophorin 43, 44, 56

M

male-only hybrid 31
Mammalian cells 220, 221, 223, 225–229
mass spectrometry 97, 98
membrane fusion 193, 194, 197, 199–202
middle silk gland 252
Mitochondria 193, 194, 202
molting hormone 62, 70–73, 75–78, 83, 85, 86
antibody 257, 258, 260
multiple nucleopolyhedrovirus 130
Muramyldipeptide 269, 270
mutants 4–14

N

New industrial sericulture 144, 158, 159
N-linked glycosylation 181
N-linked oligosaccharide 107
non-enveloped 205–209, 212–215

O

optic atrophy 1 (OPA1) 193

P

parthenogenesis 33, 34
pierisin 236, 241, 242, 243
PODS 238, 240
polyhedra 235–240, 245–247
positional cloning 7, 8
post-translational modification 88, 89, 107
Protein Data Bank 57
protein-protein interaction 89

R

recombinant protein 251–253, 260–263
recombinant protein expression 180
Recombinant protein 143, 144, 152–154, 158, 159
regulation mechanism 71, 85

S

sericin 236, 240, 241, 244–247
silkworm 107, 108, 110–112, 114–120, 122–124, 129, 131, 133, 135, 204, 206, 208–213, 215, 225, 228, 229
silk 63, 69, 82
silkworm-baculovirus expression system 287, 291, 295, 297

T

Transduction 220–222, 225–228
Transgenic silkworm 107, 111, 112, 117, 120, 124, 143–146, 150, 152–156, 158, 159, 251–257, 260–263

V

vaccination platform 212, 213, 215
Vaccine 268–275, 279, 281, 283, 284
Virosome 269–272
Virus-like particles 204, 206, 215
Vitellogenin 41–43, 45, 56
VLP 268–275, 277–282, 284